Features of the Companion Website:

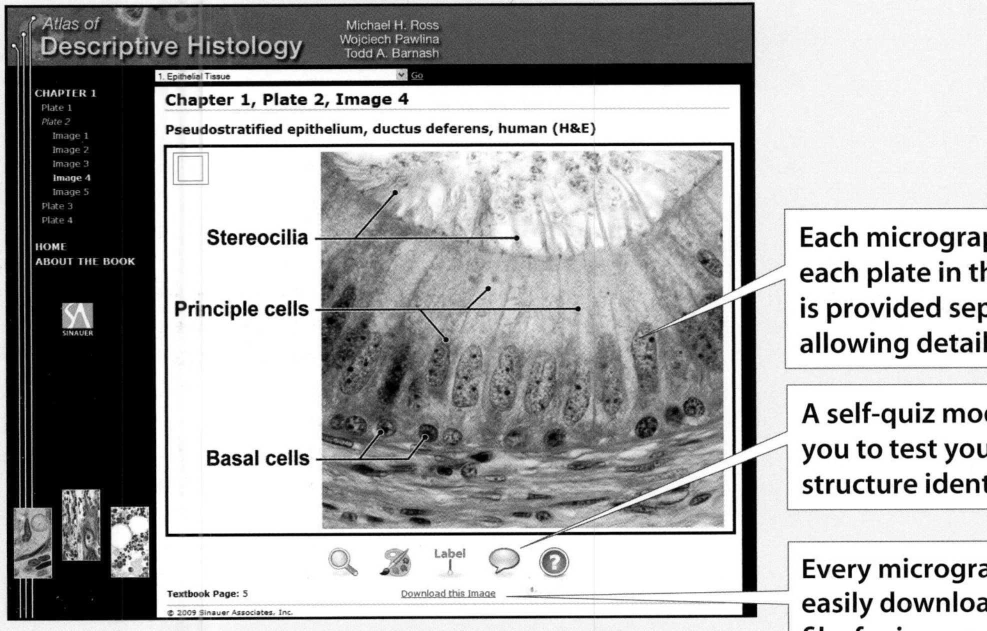

Each micrograph from each plate in the textbook is provided separately, allowing detailed viewing.

A self-quiz mode allows you to test yourself on structure identification.

Every micrograph can be easily downloaded as a JPEG file, for incorporation into your notes.

Low-resolution image, with labels and leaders visible.

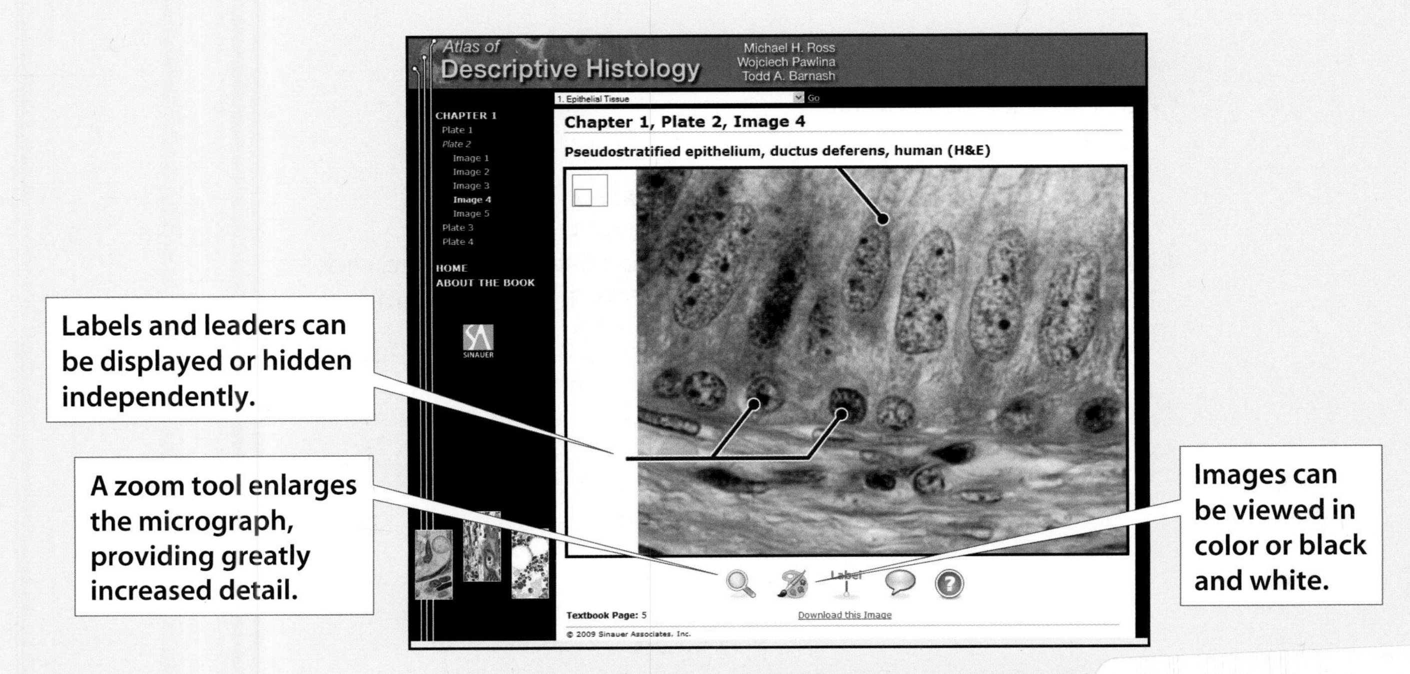

High-resolution image, with leaders visible and labels hidde

Atlas of
Descriptive Histology

Michael H. Ross
University of Florida College of Medicine
Gainesville, Florida

Wojciech Pawlina
Mayo Medical School
College of Medicine, Mayo Clinic
Rochester, Minnesota

Todd A. Barnash
University of Florida College of Medicine
Gainesville, Florida

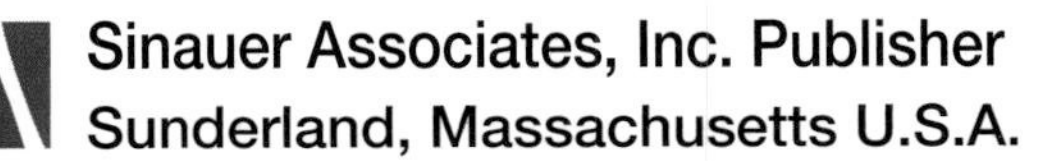
Sinauer Associates, Inc. Publisher
Sunderland, Massachusetts U.S.A.

Atlas of Descriptive Histology

Sinauer Associates, Inc.
23 Plumtree Road
Sunderland, MA 01375 U.S.A.
E-mail: publish@sinauer.com
Internet: www.sinauer.com

Library of Congress Cataloging-in-Publication Data

Ross, Michael H.
Atlas of descriptive histology / Michael H. Ross, Wojciech Pawlina, Todd A. Barnash.
p. cm.
Includes index.
ISBN 978-0-87893-696-0
1. Histology--Atlases. I. Pawlina, Wojciech. II. Barnash, Todd A. III. Title.
QM551.R669 2009
611'.018--dc22

2009015480

5 4 3 2 1

AUTHORS

Michael H. Ross, Ph.D. is Professor and Chairman Emeritus of the Department of Anatomy and Cell Biology at the University of Florida College of Medicine, Gainesville, Florida. For almost half a century, Dr. Ross has contributed to histology education not only as an author of a popular histology textbook, but an equally recognized atlas. He is also a passionate teacher and mentor of several generations of medical students, fellows, and junior faculty. His research interest was in the male reproductive system, where he pioneered studies on the blood testicular barrier and the role of the Sertoli cell in maintaining the barrier. Upon his retirement, Dr. Ross has devoted his time to new editions of his histology textbook and to bringing to fruition this color *Atlas of Descriptive Histology* designed for today's student.

Wojciech Pawlina, M.D. is Professor and Chair of the Department of Anatomy at the Mayo Medical School, College of Medicine, Mayo Clinic in Rochester, Minnesota. He serves as the Assistant Dean for Curriculum Development and Innovation at Mayo Medical School and as the Medical Director of Procedural Skills Laboratory. Dr. Pawlina teaches histology, gross anatomy, and embryology to medical students, residents, fellows, and other health care professionals. His research interest in medical education is directed towards strategies to implement professionalism, leadership, and teamwork curriculum in early medical education. He is the Editor-in-Chief of the journal *Anatomical Sciences Education*.

Todd A. Barnash is the Computer Support Specialist for the Department of Anatomy and Cell Biology at the University of Florida College of Medicine in Gainesville, Florida. Since 1993 he has been the primary information technology, information security, and digital imaging resource for the department and has collaborated with Dr. Ross, as well as other faculty members, on numerous projects throughout his tenure at the University.

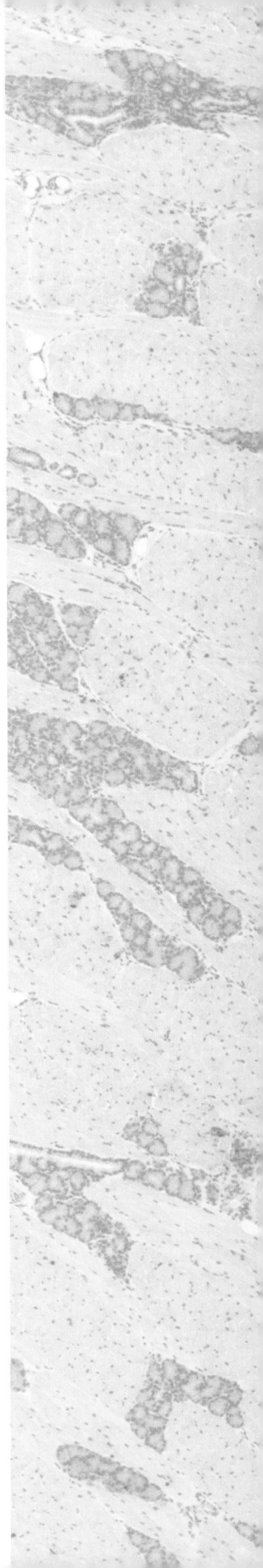

Media and Supplements to Accompany *Atlas of Descriptive Histology*

For the Student

Companion Website (www.sinauer.com/histology)
Each copy of the *Atlas of Descriptive Histology* includes an access code that gives the student a subscription to the textbook's Companion Website. (See the inside front cover for subscription details.) The site features a flexible image viewer that allows the student to view any of the micrographs in the book. Features of the image viewer include:

- Each micrograph from each plate in the textbook is provided separately.
- A zoom tool allows the student to view the micrograph at either low-resolution or high resolution—providing greatly increased detail.
- Labels and leaders can each be displayed or hidden independently.
- A self-quiz mode allows the student to test themselves on structure identification.
- Each micrograph can be downloaded for incorporation into student notes or presentations.

For the Instructor

The *Atlas of Descriptive Histology* Instructor's Resource Library includes multiple versions of every micrograph from every plate in the textbook, as well as an instructor version of the image viewer from the Companion Website. The IRL includes the following:

- Micrographs provided in three formats: Without leaders or labels, with leaders only, and with leaders and labels.
- All images provided as both low-resolution and high-resolution JPEGs.
- A PowerPoint presentation of all images for each chapter.
- The instructor version of the image viewer, which includes all the features described above, with the addition of a pen tool that allows drawing on the image during lecture.

Table of Contents

CHAPTER 4 Cartilage 31

CHAPTER 5 Bone 41

CHAPTER 6 Blood and Bone Marrow 61

CHAPTER 7 Muscle Tissue 71

CHAPTER 12 Digestive System I: Oral Cavity 169

CHAPTER 13 Digestive System II: Esophagus and Gastrointestinal Tract 191

CHAPTER 14 Digestive System III: Liver and Pancreas 215

CHAPTER 19 Female Reproductive System 303

CHAPTER 20 Eye 341

CHAPTER 21 Ear 355

INDEX 361

PREFACE

"*A picture is worth a thousand words.*" People have long quoted this sentence, often attributing it to an old Chinese proverb. Professors of histology use it when complex and difficult concepts of human tissue can be explained with a single image. Historically, students in histology classes were required to draw the image seen under the microscope, providing the histology teacher an opportunity to discuss and correlate the drawing to the students' knowledge. However, in this age of rapid advances in cell and molecular biology, genetics, new educational technologies, fast-progressing, integrated curricula, and reduced teacher–student interactions, this has become a forgotten practice. We repeatedly encounter students who struggle with information overload; this is when visualization becomes increasingly important. It is a well known fact that cognition, according to dual coding theory, involves the activity of two distinct systems: a verbal system, represented by text or auditory input, and a nonverbal system (imagery input). When these inputs are processed simultaneously, they have an additive effect on learning and recall of learned information. We frequently observe this theory in practice during histology laboratory where students reflecting on a specific histology problem resolve their issues with the assistance of an image and a simple explanation.

The idea of linking histology images (nowadays obtained either with the microscope or the computer, using virtual microscopy systems) with a simple interpretation was behind the creation of the *Atlas of Descriptive Histology*. The First Edition of the *Atlas of Descriptive Histology*, by Edward J. Reith and Michael H. Ross, was published in 1965 with two subsequent editions in 1970 and 1977. This atlas consisted of 123 black and white plates of photomicrographs with a general description of the histological structures on the opposing pages. This atlas also became the nucleus for the development of a textbook entitled *Histology, A Text and Atlas.* Currently authored by Michael H. Ross and Wojciech Pawlina, *Histology, A Text and Atlas with Correlated Cell and Molecular Biology* has provided guidance for students taking histology courses since its First Edition in 1985.

As the textbook developed into a comprehensive text of clinically oriented histology with correlated cell and molecular biology, we received many inquiries from colleagues and histology teachers about publishing a concise color atlas of descriptive histology, similar to the straightforward, black and white atlas from the 70's. Owners of this *Atlas* often proudly presented to us their copy of the book, often times showing the tell-tale signs of heavy use over the years. After careful consideration, Dr. Ross decided to digitize his histology slides, collected from around the world throughout his entire professional life, and proceeded with this project. Certainly, the addition of color and the improvement

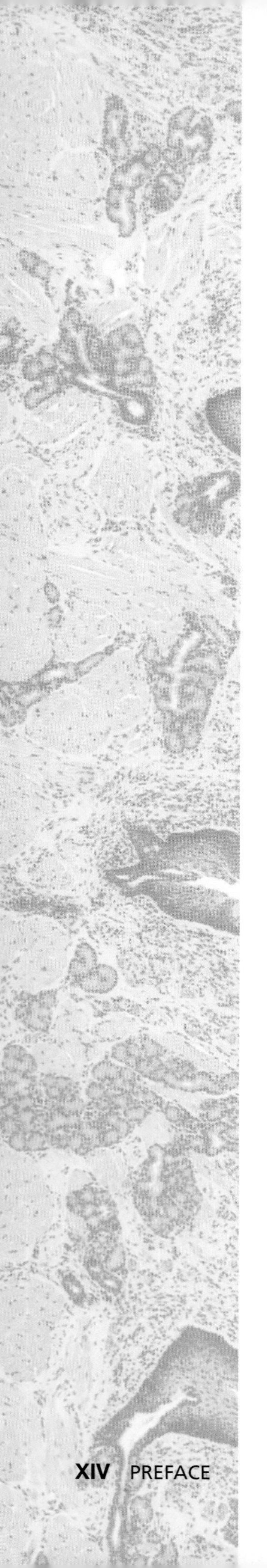

in technology for obtaining high-resolution digital images from old glass slides helps the reader to reach this original objective.

This book represents a well-designed combination of carefully selected, exceptional quality, color images with simple explanatory legends. The blueprint of the original atlas, which served many students so well in the past, is faithfully maintained in the current edition. The presentation of pages, however, is more modern and incorporates new design elements—yellow text boxes, orientation micrographs, chapter openers—provided by an experienced publisher in the field of science education: Sinauer Associates.

We anticipate that this *Atlas of Descriptive Histology* will follow the success of the previous editions and will become a helpful companion to any histology textbook or laboratory manual as it finds its place near every microscope or computer with virtual microscopy slides—or even an iPod with recorded histology lectures or presentations.

Let us finish with a quote from the opening to Professor Pio del Rio-Hortega's 1933 classic histology paper entitled "Art and artifice in the science of histology":

> *Histology is an exotic meal, but can be as repulsive as a dose of medicine for students who are obliged to study it, and little loved by doctors who have finished their study of it all too hastily. Taken compulsorily in large doses it is impossible to digest, but after repeated tastings in small draughts it becomes completely agreeable and even addictive. Whoever possesses a refined sensitivity for artistic manifestations will appreciate that, in the science of histology, there exists an inherent focus of aesthetic emotions.*

We hope that repeated "tasting" of this atlas becomes addictive to its users.

Michael H. Ross, Ph.D.
Wojciech Pawlina, M.D.
Todd A. Barnash

Acknowledgments

Producing a modern atlas such as this one requires the combined efforts of a much larger team of professionals than the three of us who are privileged to have our names on the front cover. The staff members of Sinauer Associates have produced, with great efficiency and good humor, what we consider an atlas of outstanding visual quality and educational value. Those with whom we have had the most enduring contacts are Editor Andy Sinauer and Project Editor Julie HawkOwl, but many others labored behind the scenes to ensure the book's high quality and timely production. They include Joan Gemme, Joanne Delphia, and Christopher Small. We are very grateful to all of them. We thank Julie HawkOwl for her skillful copyediting; Jason Dirks, Mara Silver, Suzanne Carter, Nate Nolet, Ann Chiara, and Tom Friedmann for their work on the media and supplements package; Marie Scavotto and Linda VandenDolder for their effective work promoting the book.

To
Jan S. Moreb, M.D.
Clinician, Researcher, Teacher, and Scholar
With Great Admiration
M.H.R.

and
Our Past, Present, and Future Students

CHAPTER 1
Epithelial Tissue

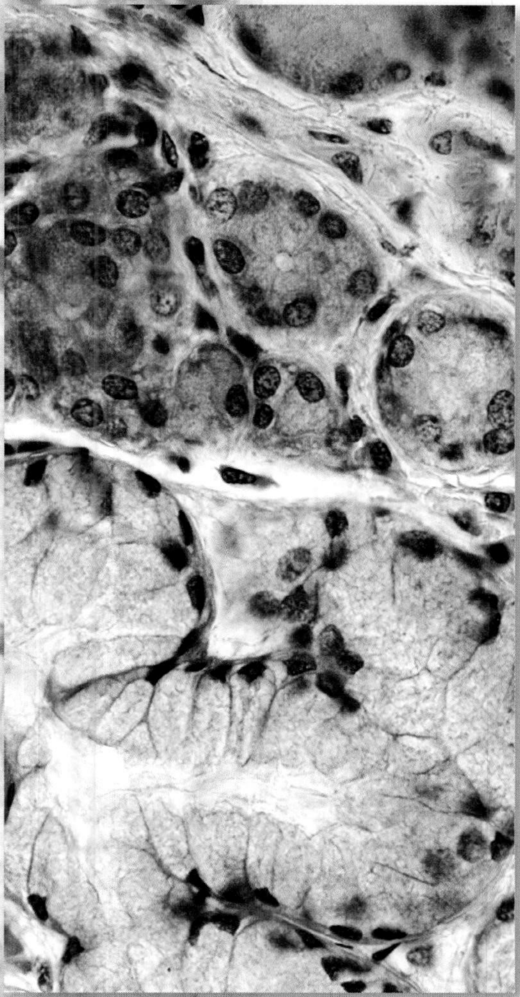

PLATE 1. SIMPLE SQUAMOUS AND SIMPLE CUBOIDAL EPITHELIA

Epithelium consists of a diverse group of cell types, each of which possesses specific functional characteristics. The cells that make up a given epithelium are arranged in close apposition with one another and typically are located at what may be described as the free surfaces of the body. Such free surfaces include the exterior of the body, the outer surface of many internal organs, and the lining of body cavities, tubes, and ducts.

Epithelium is classified on the basis of the arrangement of the cells that it contains and their shape. If the cells are present in a single layer, they constitute a simple epithelium. If they are present in multiple layers, they constitute a stratified epithelium. The shape of the cells is typically described as squamous if the cell is wider than it is tall; cuboidal if its height and width are approximately the same; or columnar if the cell is taller than it is wide. The micrographs in this Plate provide examples of simple squamous and simple cuboidal epithelia.

Simple squamous epithelium, mesovarium, human, H&E, x350; inset x875.

This specimen shows the surface epithelium of the mesovarium. The mesovarium is covered by mesothelium, a special name given to the simple squamous epithelium that lines the internal closed cavities of the body. The **mesothelial cells** (1) are recognized by their nuclei, as seen at this low magnification. Beneath the squamous mesothelial cells is a thin supporting layer of loose **connective tissue** (2), and below that are typically arranged **adipose cells** (3). The **inset** reveals at higher magnification the **mesothelial cell nuclei** (1). Note their broad width as compared to their height.

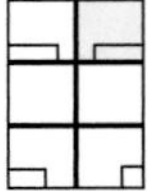

Simple squamous epithelium, mesentery, rat, silver impregnation, x350; inset x700.

This is an intermediate magnification of a whole mount of a piece of mesentery. The thin mesentery specimen was placed on the slide and prepared for microscopic examination. The microscope was focused at the surface of the mesentery. By this method, the **boundaries** (4) of the surface mesothelial cells appear as black lines from the precipitated silver. Note that the cells are in close apposition to one another, and that they have a polygonal shape. The **inset** reveals several mesothelial cells, each of which exhibits a **nucleus** (5) that has a round or oval profile. Because of the squamous shape of the mesothelial cells, their nuclei are not spherical but rather are disc-like.

Simple squamous epithelium, kidney, human, H&E, x350.

This specimen shows a kidney renal corpuscle. The wall of the renal corpuscle, known as the parietal layer of Bowman's capsule, is a spherical structure that consists of a **simple squamous epithelium** (6). The interior of the renal corpuscle contains a capillary network from which fluid is filtered to enter the **urinary space** (7) and then into the **proximal convoluted tubule** (8). **Nuclei** (9) of the squamous cells of the parietal layer of Bowman's capsule are disc-shaped and appear to protrude slightly into the urinary space. Their irregular distribution is a reflection of the probability of sectioning the nucleus in any given cell. The free surface of this simple squamous epithelium faces the urinary space, whereas the basal surface of the epithelial cells rests on a thin layer of **connective tissue** (10).

Simple cuboidal epithelium, pancreas, human, H&E, x700.

This specimen shows two **pancreatic ducts** (11) that are lined by a simple cuboidal epithelium. The **duct cell nuclei** (12) tend to be spherical, a feature consistent with the cuboidal shape of the cell. The **free surface** (13) of the epithelial cells faces the lumen of the duct, and the basal surface rests on **connective tissue** (14). Careful examination of the free surface of the epithelial cells reveals some of the **terminal bars** (15) between adjacent cells.

Simple cuboidal epithelium, lung, human, H&E, x175; inset x525.

This specimen shows the epithelium of the smallest conducting bronchioles of the lung. The epithelium in this distal part of the bronchial tree consists of **simple cuboidal epithelial cells** (16). The **inset** shows a higher magnification of the cuboidal cells. Note the spherical nuclei. These are small cells with relatively little cytoplasm, thus the nuclei appear close to one another. The free surface of the epithelial cells faces the **airway** (17), whereas the basal surface of these cells rests on its basement membrane and underlying dense **connective tissue** (18).

Simple cuboidal epithelium, liver, human, H&E, x450; inset x950.

The liver specimen shown here reveals the cords of **cuboidal cells** (19), known as hepatocytes, that make up the liver parenchyma. The hepatic cords are mostly separated from one another by blood **sinusoids** (20). Between the hepatocytes and the endothelium of the sinusoids is an extremely sparse layer of connective tissue. The **inset** shows a higher magnification of a hepatocyte and reveals an unusual feature in that several surfaces of these cells posses a groove, representing free surfaces of the cell. Where the groove of one cell faces the groove of an adjacent cell, a small channel, the **bile canaliculus** (21), is formed. Bile is secreted into the canaliculi.

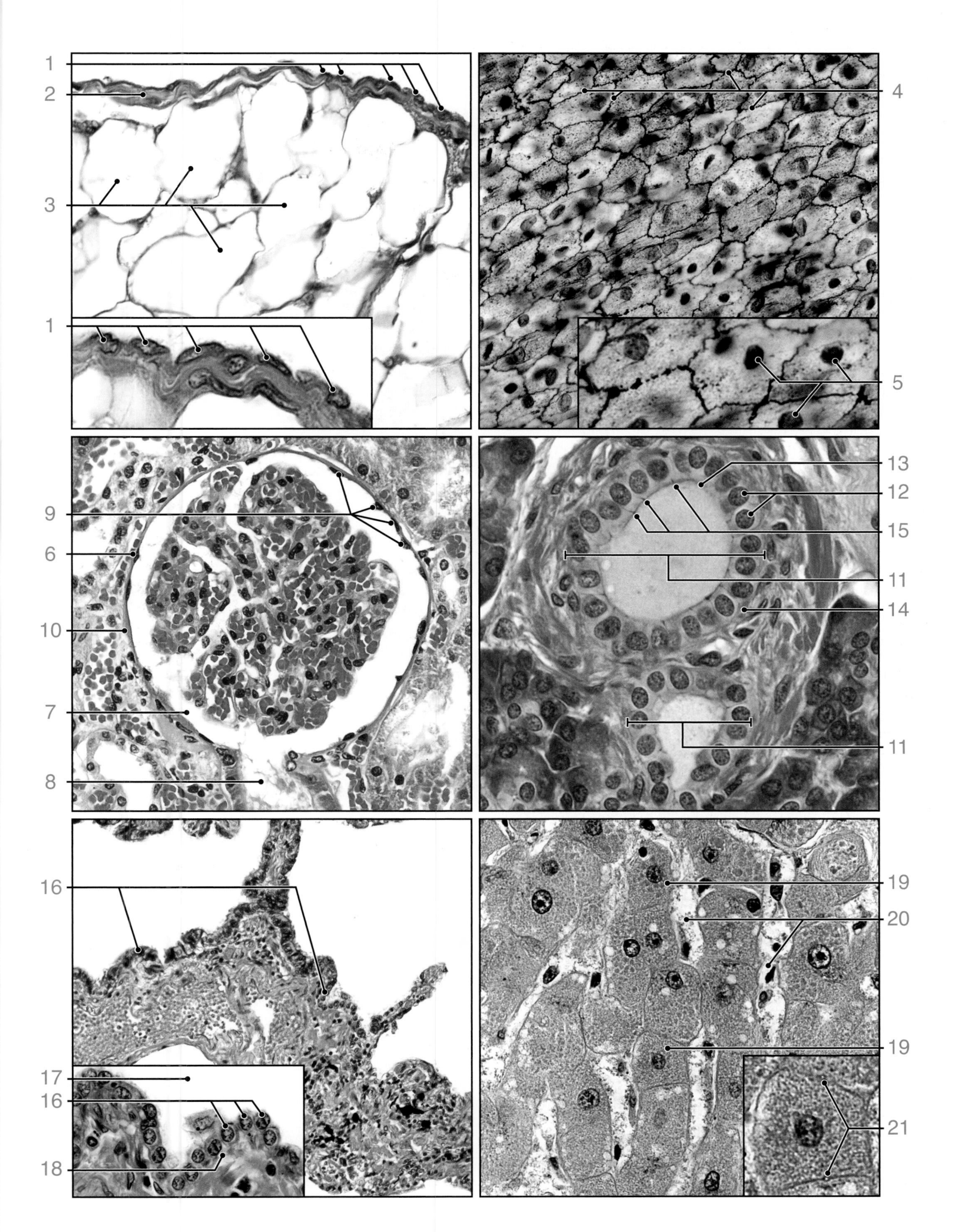

PLATE 1. SIMPLE SQUAMOUS AND SIMPLE CUBOIDAL EPITHELIA

PLATE 2. SIMPLE COLUMNAR AND PSEUDOSTRATIFIED EPITHELIA

Pseudostratified epithelium is epithelium in which the cells appear to be stratified because their nuclei appear to be arranged in more than a single layer. Really, it is comparable to simple epithelium because all of the cells rest on the basement membrane. Short cells are interspersed among taller cells, so their nuclei may appear at different levels within the thickness of the epithelium. Also, cell boundaries may not be evident, making it difficult to distinguish pseudostratified epithelia from truly stratified epithelia. Fortunately, there are relatively few instances where pseudostratified epithelium is present. It is found in the large excretory ducts of certain exocrine glands, the male urethra, the excretory passages of the male reproductive system, the eustachian tube, the tympanic cavity, the lacrimal sac, and much of the mucous membranes of the respiratory passages. In all but the male urethra and large excretory ducts of the exocrine glands, the tall cells in the epithelium exhibit cilia or stereocilia.

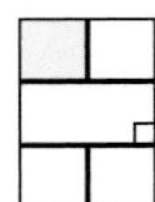

Simple columnar epithelium, jejunum, human, H&E, x525.

This micrograph shows the tip of an intestinal villus with its surface covered by a simple columnar epithelium. The epithelium consists of two types of cells—the intestinal absorptive cell, or **enterocyte** (1), and, in fewer numbers, the mucus-secreting **goblet cell** (2). Both cell types are tall, thus the columnar designation, and both are arranged as a single layer, thus it is a simple epithelium. The **nuclei** (3) of both cell types are elongate, a feature consistent with the shape of the cells. Note that the mucus cup of the goblet cell does not stain with H&E, thus its empty appearance. Several **lymphocytes** (4), which have migrated into the epithelium from the **connective tissue** (5) of the villus, can be identified by their dense, round nuclei. They are non-epithelial cells and have a transient presence in the epithelial compartment.

Simple columnar epithelium, colon, human, H&E, x440.

This specimen reveals the simple columnar epithelium that lines the luminal surface of the **colon** (6) and the **intestinal glands** (7) (crypts of Lieberkühn), which are continuous with the surface cells. These cells possess a large mucus cup that develops as the cells mature and migrate from the lower part of the crypts to the luminal surface. Note the overall height of the cells, indicated by the *brackets*. They are clearly taller than they are wide, thus columnar, and are arranged in a single layer.

Simple columnar epithelium, tongue, salivary glands, human, H&E, x725; inset x1450.

This specimen shows mucus-secreting glandular tissue on the left, and serous secreting glandular tissue on the right. The cells of both types of glands are more tall than wide, thus they are regarded as columnar. Note that the mucus-secreting cells mostly have **flattened nuclei** (8), whereas the serous secreting cells have **round nuclei** (9). Also, the serous cells have a conical or pyramidal shape. The apical surface is relatively small compared to the basal surface. The **inset** reveals the cells' **junctional complexes** (10), the dark, red-staining bodies; thus, the space between a pair of complexes represents the apical surface of one cell. These cells form a sphere-like secretory lobule. In contrast, the mucus-secreting cells form elongate, branched lobules with a large lumen (*asterisks*). Nevertheless, in both examples, the gland cells are columnar and reside in a single layer.

Pseudostratified epithelium, ductus deferens, human, H&E, x700.

The tall cells shown here are the **principle cells** (11) lining the ductus deferens. Note their tall elongate nuclei and the **stereocilia** (12) (actually long microvilli) at their apical cell surface. Also present are small **basal cells** (13). The small, round nuclei of the basal cells are surrounded by a thin rim of cytoplasm. These small cells will differentiate and replace the tall, principle cells. Both the tall principle cells and the basal cells rest on the basement membrane. While their appearance might suggest two cell layers, this is actually a simple epithelium; thus, it is designated as pseudostratified epithelium.

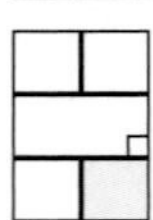

Pseudostratified epithelium, epiglottis, human, Mallory-Azan, x700.

The pseudostratified epithelium in this micrograph also gives the impression of a stratified epithelium, based on the location and appearance of the epithelial cell nuclei. Three types of cells constitute this epithelium, all of which rest on the basement membrane. Most of the nuclei that are immediately adjacent to the basement membrane belong to undifferentiated cells designated as **basal cells** (14). These cells give rise to the other two cell types, namely, mucus-secreting **goblet cells** (15) and **ciliated cells** (16). Note that only the mucus cup of the goblet cell is clearly apparent. The **cilia** (17) on the ciliated cells extend from their **basal bodies** (18), which collectively appear as a densely stained dark line.

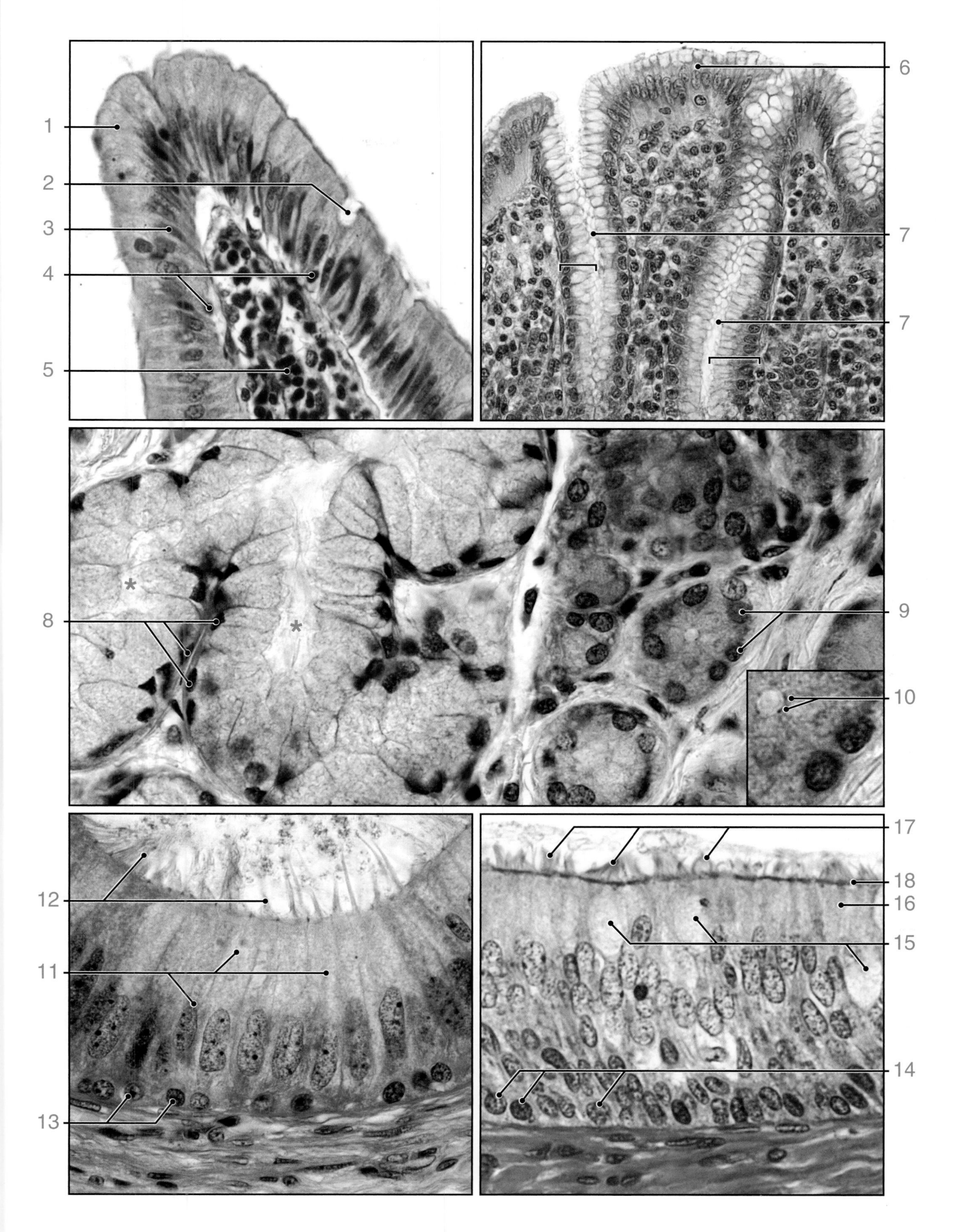

PLATE 2. SIMPLE COLUMNAR AND PSEUDOSTRATIFIED EPITHELIA

PLATE 3. STRATIFIED EPITHELIA

Stratified epithelia occur in a variety of locations. The number of cell layers and the thickness of this epithelium likewise vary markedly. For example, the epidermis has the largest number of cell layers and can reach a thickness of almost 1.5 mm. In contrast, there are many sites where only two cell layers occur—in small (though not the smallest) exocrine ducts, for example. Further variation is seen in the shape of the cells that make up a stratified epithelium. Typically, the cells may vary in shape, from squamous to cuboidal to columnar. In the case of epidermis, the basal cells (those that rest on the basement membrane) are cuboidal in shape and those that are near the surface are squamous. Thus, the epidermis would be described as stratified squamous epithelium. In the bottom two micrographs, two ducts are shown, one exhibits two layers of cells that are cuboidal, though of slightly different size. This would be classified as a stratified cuboidal epithelium, whereas the other duct exhibits basal squamous cells and columnar surface cells. This epithelium would be classified as stratified columnar. It is always the surface cell that provides the designation for the classification of a stratified epithelium.

Stratified squamous (nonkeratinized) epithelium, esophagus, human, H&E, x140; inset x350.

This micrograph shows the **stratified squamous epithelium** (1) lining the esophagus. It is a multilayered epithelium with only the **basal cells** (2) resting on the basement membrane. Through the mitotic activity of these basal cells, stratification of the epithelium is maintained. Ultimately, the cells at the surface are desquamated, sloughing off into the lumen. Note that the basal cells are small, cuboidal and have little cytoplasm. As the cells move toward the surface, their shape changes from cuboidal to squamous. Note the shape of the nuclei at different levels. The more **superficial cells** (3) have elongate, or disc-shaped nuclei, a reflection of the shape that the cells have acquired, namely squamous. The **inset** shows a surface squamous cell and, next to it, one that is desquamating.

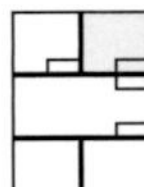

Stratified squamous (keratinized) epithelium, skin, human, H&E, x140; inset x350.

This specimen, like the previous specimen, is a **stratified squamous epithelium** (4). It differs mainly in that it is keratinized. Note that the **basal cells** (5) are small and cuboidal. As new cells are pushed towards the surface, they will change to a **squamous shape** (6). As shown in the **inset**, those cells approaching the surface undergo a keratinizing process in which their cytoplasm becomes filled with the protein keratin and the nucleus is lost. This process is signified by the production of **keratohyalin granules** (7) in the cytoplasm, which is reflected by the dark blue staining of the cytoplasm. The **mature, fully keratinized cells** (8), which eventually slough off from the body's surface, stain with eosin.

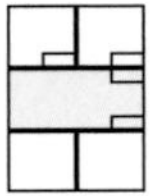

Stratified squamous and stratified cuboidal epithelium, mammary gland, human, Mallory, x120; insets x350.

This micrograph shows the terminal portion of an excretory duct of a female nipple. The more distal portion of the duct reveals a **stratified squamous keratinized epithelium** (9). To the right, where two smaller ducts are joining to form the larger duct, a **stratified cuboidal epithelium** (10) is seen in one duct and a **stratified squamous epithelium** (11) is present in the other duct. The higher magnification of the **lower inset** reveals the stratified cuboidal epithelium of the lower duct. Note that there are two layers of cells, the surface layer composed of cuboidal shaped cells. In the **upper inset**, showing the stratified squamous epithelium, note that there is a layer of cuboidal basal cells, and above that are one or two layers of squamous cells, evidenced by the shape of their nuclei. The surface cells are clearly squamous, thus it is a stratified squamous epithelium.

Stratified cuboidal epithelium, tongue, human, H&E, x275.

This micrograph shows a cross-sectioned duct from a salivary gland. The **duct epithelium** (12) consists of two cell layers. The cells of the basal layer are cuboidal. The cells of the surface layer are also generally cuboidal, but are larger. Some of the cells in this layer appear to have **elongate nuclei** (13) rather than spherical nuclei, suggesting that these few cells are columnar. Given that the majority of the surface cells are cuboidal, the epithelium in this duct is identified as stratified cuboidal.

Stratified columnar epithelium, tongue, human, H&E, x425.

The duct in this micrograph is from the same specimen as the previous micrograph. Note that most of the **basal cells** (14) appear to be squamous, based on nuclear shape. Most of the cells in the surface layer exhibit **tall, elongate nuclei** (15) indicative of a columnar cell type. Thus, in this instance, the duct epithelium is described as stratified columnar.

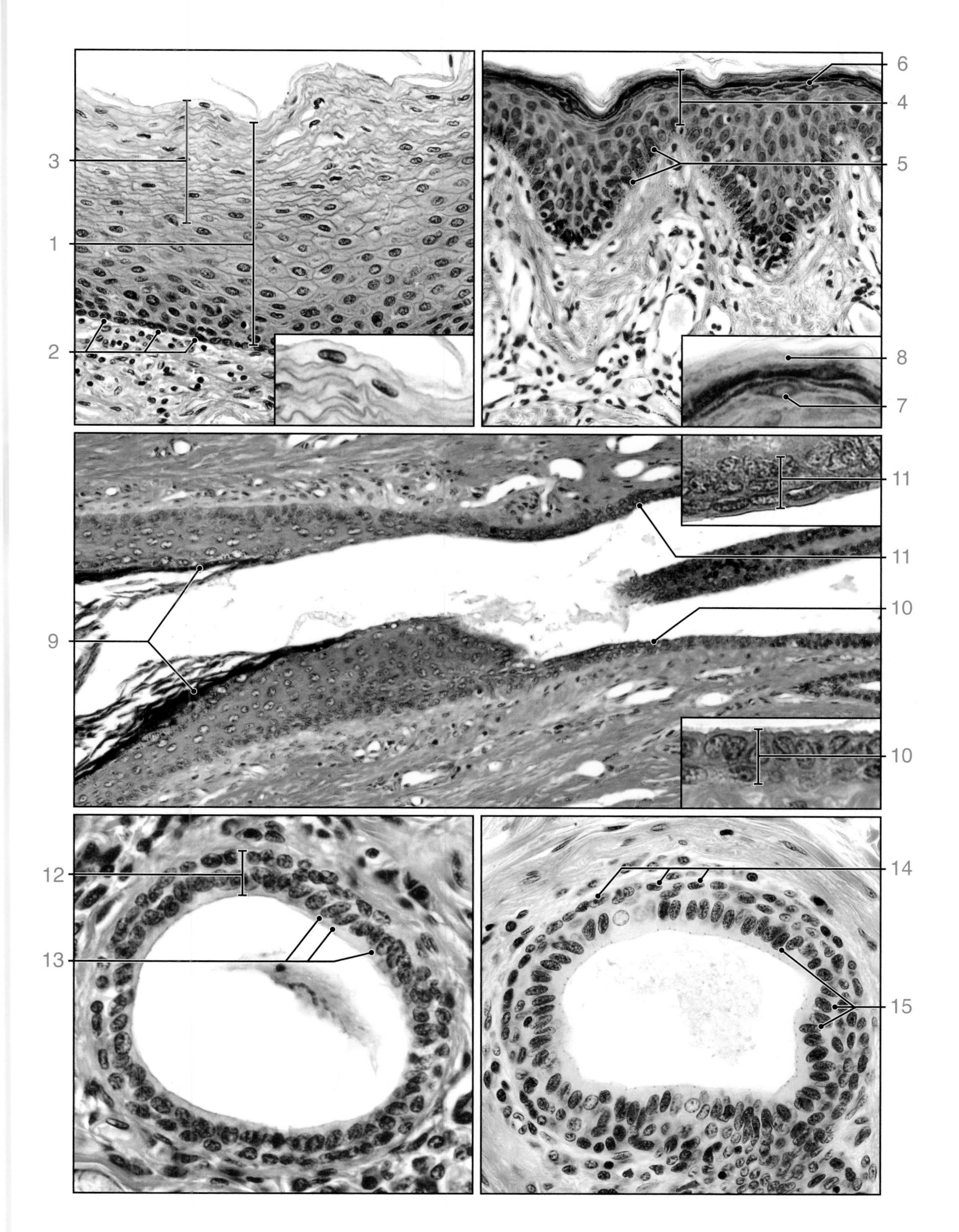

6
4
3
5
1
2
8
7
11
11
10
9
10
12
14
13
15

PLATE 4. TRANSITIONAL EPITHELIUM AND EPITHELIOID TISSUES

Transitional epithelium is a term originally assigned to epithelium lining organs that undergo considerable distention, such as the urinary bladder. The cells in the upper layer of the epithelium that lines this organ undergo a transition from a cuboidal to a squamous shape as the bladder fills. The epithelium is stratified, but the term for this epithelium would change based on the shape of the surface cells at a given time. Thus, it is called transitional epithelium for consistency.

Epithelioid epithelium is so named because the close apposition of the cells is similar to typical epithelium. They are classified in a separate category because these cells lack other epithelial characteristics. A major difference is that they do not exhibit a free surface. The epithelium of endocrine organs typically develops at a free surface, then migrates from that site. In other instances, such as the perineurium of nerves, the cells arise from a mesodermal site rather than from ectoderm or endoderm.

Transitional epithelium, urinary bladder, human, H&E, x140.

The **transitional epithelium** (1), seen in a nondistended urinary bladder, is shown in this micrograph. It consists of four to five layers of epithelial cells. The **surface cells** (2), also shown in the **inset**, are relatively large and often exhibit a slightly rounded or dome-like surface. The cells resting on the basement membrane are the smallest, and those between the basal cells and surface cells tend to be intermediate in size. When the bladder is distended, the more superficial cells are stretched and have the appearance of a squamous cell. In this state, the epithelium appears to have reduced in thickness to about three cells deep. The bladder wall usually contracts when it is removed, unless special steps are taken to preserve it in a distended state. Thus, its appearance is usually like that seen here.

Transitional epithelium, urethra, human, H&E, x140, inset x350.

This micrograph shows the transitional epithelium of the urethra in the distal part of the penis. The **transitional epithelium** (3) seen here is similar to that in the top left micrograph. Again, the **superficial cells** (4) are large and show a curvature on the apical surface. The **inset** shows the surface cells at higher magnification (note their columnar shape). A **lymphocyte** (5) can be seen between the surface epithelial cells. Also present in this part of the urethral epithelium are islands of **mucus-secreting cells** (6). They exhibit a clear cytoplasm, and they may be arranged as a simple epithelium or in more extensive groupings where they form alveolar structures.

Interstitial (Leydig) cells, testis, human, H&E, x200.

This micrograph shows a large group of **interstitial cells** (7) lying between two **seminiferous tubules** (8). In the upper left is a **blood vessel** (9). The interstitial cells possess certain epithelial characteristics but they do not possess a free surface. They are epithelioid because they are in close contact, like epithelial cells of a true epithelium. Interstitial cells of the testis are an endocrine tissue and develop from mesenchyme and embryonic nonsurface tissue.

Islet of Langerhans, pancreas, H&E, x400.

This micrograph shows the endocrine **islet of Langerhans** (10) of the pancreas. These cells also have an epithelioid arrangement. The cells are in contact, but lack a free surface. In this instance, they have developed from an epithelial surface by invagination. In contrast, the surrounding alveolar structures of the **exocrine pancreas** (11), which develop from the same epithelial surface, are made up of cells with a free surface onto which their secretory product is discharged. Similar examples of epithelioid tissue are seen in the adrenal, parathyroid, and pituitary glands, all of which are endocrine glands.

Myelinated nerve, human, trichrome, x350.

This micrograph shows a portion of a myelinated nerve. The **nerve fibers** (12) are seen in cross-section. Surrounding this bundle of nerve fibers is the **perineurium** (13) of the nerve. It consists of multiple layers of flattened, squamous-like cells. The nuclei of these cells are stained red. The perineurial cells within each layer are arranged in close apposition to one another. Unlike the cells of a true epithelium, perineurial cells do not have a free surface. In addition, they exhibit cytoplasmic features similar to smooth muscle cells, namely that they are contractile. The cells within each layer create a semipermeable, layered, sheathlike structure. Thus, their arrangement is epithelioid.

Thymus, human, H&E, x500.

Another example of an epithelioid tissue is seen in the thymus. A supporting reticular stroma is formed from endodermal epithelium. Lymphocytes come to lie between these epithelial cells, widely separating them, to form a cellular reticulum. These epithelial cells are referred to as **epithelioreticular cells** (14). Note how the lymphocytes are massed between the epithelioreticular cells. The epithelioreticular cells sometimes appear in small clusters of cells and sometimes as individual cells, totally isolated from other epithelioreticular cells. Though not visible in this section, they are connected, forming the stroma of the organ. Because the epithelioreticular cells are no longer at the surface from which they originated, they are regarded as an epithelioid tissue.

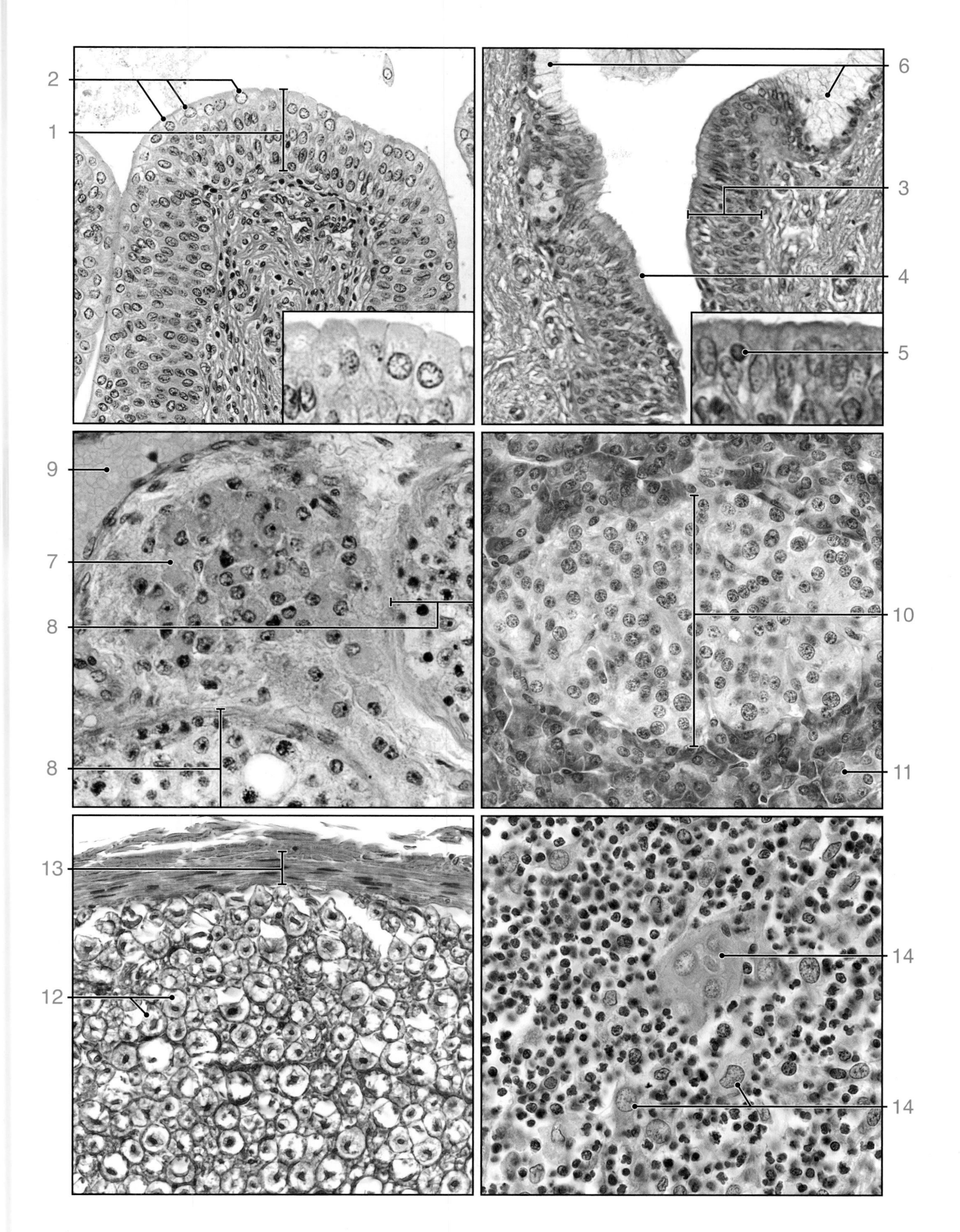

PLATE 4. TRANSITIONAL EPITHELIUM AND EPITHELIOID TISSUES

CHAPTER 2
Connective Tissue

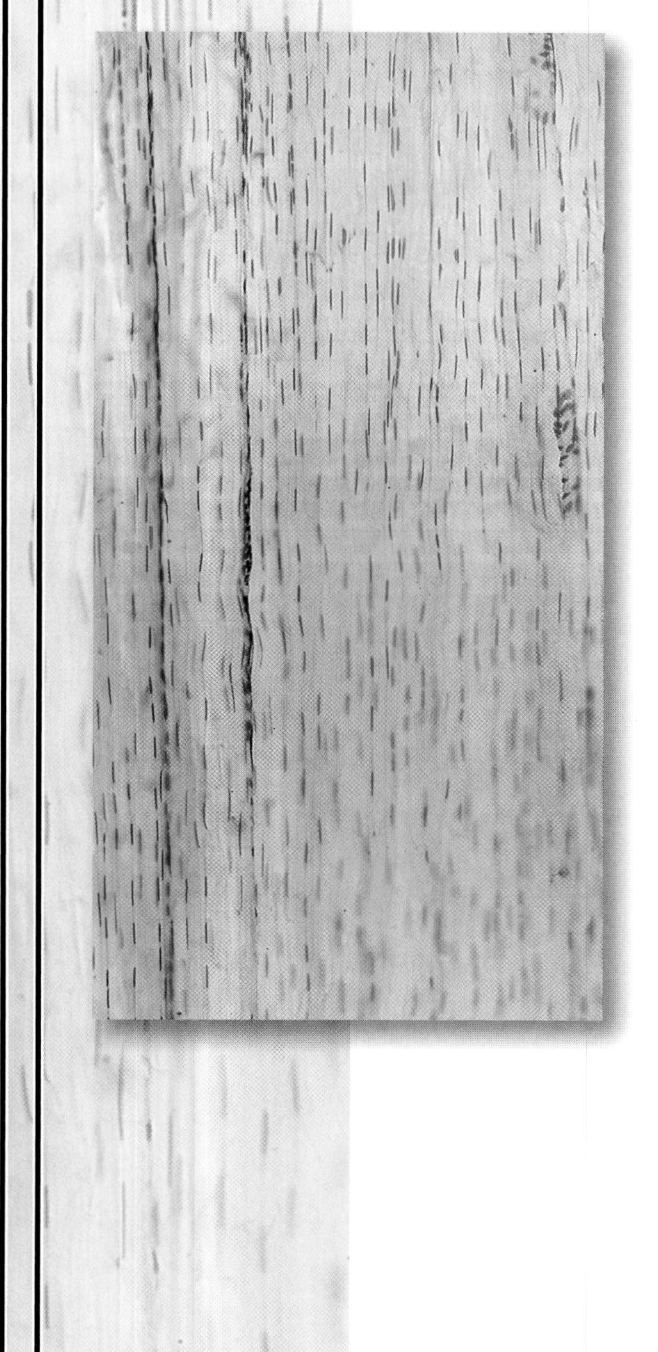

PLATE 5. LOOSE AND DENSE IRREGULAR CONNECTIVE TISSUE

Loose and dense connective tissues represent two of the several types of connective tissue. The others are cartilage, bone, blood, adipose tissue, and reticular tissue. Loose connective tissue is characterized by a relatively high proportion of cells within a matrix of thin and sparse collagen fibers. In contrast, dense connective tissue contains few cells, almost all of which are fibroblasts responsible for the formation and maintenance of the abundant collagen fibers that form the matrix of this tissue. The cells that are typically associated with loose connective tissue are fibroblasts (the collagen forming cells) and cells that function in the immune system. Thus, in loose connective tissue, there are varying quantities of lymphocytes, macrophages, eosinophils, plasma cells, and mast cells.

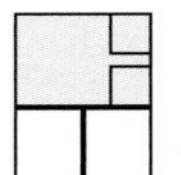

Mammary gland, human, H&E, x175; insets x350.

This micrograph shows, at low magnification, both **loose connective tissue** (1) and **dense irregular connective tissue** (2) for comparative purposes. The loose connective tissue surrounds the **glandular epithelium** (3). The dense connective tissue consists mainly of thick bundles of collagen fibers with few cells present, whereas the loose connective tissue has a relative paucity of fibers and a considerable number of cells. The **upper inset** is a higher magnification of the dense connective tissue. Note that only a few cell nuclei are present relative to the large expanse of collagen fibers. The **lower inset**, revealing the glandular epithelium and surrounding loose connective tissue, shows very few fibers but large numbers of cells. Typically, the cellular component of loose connective tissue contains a relatively small proportion of fibroblasts and a large proportion of lymphocytes, plasma cells, and other connective tissue cell types.

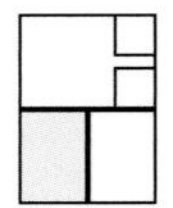

Colon, monkey, Mallory, x250.

This micrograph reveals a highly cellular **loose irregular connective tissue** (4), also called lamina propria, which is located between the intestinal glands of the colon. The simple columnar, mucous secreting epithelial cells seen here represent the glandular tissue. The Mallory stain colors cell nuclei red, and collagen blue. Note how the cells are surrounded by a framework of the blue stained collagen fibers. Also shown in this micrograph is a band of smooth muscle, the **muscularis mucosa** (5) of the colon, and below that, seen in part, is **dense irregular connective tissue** (6) that forms the submucosa of the colon. Typically, the **collagen fibers** (7) that lie just below the **epithelial cells** (8) at the luminal surface are more concentrated and thus appear prominently in the micrograph.

Colon, monkey, Mallory, x700.

This micrograph is a higher magnification of the boxed area on the previous image. The bases of the epithelial cells are seen on each side of the micrograph. The **collagen fibers** (9) appear as thin threads that form a stroma surrounding the cells. Most of the cells present here are lymphocytes and **plasma cells** (10). Other cells present within the stromal framework are fibroblasts, smooth muscle cells, macrophages, and occasional mast cells.

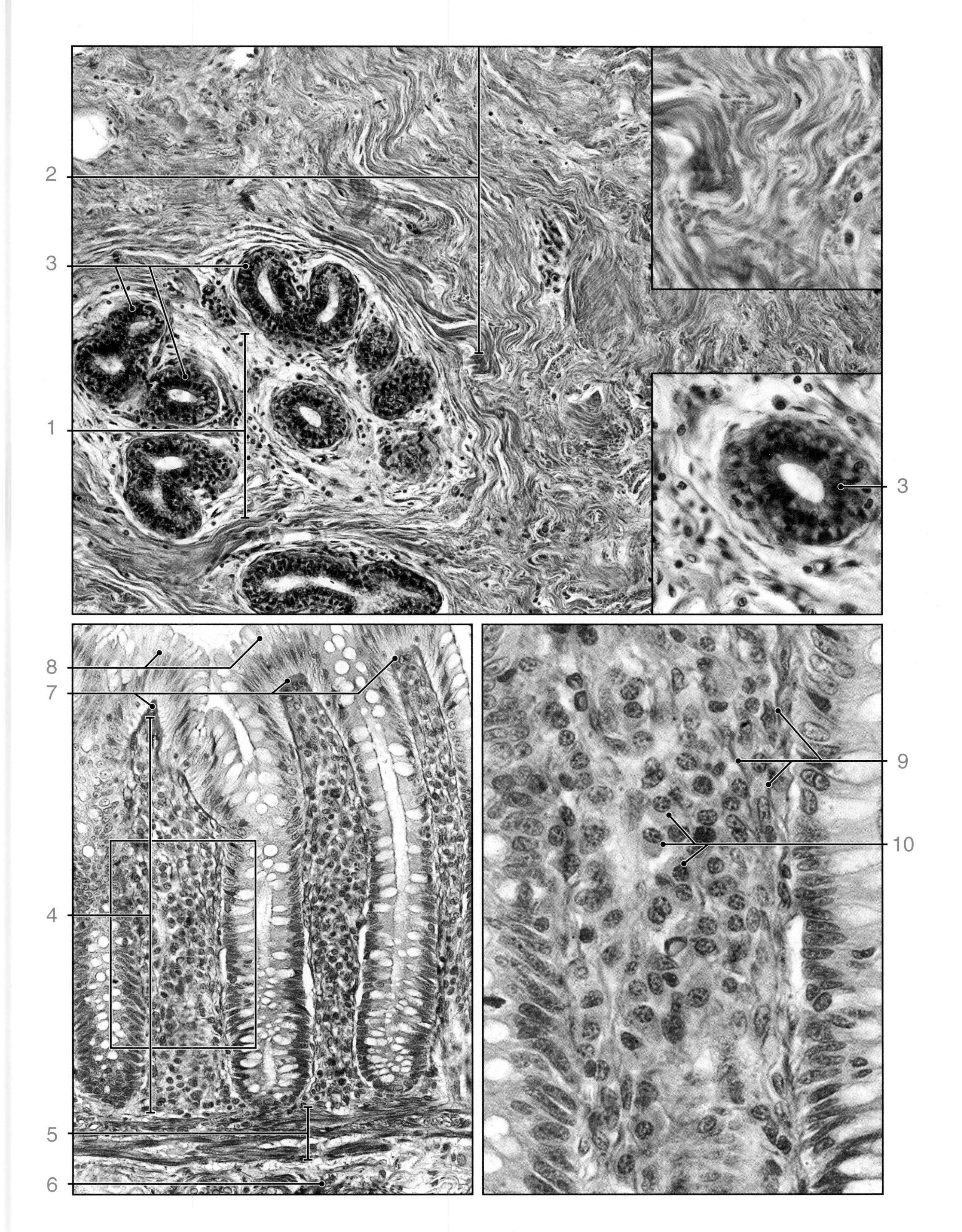

PLATE 5. LOOSE AND DENSE IRREGULAR CONNECTIVE TISSUE

PLATE 6. **DENSE IRREGULAR CONNECTIVE TISSUE**

Dense irregular connective tissue is characterized by an abundance of tightly packed collagen fibers and few cells. The only cells present are, with few exceptions, fibroblasts. They are responsible for the production and maintenance of the collagen. In some instances, occasional macrophages may also be present. Because the collagen fibers are arranged in an irregular pattern, that is, they appear to course in different directions, the term "irregular" is applied. Dense irregular connective tissue is found at sites where strength and integrity of a structure are required. It is organized to resist physical forces to which an organ might be exposed. To illustrate examples of this type of connective tissue, the dermis of the scalp and a meniscus of the knee joint have been selected.

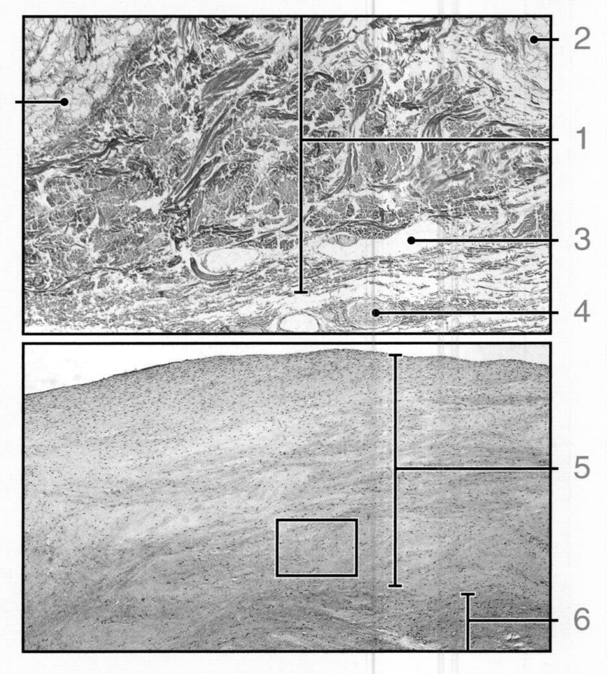

ORIENTATION MICROGRAPHS: The *upper micrograph* shows **dense irregular connective tissue** (1) that lies below the integument of the scalp. Also seen in this view are areas containing **adipose tissue** (2), which is another form of connective tissue, as well as a **vein** (3), and a small **nerve** (4).

The *lower micrograph* shows, at low power, the superficial portion of one of the menisci from the knee joint. The surface of the meniscus is seen at the top of the micrograph. The major portion of the meniscus shown in this micrograph, namely the lighter staining area, consists of **dense irregular connective tissue** (5). The deeper portion of the meniscus, which is more densely stained, consists of **fibrocartilage** (6).

Scalp, dense irregular connective tissue, human, H&E, x375; inset x700.

This micrograph shows an area of dense irregular connective tissue from the human scalp. The most significant feature is the presence of numerous, thick bundles of collagen fibers, which occupy most of the field. Some of the bundles have been **longitudinally sectioned** (1), whereas the majority of the **collagen bundles** (2) appear to be cross-sectioned. Dense connective tissue, as noted in the previous Plate, has a paucity of cells, virtually all of which are **fibroblasts** (3). In most preparations using H&E stain, only the nucleus of the fibroblast is apparent. The cytoplasm of the cell typically blends in with the adjoining collagen fibers, as both are stained with eosin. The **inset** shows, at higher magnification, several fibroblasts. The cytoplasm of the uppermost fibroblast is visible in this view because of this cellís separation from adjoining fibers. The other fibroblasts clearly exhibit nuclei, but their cytoplasm is not distinguishable. Dense irregular connective tissue is not highly vascularized; however, in this specimen, several **lymphatic vessels** (4) can be seen as well as several small **blood vessels** (5).

Meniscus, dense irregular connective tissue, human, Masson, x375; inset x700.

Another example of a dense irregular connective tissue is shown in this specimen. This image, tissue from the more superficial portion of the meniscus, is an magnified view of the region within the rectangle on the lower orientation micrograph above. Collagen fibers comprise the bulk of the tissue in this area and are arranged as closely packed irregular arrays of collagen bundles. The cellular component consists of **fibroblasts** (6). The fibroblast nuclei appear either as elongate profiles or oval profiles depending on the plane of section in which the cell was cut. The area from the square is magnified in the **inset**, where several fibroblasts are visible. The collagen fibers appear blue-green in color whereas the cells have a contrasting black coloration. An advantage in using the Masson stain is that it readily distinguishes the thin elongate **cytoplasm of the fibroblasts** (7) from the collagen fibers. Typically, as noted in the H&E section above, the nucleus of the fibroblast is readily discerned, but the cytoplasm, which stains with eosin, blends with the collagen fibers.

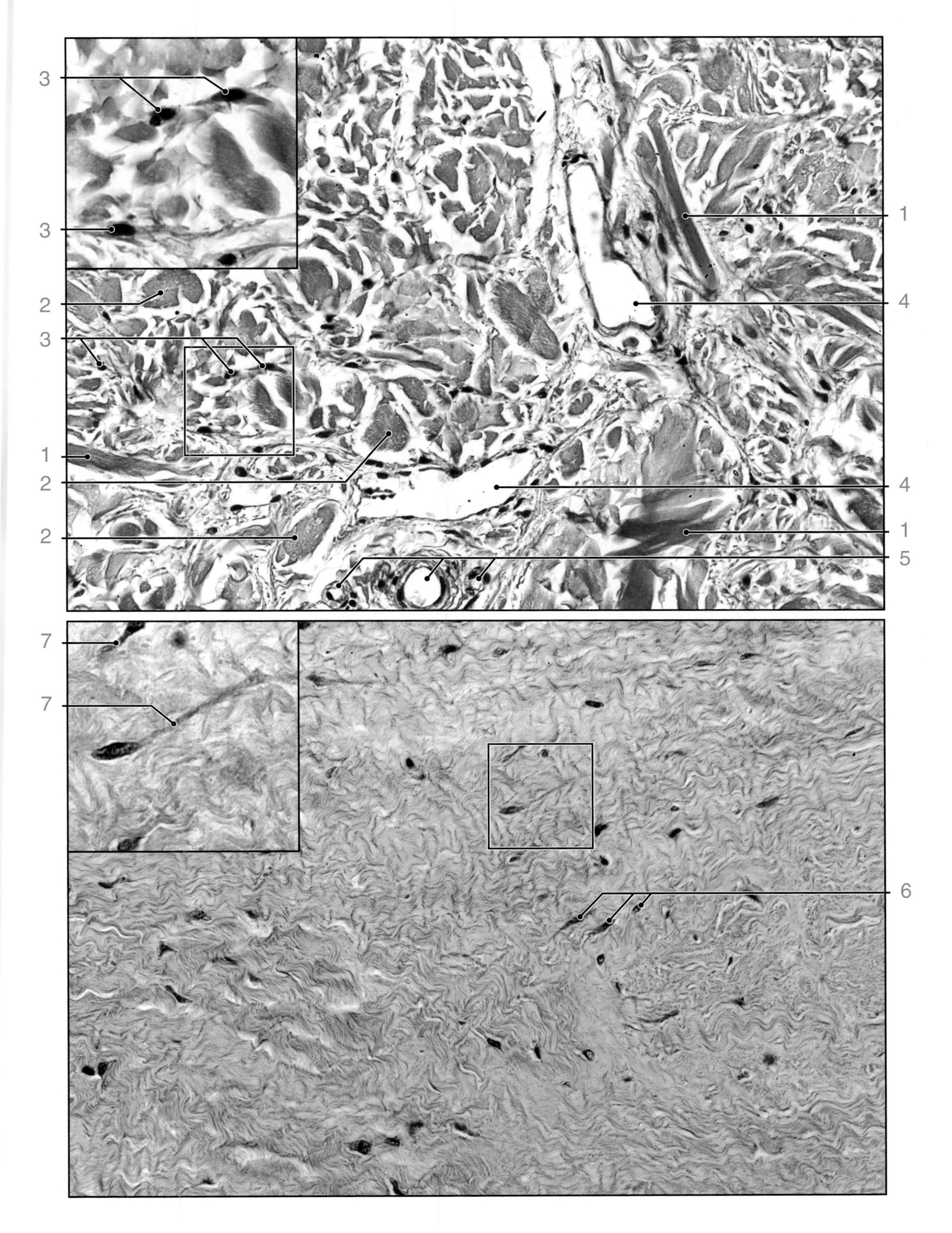

PLATE 6. DENSE IRREGULAR CONNECTIVE TISSUE

PLATE 7. CONNECTIVE TISSUE, ELECTRON MICROSCOPY

Uterine tube, human, electron micrograph, x4100.

The elements that constitute a connective tissue, namely, the fibers and cells, are readily distinguished by transmission electron microscopy. The specimen shown here is from a section through the wall of a uterine tube. The tissue is rather cellular, but it also contains a considerable amount of fibrous material. In terms of its constituent elements, it is comparable to the connective tissue shown in the boxed area of the lower micrograph on Plate 6. The **fibroblasts** (1), which constitute the bulk of the connective tissue cell population, usually exhibit long cytoplasmic processes that pass between the collagen bundles. The processes extend for indeterminate distances and may become so attenuated (2) that their thinness precludes the possibility of being visualized with the light microscope. The **inset** shows a part of two fibroblasts, including the nucleus of one. These cells can be immediately identified as fibroblasts based on the presence of a relatively extensive amount of granular **endoplasmic reticulum** (3). Note that the cisternae of the endoplasmic reticulum are dilated, particularly in the cell on the right, and contain a homogeneous substance of moderate density. This substance is a product of the synthetic activity of the ribosomes on the surface of the reticulum, and primarily represents a precursor of the collagen to be produced. In the same cell, a portion of the **Golgi apparatus** (4) is visible. The Golgi consists of a multitude of flattened saclike profiles and small vesicles.

In addition to fibroblasts, there are at least two other connective tissue cell types present in this specimen. One is a cell with both myoid and fibroblast-like features. This we refer to as a **myofibroblast** (5). It differs from the fibroblast in that its cytoplasm possesses an extensive filamentous component, which would be visible at higher magnification. These cells also exhibit cytoplasmic densities. The combination of the filaments and the cytoplasmic densities is a feature characteristic of smooth muscle cells. The myofibroblast functions both as a fibroblast and as a contractile cell. The other is a cell that has certain features that suggest it is a **monocyte** (6). In the light microscope, this cell might appear to be a fibroblast. The nucleus is elongate, but there is little cytoplasm surrounding the nucleus. Based on its appearance here, it is cytologically different from the fibroblast. Presumably, this cell is a monocyte that is in transition to a tissue macrophage.

In this specimen, almost all of the **collagen fibers** (7), have been cross-sectioned, resulting in end-on profiles of the individual collagen fibrils. Consequently, it is possible to discern their various sizes and shapes. These threadlike fibrils aggregate into bundles, forming the fiber that is seen at the light microscope level.

During the fixation and dehydration stages in the preparation of a routinely prepared light microscopic specimen, there may be considerable shrinkage of initially hydrated tissue. One consequence of this process is the artificial separation of the collagen fibers. Thus, in the light microscope, the fibers appear as wispy, isolated, threadlike elements rather than the more evenly distributed fibers seen in the electron microscope.

Typically, one finds small blood vessels passing through connective tissue. In this view, there is a **capillary** (8) as well as a longitudinally sectioned **venule** (9). The venule contains several **red blood cells** (10).

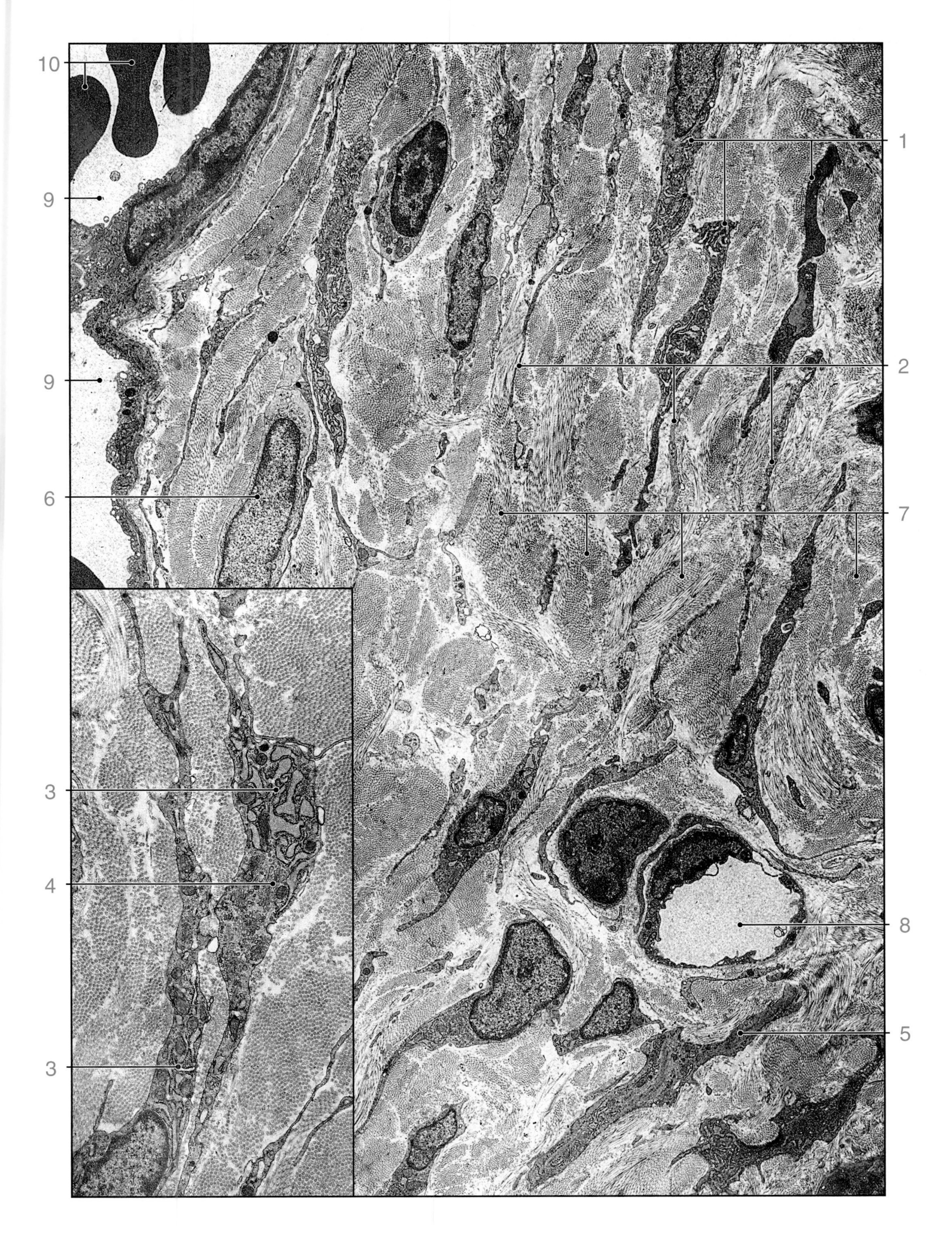

PLATE 7. CONNECTIVE TISSUE, ELECTRON MICROSCOPY

PLATE 8. DENSE REGULAR CONNECTIVE TISSUE

Dense regular connective tissue is characterized by an orderly arrangement of collagen fibers. The cellular constituent of this tissue consists only of fibroblasts. The collagen fibers are made up of very closely packed collagen fibrils. The fibroblasts occur in linear rows, interspersed between the collagen fibers. Tendons, which attach muscle to bone, and ligaments, which attach bone to bone, are examples of dense regular connective tissue. The difference between these two tissue structures is mainly that tendons exhibit a more orderly array of collagen fibers than ligaments. Typically, ligaments and tendons are covered by a dense irregular connective tissue. In tendons, the surrounding connective tissue may be referred to as epitendineum. Epitendineum also penetrates the tendon, creating fascicles, or bundles, of collagen fibers. The connective tissue is referred to as endotendineum. Both the epitendineum and endotendineum contain vessels and nerves. The fibroblasts in tendons may be referred to as tendinocytes. They are elongated cells whose cytoplasm extends into exceedingly thin processes between the collagen fibers. When seen in longitudinal section, the nuclei appear as uniformly elongate structures. The cytoplasm tends to blend in with the collagen in H&E preparations, making it difficult to detect. The cytoplasmic processes of adjacent fibroblasts tend to be in contact at their edges, forming a syncytial-like network.

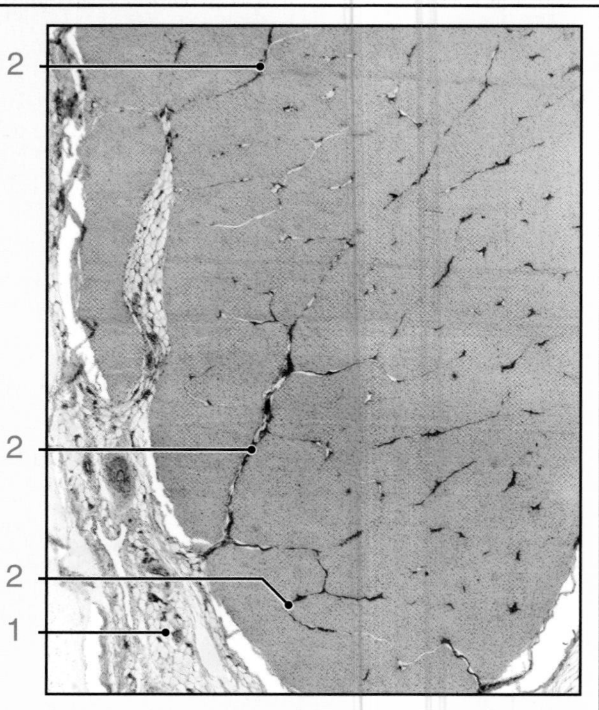

ORIENTATION MICROGRAPH: The orientation micrograph reveals part of a tendon cut in cross-section. A layer of irregular dense connective tissue, the **epitendineum** (1), is seen on one side of the tendon. Within the tendon are connective tissue septa, the **endotendineum** (2), that extend from the epitendineum and subdivide the tendon into fascicles of various sizes.

Tendon, human, H&E, x175; inset x350.

This is a longitudinal section of a tendon that is comparable in size to the cross-sectioned tendon shown in the orientation micrograph. The direction of the section of the tendon is revealed by the appearance of the **fibroblast (tendinocyte) nuclei** (1), which are arranged in rows and have thin, elongate profiles. Visually, the cytoplasm of these cells typically blends in with the collagen fibrils that make up the tendon fibers. The **endotendineum** (2), which separates the tendon fascicles, consists of dense irregular connective tissue. Where a blood vessel is seen within the endotendineum, the endotendineum appears more cellular. The **inset** shows the elongate tendinocyte nuclei at higher magnification. At this magnification, the **cytoplasm** (3) of some of the tendinocytes is just barely evident. A short segment of a slightly oblique-sectioned **arteriole** (4) is present in the micrograph.

Tendon, human, H&E, x180; inset x350.

The micrograph shown here magnifies part of the cross-sectioned tendon in the orientation micrograph. The dense irregular connective tissue in the upper part of the micrograph is the **epitendineum** (5). Associated with the epitendineum are a number of blood vessels that are coursing between the epitendineum and the underlying tendon. The numerous nuclei in this view of the epitendineum are associated with blood vessels and give this part of the epitendineum a densely stained appearance. **Endotendineum** (6) is seen at several sites within the tendon. Again note that the more dense staining areas in this region are also due to numerous nuclei associated with the blood vessels. Note that the fibroblast nuclei in the tendon, when cut in cross-section, appear as irregular, dot-like structures at this magnification. Compare their shape with the nuclei of the longitudinally sectioned tendon in the above micrograph. The **inset** shows a higher magnification of the area within the rectangle on this micrograph. It reveals the general shape of the tendon cell nuclei. Note that they actually have an angular shape, due to the fact that the tendinocytes are compressed between bundles of adjoining collagen fibers. One of the **blood vessels** (7) is also seen in this view. Note the surrounding **nuclei** (8) that belong to the dense regular connective tissue of the endotendineum.

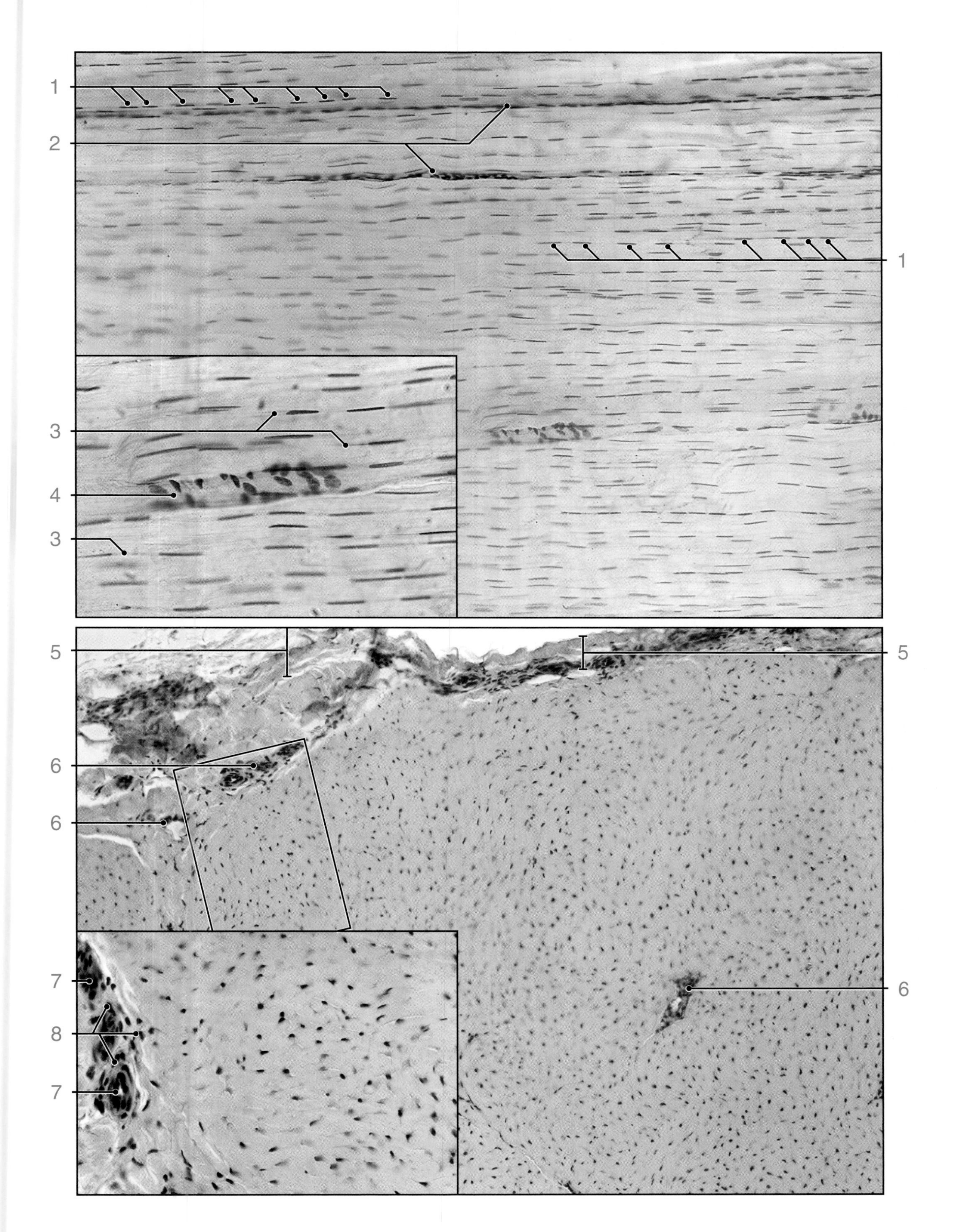

PLATE 8. DENSE REGULAR CONNECTIVE TISSUE

PLATE 9. **ELASTIC FIBERS**

Elastic fibers are present in a number of organs. They provide the ability for tissues to respond to distention and stretch. They are thinner than collagen fibers and are arranged in a three-dimensional branching pattern. The fibers are interlaced with collagen fibers, which limit the degree of distention of the tissue. Generally, elastic fibers stain very poorly with eosin and require special stains, such as resorcin-fuchsin, Weigert's stain, or Verhoeff's stain.

Elastic fibers are produced by fibroblasts and, in certain sites, by smooth muscle cells. Structurally, elastic fibers are made up of two components. They have a central core of elastin and fibrillin microfibrils surround the elastin. The elastin component is a protein that is rich in proline and glycine, but unlike collagen, it is poor in hydroxyproline and lacks hydroxylysine. Fibrillin is a glycoprotein that forms the microfibrils. They are initially deposited during elastic fiber formation and then followed by deposition of the elastin. The Fibrillin microfibrils help organize the elastin into fibers.

Elastic tissue is found in several sites in a non-fiber form. For example, in blood vessels, the elastin occurs in sheets or lamellae. In this case, the elastin is produced by smooth muscle cells. Also, the Fibrillin microfibrils are not associated with lamellar elastin formation.

Skin, elastic fibers, monkey, Weigert's stain, eosin counterstain, x125.

This specimen shows, at low magnification, the **epithelium** (1) of the skin and the underlying **dense connective tissue** (2). A special feature revealed by the Weigert's stain is the **elastic fiber** (3). The elastic fibers are the blue/black threadlike strands dispersed as a network of branching elements.

Skin, elastic fibers, monkey, Weigert's stain, x200.

This micrograph shows a higher magnification of the dense connective tissue and elastic fibers from the same specimen as the previous micrograph. At this magnification, it is evident that the threadlike elastic fibers are cut in varying planes. Some **elastic fibers** (4) are longitudinally sectioned, whereas other **elastic fibers** (5) appear in cross-section, thus revealing their threadlike structure. Thick eosin-stained bundles of **collagen fibers** (6) occupy the bulk of the connective tissue. The **nuclei** (7) in the lower half of the micrograph belong to epithelial cells of sweat glands that have been sectioned in various planes.

Mesentery, whole mount, rat, Verhoeff's elastic tissue stain and orange safranin counterstain, x125.

This low power micrograph reveals a **vein and its tributaries** (8) lying within the thickness of the mesentery. The **elastic fibers** (9) appear as fine, black lines dispersed throughout the specimen. Numerous faintly orange stained **collagen fibers** (10) are also present. The small, dark, round structures are nuclei. They represent nuclei of the surface mesothelial (epithelial) and connective tissue cells.

Mesentery, whole mount, rat, Verhoeff's elastic tissue stain and orange safranin counterstain, x200.

The higher magnification of this micrograph accentuates the **elastic fibers** (11), which appear as well defined black lines. Some of the fibers are relatively thick compared to thin fibers, which have a lighter or faint appearance. Note the branching character of the thicker fibers. The **collagen fibers** (12) are the broader, orange bands that crisscross the specimen. Unlike the elastic fibers, they do not branch. The nuclei, as noted in the previous micrograph, belong to connective tissue and mesothelial cells. In this specimen, many of the mesothelial cells were lost during the tissue preparation. The nuclei that have a lighter or less dense appearance probably belong to **mesothelial cells** (13). In this micrograph, **mast cells** (14) are readily identified by the staining of their cytoplasm. The **mitotic figure** (15) probably represents a dividing mesothelial cell.

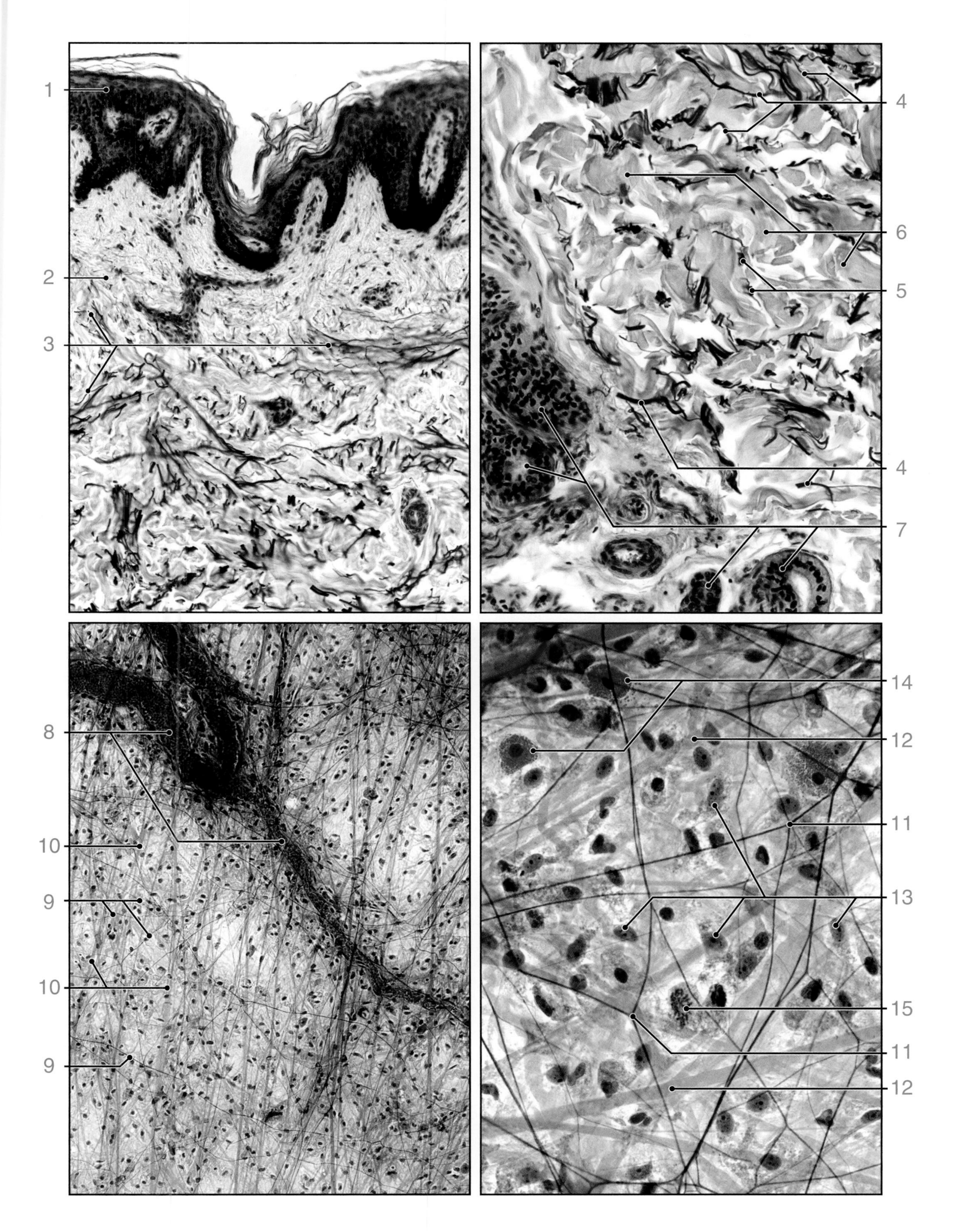

PLATE 9. ELASTIC FIBERS

PLATE 10. **RETICULAR FIBERS**

Reticular fibers are similar to collagen fibers in that they are both made up of collagen fibrils and both have a 68 nm banding pattern. Reticular fibrils, however, consist of type III collagen rather than the type I collagen associated with collagen fibrils. In general, reticular fibers provide a fine, supporting framework for the cellular constituents of various tissues. They cannot be recognized in routinely stained H&E preparations but require special staining techniques for visualization. The Gomori and Wilder staining methods, which are silver staining procedures, enable reticular fibril visualization. When stained by these methods, the reticular fibers appear as thin, black, threadlike networks. In these preparations, collagen fibers are substantially thicker in appearance and stain a brownish color.

Reticular fibers are produced by fibroblasts in most sites. Exceptions are in hemopoietic and lymphatic tissues, where reticular fibers are produced by a special cell type, namely, the reticular cell. In these organs, the reticular cell maintains a special relationship to the fibers that it has produced; the cytoplasm of these cells forms a sheath around the fiber, much like the insulation that covers an electrical wire, thus isolating it from other tissue constituents. Other exceptions include peripheral nerves, where Schwann cells produce the reticular fibers, and smooth muscle cells that produce reticular fibers in the tunica media of blood vessels and the muscularis of the alimentary canal.

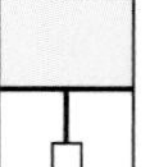

Heart, reticular fibers, human, Wilder stain, x500.

This micrograph of cardiac muscle has been stained to show the reticular fibers that bind together the **muscle fibers** (1) (cells) of the heart. The muscle fibers have been cut longitudinally and are horizontally oriented in the micrograph. The multiple **nuclei** (2) of each muscle fiber are clearly seen. Another feature characteristic of cardiac muscle fibers is their **cross-striations** (3). The **reticular fibers** (4) appear as the dark, black lines, many of which give the appearance of wrapping around the muscle fibers.

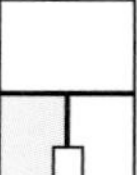

Lymph node, reticular fibers, human, Wilder stain, x125.

This specimen is shown at low magnification for orientation. It reveals the outer portion of a lymph node, including the **capsule of the node** (5), the underlying **subcapsular sinus** (6) through which lymph fluid flows, and part of the **lymph node cortex** (7). This specimen can be compared with a lymph node stained with H&E in Plate 60 for orientation purposes.

Lymph node, reticular fibers, human, Wilder stain, x500; inset x1000.

This is a high magnification of the boxed area in the previous micrograph. The lymph node **capsule** (8), **subcapsular sinus** (9), and **cortical tissue** (10) exhibit **reticular fibers** (11), which appear as thin, black lines. They form a supporting stroma throughout the organ. The reticular fibers are most prominent in the capsule and the underlying sinus. Careful examination also reveals these fibers within the substance of the cortex. The **reticular cells** (12), which produce the reticular fibers, are best seen in the sinus where they are unobstructed by other cells, namely, the lymphocytes in the cortical tissue or the dense collagen fibers in the capsule. The reticular cells within the sinus are distinguished by their elongate nuclei, in contrast to the relatively few **lymphocytes** (13) whose nuclei are spherical. The **inset** shows the nuclei of three reticular cells at high magnification. Note how the fibers appear to extend from the nuclear region of the reticular cell. Not apparent in this preparation is the cytoplasm of the reticular cell. In the case of the capsule of the lymph node, the reticular fibers are produced by fibroblasts and do not have a covering of cytoplasm.

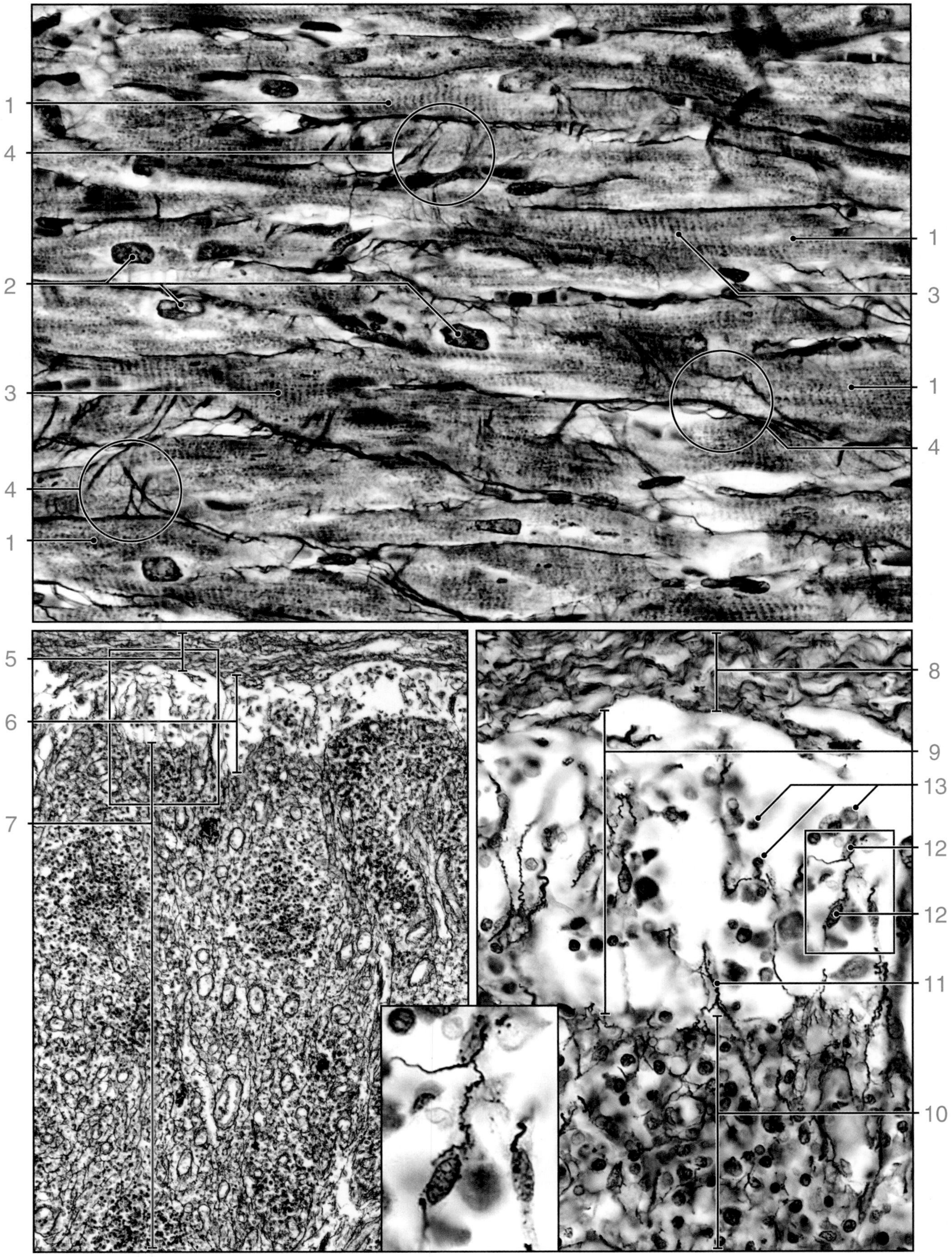

PLATE 10. RETICULAR FIBERS

CHAPTER 3
Adipose Tissue

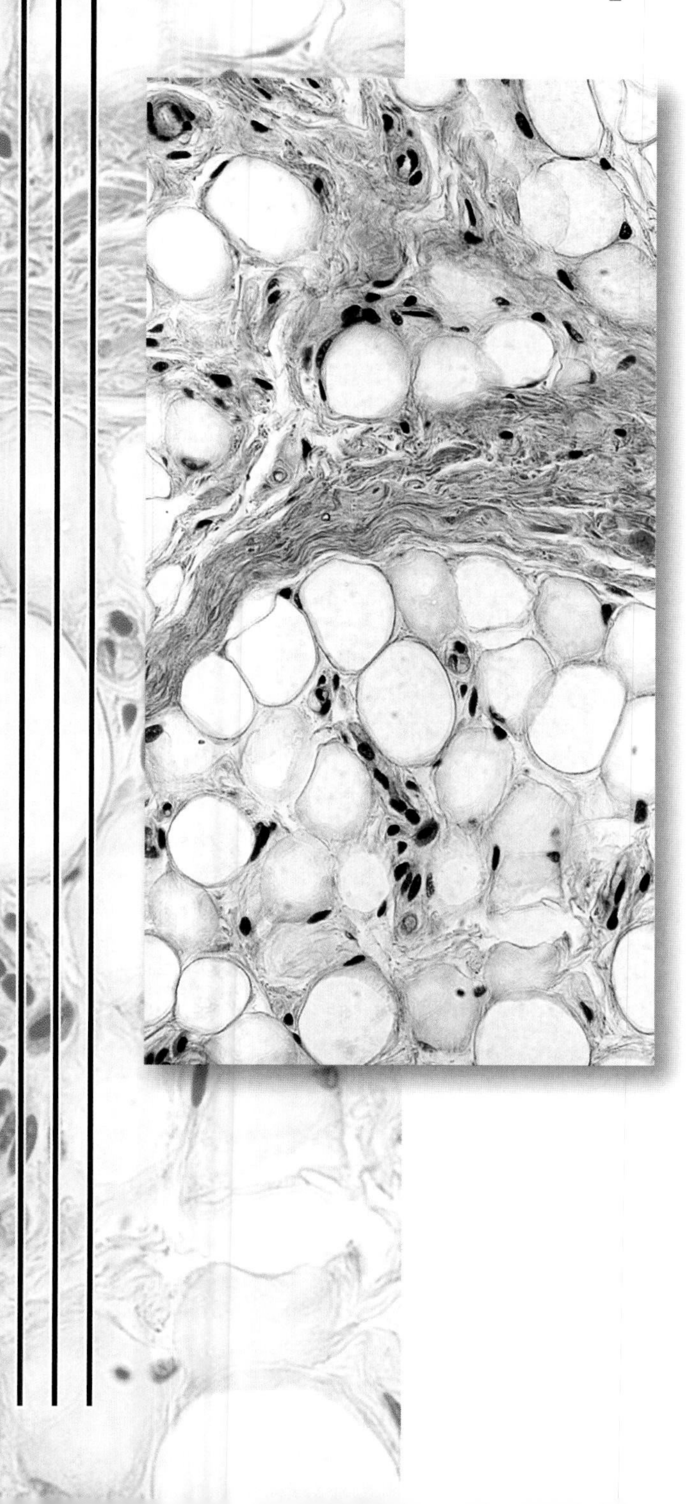

PLATE 11. ADIPOSE TISSUE I

Adipose tissue is a specialized connective tissue that is widely distributed throughout the body. Two types of adipose tissue are recognized: white adipose tissue and brown adipose tissue. White adipose tissue is more common. Its distribution varies between different individuals, depending on sex, nutritional, and hormonal status. Principal cells in white adipose tissue are fat storing. They are very large cells whose cytoplasm contains a single, large accumulation of lipid in the form of triglycerides. Brown adipose tissue consists of smaller cells whose cytoplasm is characterized by numerous lipid droplets that occupy much of each cell's volume. Both adipose tissues are very richly vascularized.

ORIENTATION MICROGRAPH: This micrograph shows white adipose tissue from the hypodermis of skin. It consists of numerous adipocytes closely packed in lobules. **Dense irregular connective tissue** (1) surrounds the adipose tissue. When observed in a typical H&E section, the loss of the lipid from within the cells gives adipose tissue a mesh-like appearance. Note the small **blood vessels** (2) observed at the periphery of the tissue. They provide a rich capillary network within the adipose tissue. Several **sweat gland ducts** (3) are also present in the connective tissue between the fat lobules.

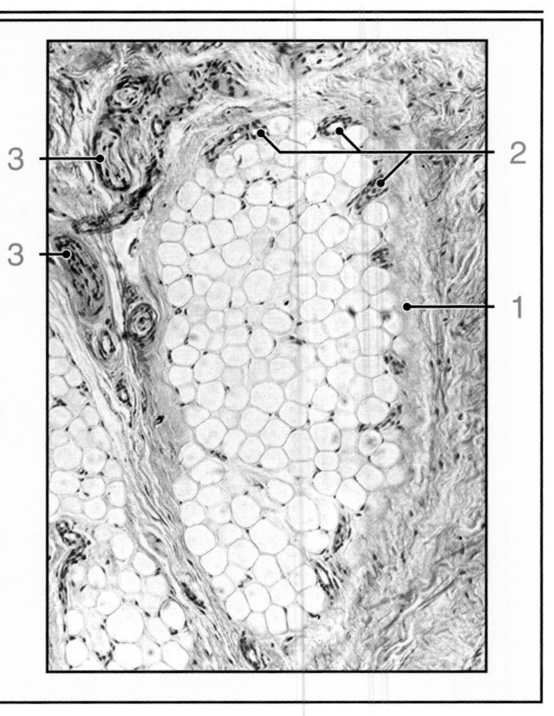

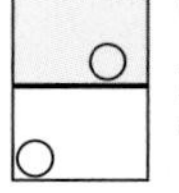

White adipose tissue, human, H&E, x363; inset x700.

This is a higher magnification micrograph of white adipose tissue from the specimen shown in the orientation micrograph. It reveals portions of several lobules of adipose cells. **Dense irregular connective tissue** (1) separates the lobules from surrounding structures. In well-preserved specimens, the **adipocytes** (2) have a spherical profile in which they exhibit a very thin rim of cytoplasm surrounding a single, large, lipid droplet. Because the lipid is lost during tissue preparation, one only sees the rim of cytoplasm and an almost clear space. Between the cells, there is an extremely thin, delicate connective tissue stroma holding the adipocytes together. Within this stroma are small **blood vessels** (3)—mostly capillaries and venules. The majority of nuclei observed in white adipose tissue belong to fibroblasts, adipocytes, or cells of small blood vessels. Distinguishing between fibroblast nuclei and adipocyte nuclei is often difficult. The **inset** shows an adipocyte whose **nucleus** (4) is relatively easy to identify. It appears to reside within the rim of **cytoplasm** (5), giving the adipocyte the classic "signet ring" appearance. A second **nucleus** (6), partially out of the plane of section, appears to reside between the cytoplasmic rim of two adjacent cells. This is probably the nucleus of a fibroblast. Because of the relatively large size of adipocytes, the nucleus of these cells is frequently not included in the plane of section of a given cell. Other cells that may be seen within the delicate connective tissue stroma are **mast cells** (7).

Brown adipose tissue, human, H&E, x450; inset x1100.

The brown adipose tissue shown here consists of small cells that are very closely packed, with minimal intercellular space. Because of this arrangement, it is hard to define individual cells at this magnification. One cell, whose boundary was identified at higher magnification, is circumscribed by a *dotted line*. The **nucleus** (8) of this cell is visible in this view. Each cell contains many small, lipid droplets embedded in the cytoplasm. Brown adipose tissue is highly vascularized. Note the numerous **blood vessels** (9), evidenced by the red blood cells that they contain. It is even more difficult to distinguish fibroblast nuclei from nuclei of the adipocytes within the lobule. Even at higher magnification (**inset**), it is difficult to determine which nuclei belong to which cells. A **capillary** (10) can be identified in the inset. Again, it is recognized by the presence of red blood cells. Where the lobules are **slightly separated** (11) from one another, small elongate nuclei can be recognized. These belong to fibroblasts in the connective tissue that forms the septa.

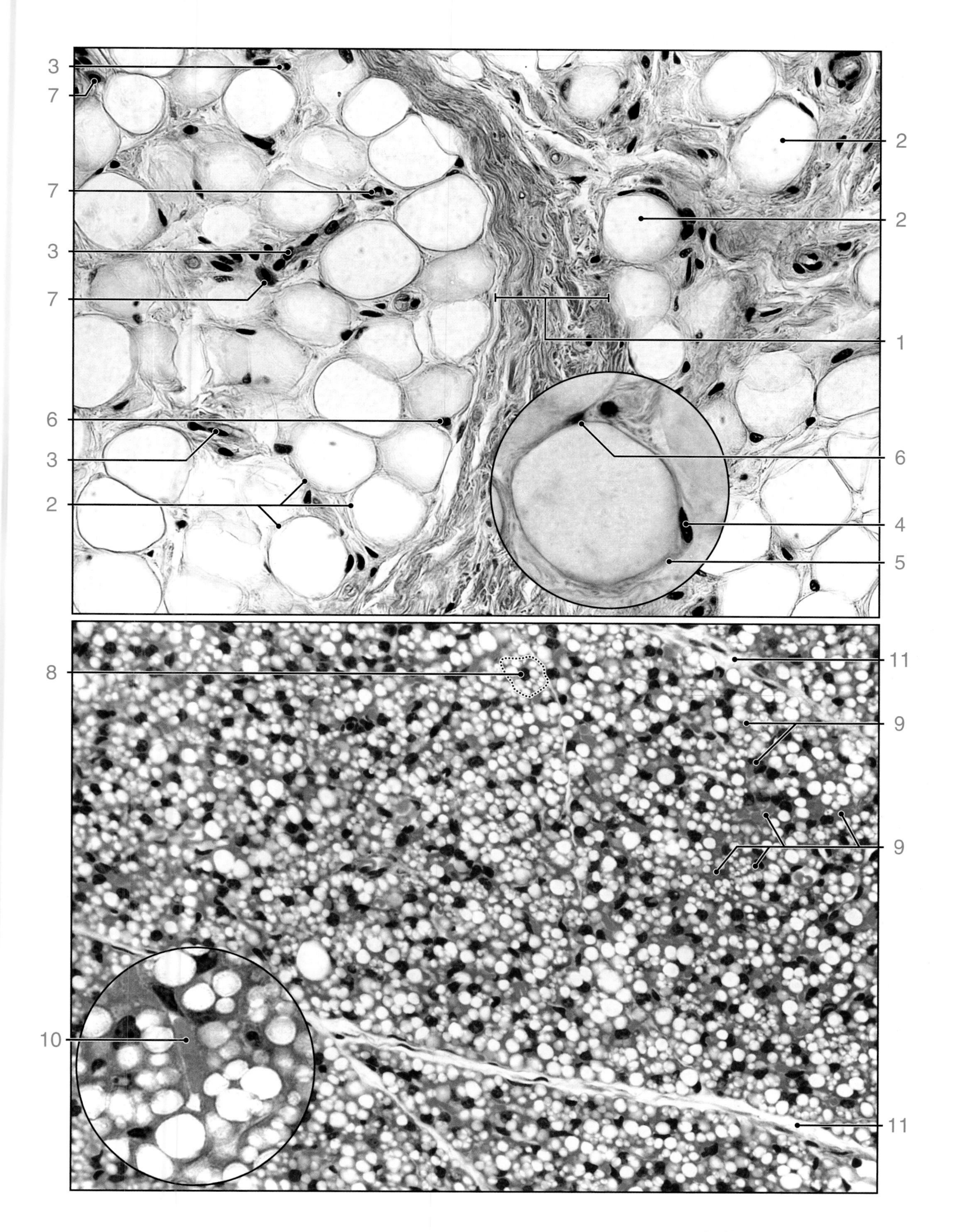
3
7
7
3
7
6
3
2
2
2
1
6
4
5
8
11
9
9
10
11

PLATE 12. ADIPOSE TISSUE II

White adipose tissue, rat, H&E, x1325.

The adipose tissue shown here is from a well-preserved, glutaraldehyde-fixed specimen embedded in plastic. The lipid content of the adipose cells, like that of formaldehyde, paraffin-embedded tissue, has been lost in preparation, thus the empty-appearing spaces. The cytoplasm is well defined appearing as a thin rim of somewhat variable thickness. In some places, exceedingly small, empty-appearing **vesicles** (1) can be seen in the cytoplasm. These vesicles contain lipid and will ultimately coalesce forming larger vesicles. Because of the extremely large size of these cells, their **nuclei** (2) are only occasionally observed. The intercellular space is occupied by a number of connective tissue cells such as **fibroblasts** (3) and occasional **mast cells** (4). Several other connective tissue cells are seen in this preparation, but are difficult to identify with certainty. **Collagen fibers** (5) are represented in the interstitium by the light staining material. Note also the presence of **capillaries** (6) and **venules** (7) within the stroma of the adipose tissue.

White adipose tissue, rat, electron micrograph, x15,000; upper inset x65,000; lower inset x30,000.

This micrograph shows portions of two adjacent adipose cells and the thin processes of several **fibroblasts** (8). In the cytoplasm of the adipocytes, **mitochondria** (9) and **glycogen** (10) are visible. The glycogen appears as fine, very dense, black granules. The **upper inset** shows the **attenuated cytoplasm** (11) of two adjoining adipose cells. Each cell is separated by a narrow space containing **basal lamina material** (12). The **lower inset** shows the **basal lamina** (13) of the adipose cells as a discrete layer at a site where the two cells are obviously separate from one another.

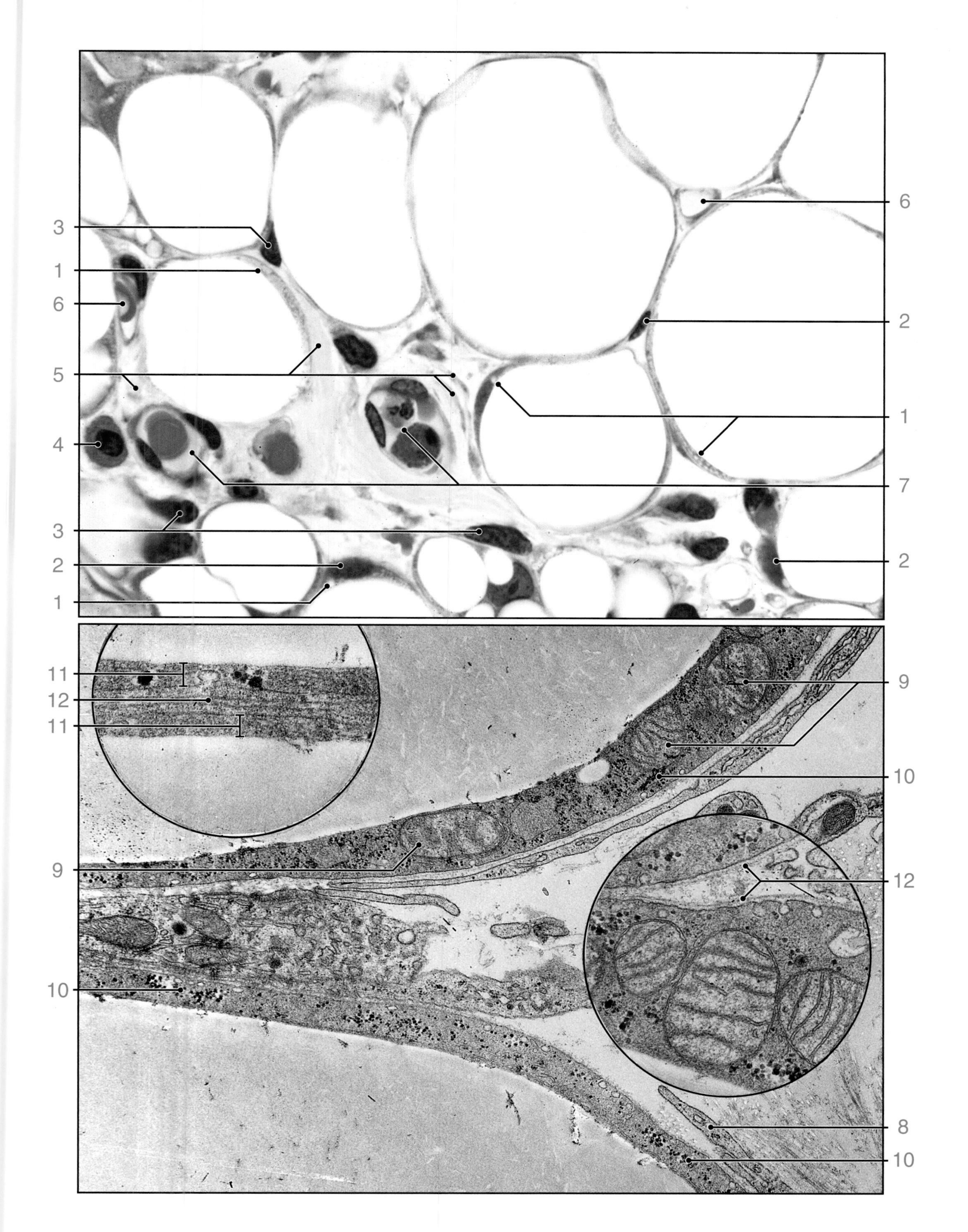
3
1
6
5
4
3
2
1
6
2
1
7
2
11
12
11
9
10
9
10
12
8
10

CHAPTER 4
Cartilage

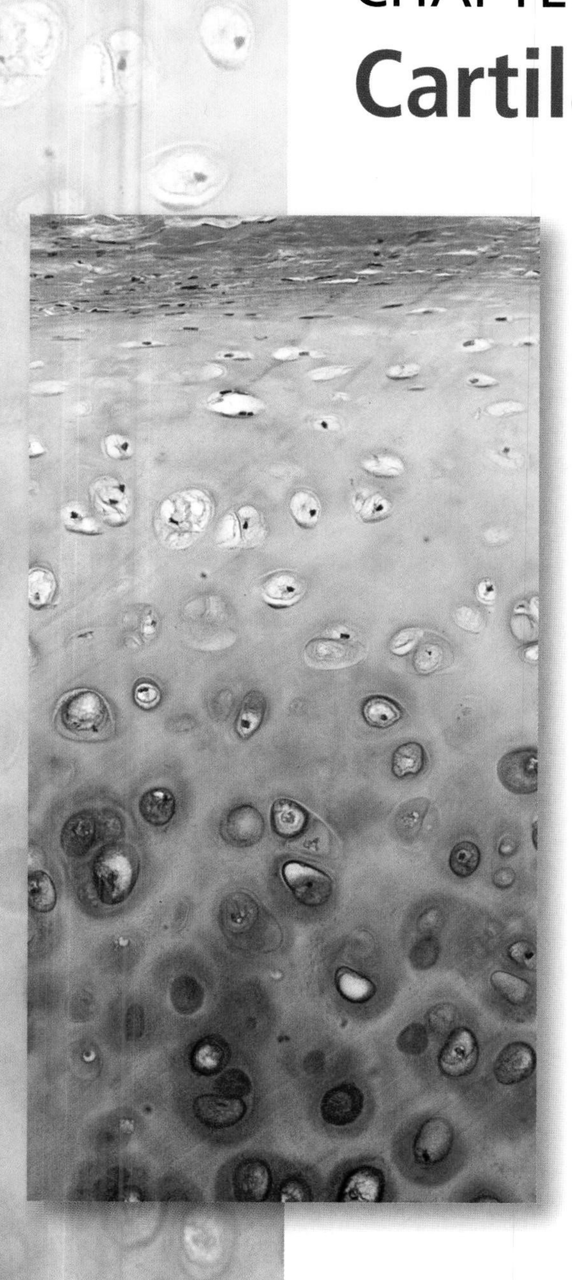

PLATE 13. HYALINE CARTILAGE I

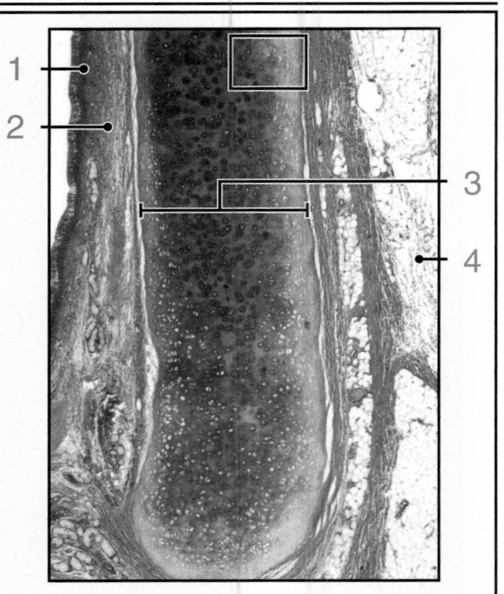

Hyaline cartilage is a bluish white, avascular tissue that is solid to the touch, but is pliable and resilient to deformation. It consists of cells called chondrocytes surrounded by an amorphous matrix that contains type II collagen fibrils; and it is rich in proteoglycan aggregates and multiadhesive glycoproteins. Also, it is highly hydrated—more than half its net weight consists of water, most of which is bound to proteoglycan aggregates. This bound water gives the cartilage its resilient character. Some of the water is not bound and provides a pathway for diffusion of metabolites to and from the chondrocytes contained within the matrix.

Hyaline cartilage, at most sites, is surrounded by dense irregular connective tissue, the perichondrium. When cartilage is actively growing, the part closest to the hyaline cartilage matrix (the inner cellular layer of the perichondrium) contains formative chondrocytes, which have the ability to form new cartilage matrix. The outer fibrous layer of the perichondrium appears similar to the dense irregular connective tissue that forms the capsule of other organs. In addition to cartilage growth from its periphery (appositional growth), hyaline cartilage has the capacity to grow from within by the division of existing chondrocytes (interstitial growth). The newly divided cells go on to produce additional cartilage matrix, thus expanding the volume of the cartilage.

ORIENTATION MICROGRAPH: This micrograph shows the end portion of one of the C-shaped rings of the trachea. Included in this view is the **epithelial lining** (1) of the trachea. Beneath that are **dense irregular connective tissue** (2) and the **hyaline cartilage** (3) of the tracheal ring. On the opposite side, beyond the connective tissue sheath, is **adipose tissue** (4).

Trachea, hyaline cartilage, human, H&E, x180; inset x 550.

This micrograph is a higher magnification of the area within the rectangle in the orientation micrograph. The **perichondrium** (1) consists of dense irregular connective tissue and stains with eosin. The remainder of the micrograph consists of cartilage, which has a greater affinity for hematoxylin. The inner layer of the perichondrium shows a transition in which **formative chondrocytes** (2) are present. They are in the early stage of matrix production. The **inset** shows a formative chondrocyte at higher magnification. The nucleus is less elongate and a hint of cytoplasm can be seen at either end. Based on the homogenous, lighter-staining extracellular material in the immediate vicinity of the cell, cartilage matrix is being produced. Slightly further into the cartilage matrix, a **chondrocyte** (3), whose nucleus has a somewhat ovoid appearance, is present. These cells are responsible for appositional growth of the cartilage. The remainder of the lighter staining region of cartilage matrix shows chondrocytes with round nuclei. The **cytoplasm** (4) of these cells is not well preserved and gives the impression of an empty space. Deeper into the cartilage, the matrix becomes increasingly basophilic, and the **chondrocytes** (5) are larger. The matrix surrounding the chondrocytes in this region, stains more intensely. This is the **capsular**, or **pericellular**, **matrix** (6). It contains the greatest concentration of sulfated proteoglycans, hyaluronan biglycans, and several multiadhesive glycoproteins. Also present is type VI collagen, which binds the matrix to the chondrocyte. The remainder of the densely staining matrix is the **territorial matrix** (7). It contains a network of randomly arranged type II collagen fibrils and some type IX collagen. Some of the chondrocytes are seen in close apposition to one another and are surrounded by a common territorial matrix. These cells are referred to as **isogenous groups** (8). They represent cells that have recently divided. As they mature and produce additional matrix, they move apart and become surrounded by their own territorial matrix. This division and production of new matrix material is responsible for interstitial growth. The remaining, lighter staining cartilage matrix that occupies the space between chondrocytes is referred to as the interterritorial matrix.

Vertebra, hyaline cartilage, human, H&E, x160.

This micrograph shows the hyaline cartilage of a **developing vertebral body** (9) and its **transverse process** (10). The chondrocytes of the vertebral body, which is rapidly growing, are small and in close proximity to one another. Relatively little matrix has been produced at this stage. The **chondrocytes** (11) in the part of the transverse process more distal from the vertebral body are larger and have produced more matrix material. This portion of the transverse process will ultimately be replaced by bone; the cartilage only serves as a model for the developing vertebrae. The same process has begun in the part of the vertebra that will form the **vertebral arch** (12).

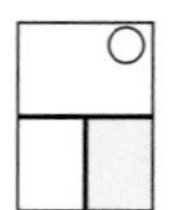

Vertebra, hyaline cartilage, H&E, x340.

The hyaline cartilage of the transverse process of the developing vertebrae is shown in this micrograph at higher magnification. Careful examination shows thin strands of **cartilage matrix** (13) surrounding the chondrocytes. It has a bluish coloration. The beginning of **bone formation** (14) is seen in the upper part of the micrograph. Endochondral bone formation will be explained further on Plate 20.

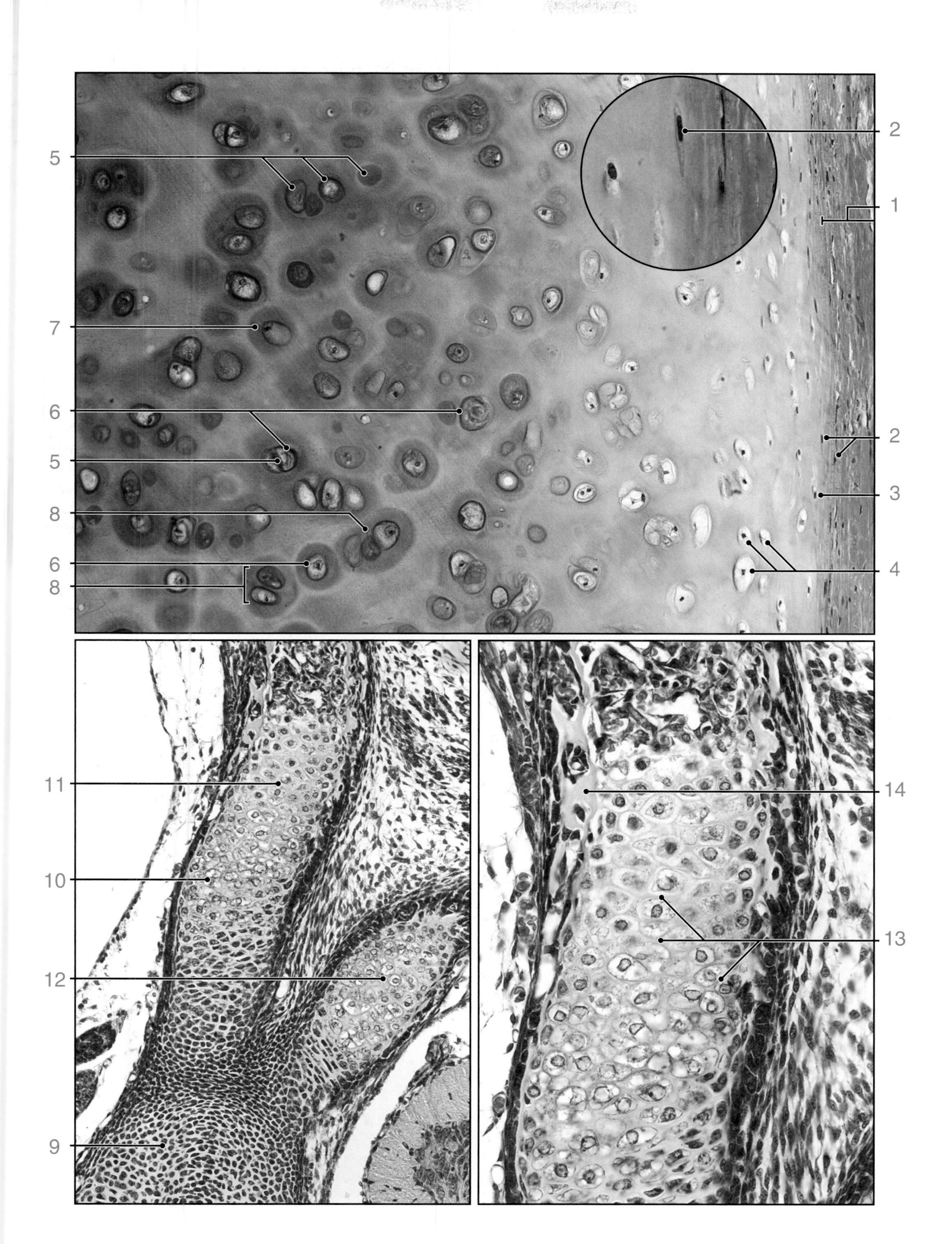
2
5
1
7
6
2
5
3
8
6
4
8
11
14
10
13
12
9

PLATE 14. HYALINE CARTILAGE II, ELECTRON MICROSCOPY

ORIENTATION MICROGRAPH: The electron micrograph shown here corresponds to the area in the inset on the upper micrograph of the previous plate. The **perichondrium** (1) consists of fibroblasts and collagen fibrils that occupy the intercellular space. The fibroblast nuclei are flattened and surrounded by scant cytoplasm, but possess long, thin, sheetlike processes that extend between the collagen fibrils. Although these cells are morphologically typical of fibroblasts in appearance, they are referred to as perichondrial cells because they constitute the cellular component of the perichondrium. In contrast, the mature **chondrocytes** (2) exhibit round or ovoid nuclei with a moderate amount of surrounding cytoplasm. The cartilage shown here is in a growing state, thus it is possible to point to the changes that occur during the process of appositional growth. Between these two well-defined layers, the perichondrium and the cartilage proper, several cells are present that indicate a transition from the morphologically identifiable **fibroblast** (3) to a **young cartilage cell** (4). A later stage, **differentiated cartilage cell** (5) is also visible. This cartilage cell still possesses an elongate nucleus, but it also possesses more cytoplasm relative to the nucleus.

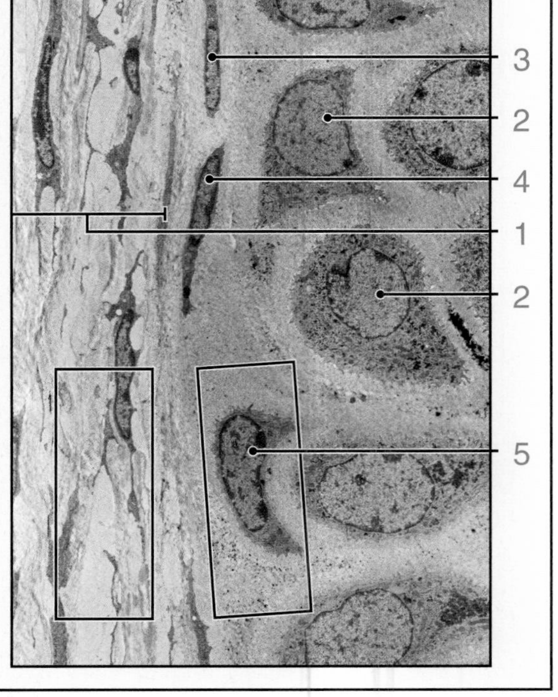

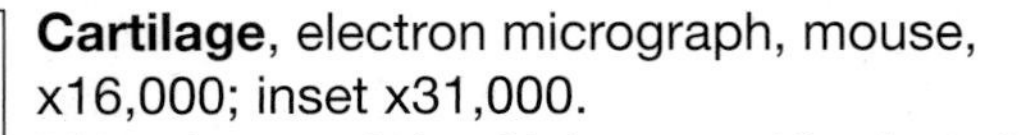

Cartilage, electron micrograph, mouse, x16,000; inset x31,000.

This micrograph is a higher magnification of the area within the rectangle on the left in the orientation micrograph. The perichondrium shown here reveals portions of several fibroblasts, or perichondrial cells. One of the cells displays multiple arrays of **rough endoplasmic reticulum** (1), a feature characteristic of fibroblasts. The surface of the cell is relatively smooth. Also characteristic of fibroblasts are the long, tenuous **cytoplasmic processes** (2) that extend between the bundles of collagen fibrils. The **lower inset** is a higher magnification of the area enclosed by the circle. It clearly reveals collagen fibrils cut in cross-section. Also present are several profiles of **elastic fibers** (3) within the matrix. They are comprised of an electron-lucent core material and numerous surrounding electron-dense microfibrils.

Cartilage, electron micrograph, mouse, x16,000; inset x31,000.

This micrograph is a higher magnification of the area within the rectangle on the right in the orientation micrograph. The cell shown here is a young chondrocyte. Small, irregular **projections** (4) on the surface of the newly differentiated chondrocyte give the cell a ruffled, or scalloped appearance. These projections first appear when cartilage matrix formation is initiated. Other than the accumulation of lipid and **glycogen** (5), the cytoplasm of the cartilage cell does not appear appreciably different from that of a fibroblast. Rough endoplasmic reticulum is particularly well developed in the active chondrocyte but, because of the particular plane of this section, it is not visible in the chondrocyte seen here. This micrograph displays a fairly prominent portion of the cell's **Golgi apparatus** (6). The area enclosed by the circle is seen at higher magnification in the **upper inset**. It reveals extremely fine (5–20 nm) matrix fibrils. These fibrils consist primarily of type II collagen. They have a smaller diameter and a less discernable periodic banding pattern than type I collagen. Note how the collagen fibrils here give a fine, stippled appearance compared to the type I collagen in the perichondrium seen in the lower inset. The relative homogeneity of the cartilage matrix, as seen here, accounts for its amorphous appearance when observed in the light microscope.

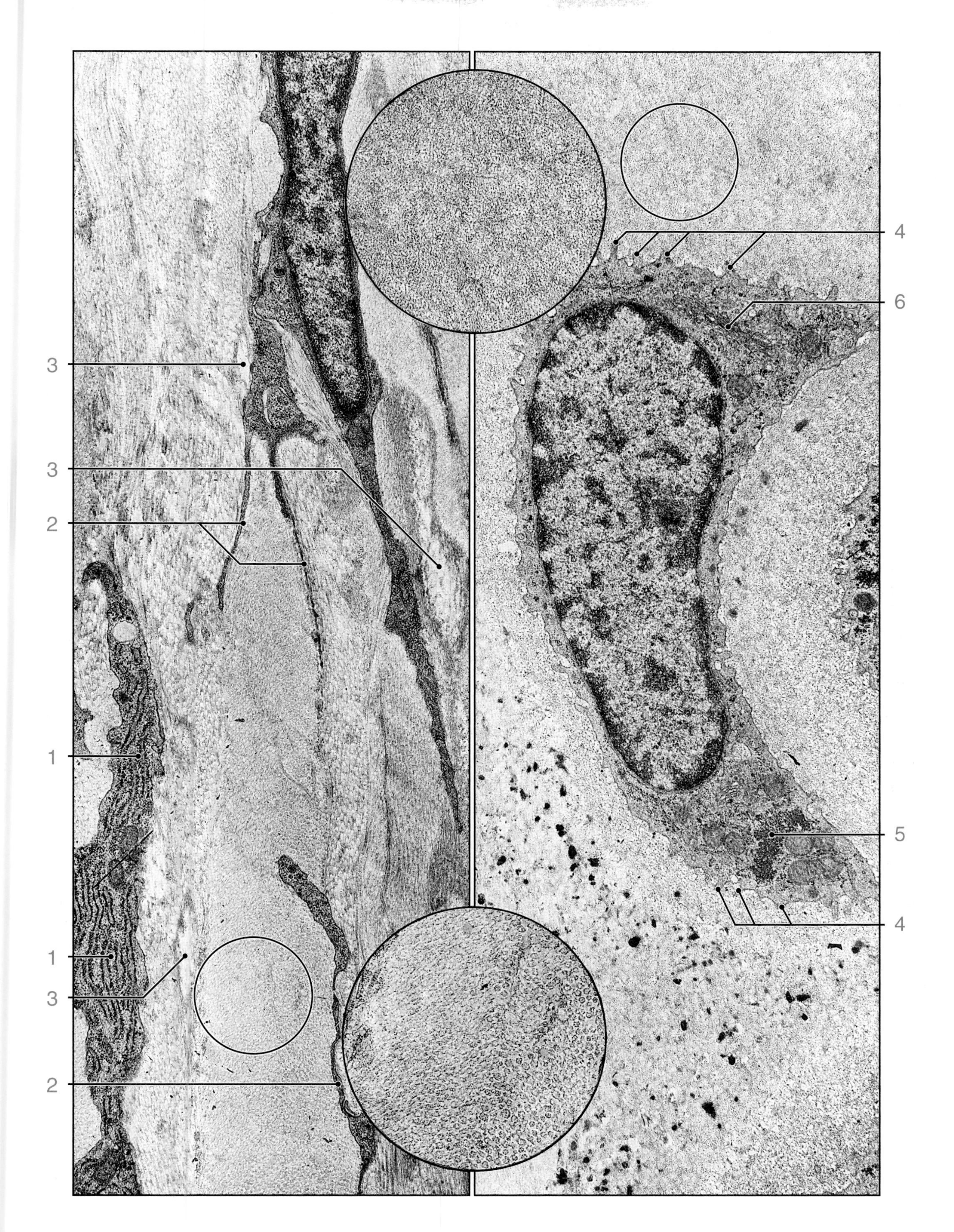

PLATE 14. HYALINE CARTILAGE II, ELECTRON MICROSCOPY

PLATE 15. HYALINE CARTILAGE III AND THE DEVELOPING SKELETON

During early fetal development, hyaline cartilage is the precursor of bones that develop by the process of endochondral bone formation (see Plate 20). The cartilage initially takes a shape resembling that of the mature bone. As development and growth of the individual proceeds, the cartilage initially serves as a model of the future bone, and most of the cartilage is replaced by bone. The cartilage that will remain throughout the growing process in long bones is the epiphyseal disk, the site where new cartilage is produced to allow growth, as well as the articular surfaces of the bone. Growth of the bone stops when the cartilage in the epiphyseal disc no longer produces new cartilage. The remaining sites of the hyaline cartilage are on the articular surfaces of the bone and within the rib cage, where it remains as the costal cartilages. Articular cartilages provide a smooth and well-lubricated surface against which the end of one bone moves against the other bone in a joint.

ORIENTATION MICROGRAPH: This micrograph is a horizontal section through a developing thoracic vertebra. The **spinal cord** (1) is partially surrounded by vertebral cartilage at this early stage. The recognizable components of the vertebra consist of the developing **vertebral body** (2) and its **costal processes** (3). Also evident in the micrograph is a portion of the **developing lung** (4).

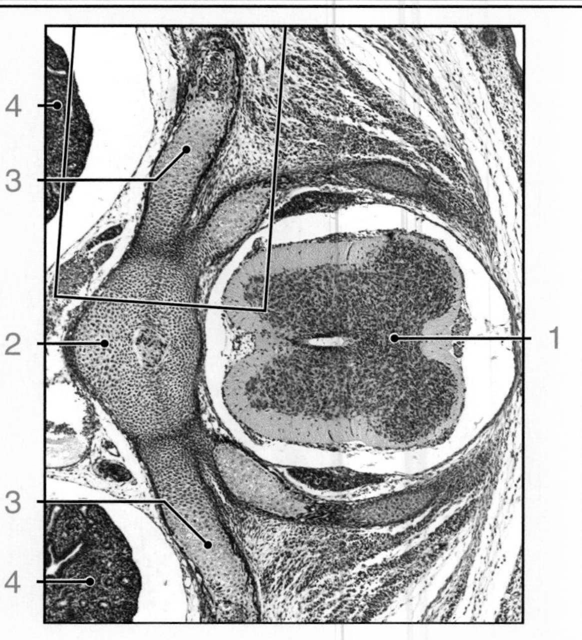

Hyaline cartilage, human fetus, H&E, x125.
This micrograph is a higher magnification of the area in the rectangle of the orientation micrograph. It shows the **transverse process** (1), part of the **vertebral body** (2), and a small portion of the **spinal cord** (3). The distal portion of the costal process shows **bone tissue** (4) in the process of being produced at the surface of the cartilage mass. The cells that produced the bone tissue arose from the **mesenchymal osteoprogenitor cells** (5) surrounding the cartilage of the costal process.

Hyaline cartilage, human fetus, H&E, x270.
This micrograph is a higher magnification of the distal part of the costal process. The amorphous-looking material at the surface of the cartilage is **bone matrix** (6). Cells that appear to be embedded in the bone matrix are **osteocytes** (7)—the cells that maintain the bone matrix. Concomitantly, in the area of bone production, the cartilage matrix is in the process of degenerating and is being replaced by blood vessels that have migrated in from the mesenchyme, along with other mesenchymal cells. The **inset** is a higher magnification of the area within the rectangle on this micrograph. It shows clearly the **bone matrix** (6) and several **blood vessels** (8) that have been incorporated into the bone.

Hyaline cartilage, human fetus, H&E, x180.
This micrograph is a cross section of a developing long bone in a fetus. At this stage of development, the bulk of the structure consists of **hyaline cartilage** (9). The earliest formed cartilage cells are relatively large, but younger cartilage cells, seen at the periphery of the cartilage mass, are small and homogeneous. At the stage shown here, **bone tissue** (10) is being produced at the surface of the cartilage. The bone matrix is readily recognized by its eosinophilic appearance. The cells that are engulfed by the bone matrix are the **osteocytes** (11). The continued accretion of bone tissue is produced by **osteoblasts** (12). These cells develop from the surrounding mesenchyme; they are called osteocytes once they have produced and become surrounded by bone matrix. The osteoblasts are recognized by the round or cuboidal shape of their nuclei and their linear distribution at the site where they are active in producing bone matrix. Another feature of interest in this micrograph is the presence of a **developing muscle** (13), seen in proximity to the developing bone. The muscle at this stage consists of a linear arrangement of cells (myoblasts) that have been cut in cross-section. For further understanding and orientation of developing muscle fibers see Plate 34.

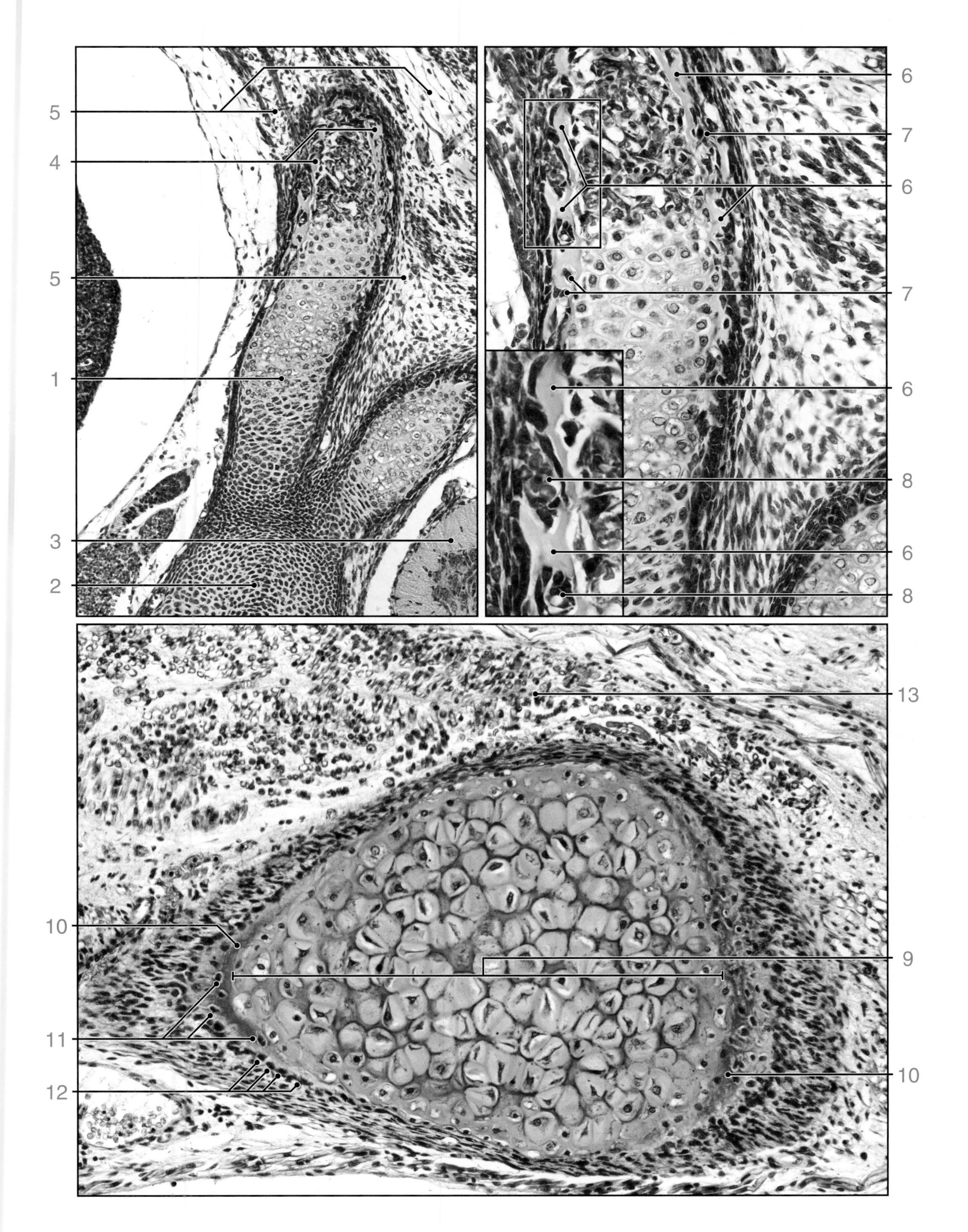

PLATE 15. HYALINE CARTILAGE III AND THE DEVELOPING SKELETON

PLATE 16. **FIBROCARTILAGE AND ELASTIC CARTILAGE**

Fibrocartilage consists of dense irregular connective tissue and, in varying amounts, cartilage. It occurs in the intervertebral disc, pubic symphysis, menisci of the knee joint, temporomandibular joint, sternoclavicular and shoulder joints, as well as at junctions of certain tendons and ligaments with bone. The cartilage cells are present in small groups (isogenous groups) interspersed among collagen fibers. The presence of cartilage matrix among the collagen fibers helps absorb sudden physical impact. The ability of the cartilage to compress and absorb shearing forces reduces excessive force on the collagen fibers. Unlike hyaline cartilage, there is no perichondrium.

Elastic cartilage contains elastic fibers and elastic lamellae, in addition to the typical components found in hyaline cartilage. The auricle of the ear, the auditory tube, the epiglottis, and part of the larynx contain elastic cartilage. The elastic fibers in elastic cartilage provide the cartilage matrix with increased resilience to shearing force, whereas in hyaline cartilage, the matrix only provides resilience to physical force. Also, hyaline cartilage normally calcifies with age, whereas elastic cartilage resists this process. Like hyaline cartilage, elastic cartilage possesses a perichondrium.

ORIENTATION MICROGRAPH: The *upper orientation micrograph* shows part of the disc-like meniscus of the knee joint, including a surface. The *lower orientation micrograph* shows a sagittal section of the epiglottis.

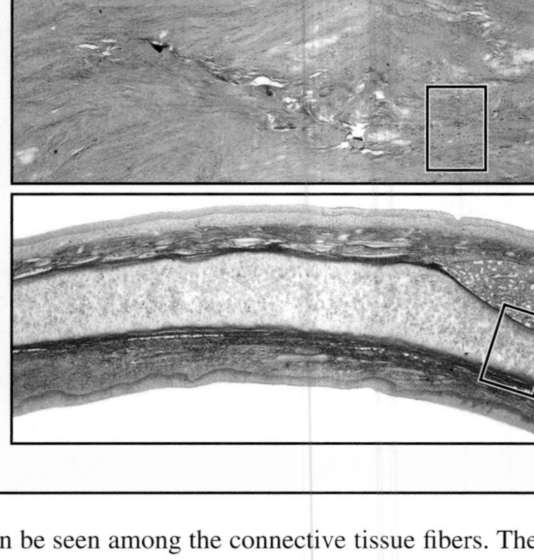

Meniscus, human, Masson stain, x114.
This low magnification micrograph is from the boxed area in the upper orientation micrograph. Note the wavy appearance of the collagen fibers that make up the bulk of the tissue. A number of **nuclei** (1), mostly in small isogenous groups, can be seen among the connective tissue fibers. These are chondrocyte nuclei. Fibroblasts are also present but are much fewer in number. When identified (preferably at higher magnification), fibroblast nuclei exhibit an elongate shape and are widely dispersed.

Meniscus, human, Masson stain, x1030.
The area within the left hand box on the top left micrograph is shown here at higher magnification. It shows a cluster of chondrocytes surrounded by **collagen fibers** (2). Most of the **chondrocyte nuclei** (3) have a round profile. The **cytoplasm** (4) of these cells is relatively scant. Because of the eccentric location of the nuclei, varying amounts of cytoplasm are observed relative to the nucleus of any given cell. Most of the nuclei exhibit both euchromatin and heterochromatin, a reflection of the activity of these cells. Interspersed between the chondrocytes is **cartilage matrix** (5). It is lightly stained.

Meniscus, human, Masson stain, x1030.
The area within the right hand box on the top left micrograph is shown here at higher magnification. In this view, it is possible to recognize fibroblasts based on the shape of their nuclei. The **fibroblast nuclei** (6) have a typical elongate profile. The chondrocyte has a **round nucleus** (7) and exhibits both euchromatin and heterochromatin. In contrast, the dark staining of the fibroblast nuclei demonstrates mostly heterochromatin, a reflection of the less active state of these cells. The chondrocyte nucleus is surrounded by a small amount of **cytoplasm** (8). In this micrograph, it is also possible to distinguish the **cytoplasmic processes** (9) of the fibroblasts, which appear as thin, blue-black strands among the collagen fibers.

Epiglottis, human, Mallory-Azan, x113.
This micrograph is a higher magnification of the boxed area within the lower orientation micrograph. The **elastic cartilage** (10) that forms the epiglottis is covered by a **perichondrium** (11), which is similar to the perichondrium found in hyaline cartilage. The Mallory-Azan stain colors the collagen fibers of the perichondrium a dark blue. The matrix of the elastic cartilage contains elastic fibers, which are stained yellow, and cartilage, which is stained pale blue.

Epiglottis, human, Mallory-Azan, x320.
This is a higher magnification of the boxed area in the bottom left micrograph. The micrograph shows the **perichondrium** (12), consisting of dense connective tissue, and several **fibroblasts** (13) with small, elongate nuclei. The nuclei seen near the interface of the cartilage matrix will give rise to chondrocytes as the cartilage grows. The remainder of the micrograph, beneath the perichondrium, consists of elastic cartilage. The elastic material stains orange and ranges from very thin fibers to thick lamellae. The **chondrocytes** (14) in the deeper portion of the cartilage are large and, with the exception of the red staining nuclei, appear empty. The "emptiness" is due to the loss of accumulated lipid in the cytoplasm and the typical, poor preservation of the cytoplasm.

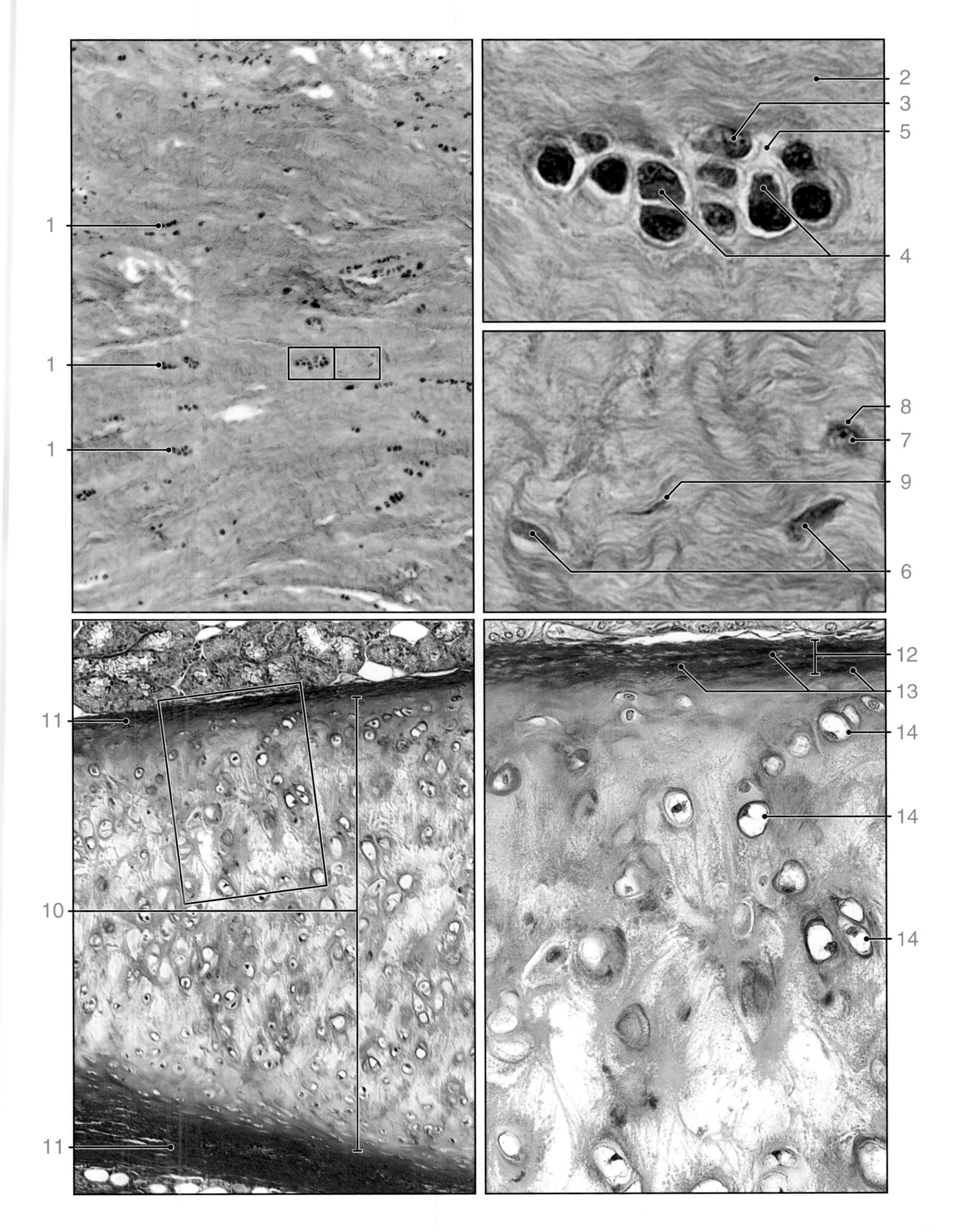

PLATE 16. FIBROCARTILAGE AND ELASTIC CARTILAGE

CHAPTER 5
Bone

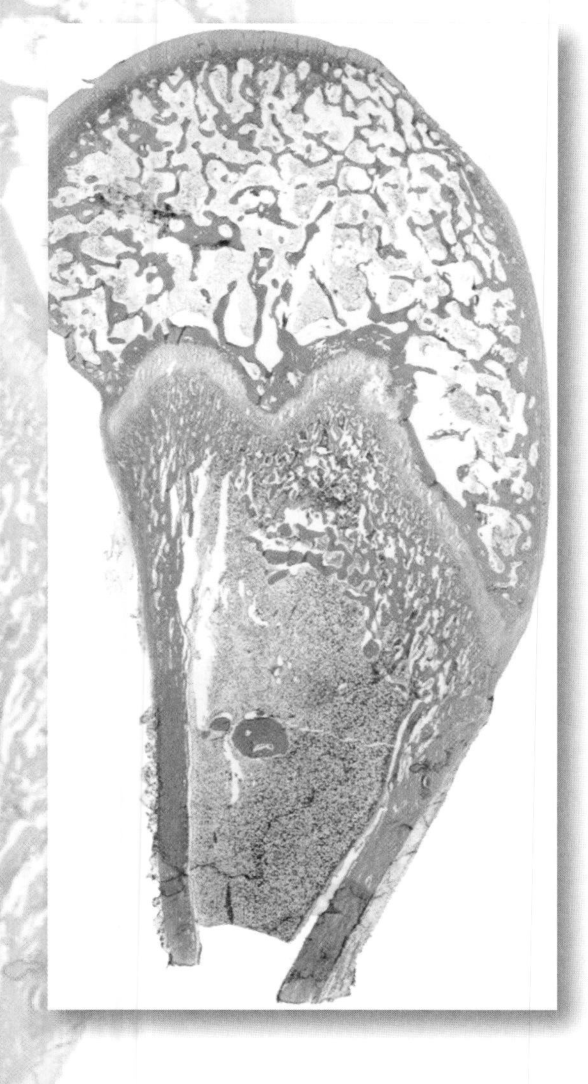

PLATE 17. BONE AND BONE TISSUE

Bone represents one of the specialized connective tissues. It is characterized by a mineralized extracellular matrix. It is the mineralization of the matrix that sets bone tissue apart from the other connective tissues and results in an extremely hard tissue that is capable of providing support and protection to the body. The mineral is calcium phosphate in the form of hydroxyapatite crystals. Bone also provides a storage site for calcium and phosphate. Both can be mobilized from the bone matrix and taken up by the blood as needed to maintain normal levels. Bone matrix contains type I collagen and, in small amounts, a number of other types of collagen (i.e., types V, III, XI, and XIII). Other matrix proteins that constitute the ground substance of bone are also present, such as proteoglycan macromolecules, multiadhesive glycoproteins, growth factors, and cytokines. Bone is typically studied in histological preparations by removing the calcium content of the bone (decalcified bone), thus allowing it to be sectioned like other soft tissues.

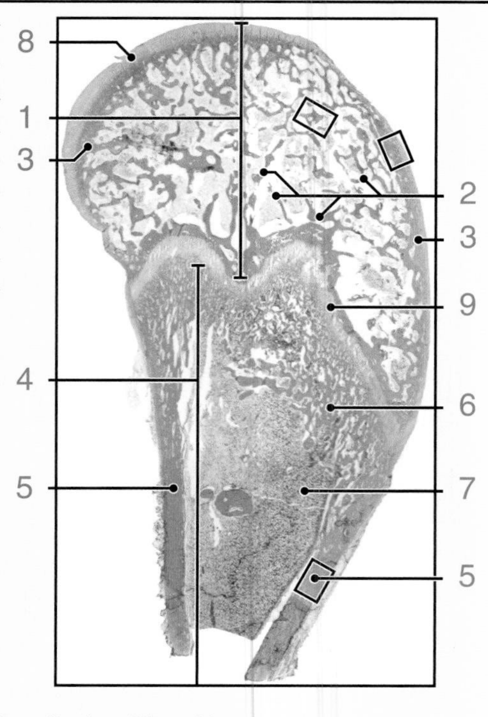

ORIENTATION MICROGRAPH: The orientation micrograph shows the proximal end of a decalcified humerus from an infant. The interior of the head of the bone, the **epiphysis** (1), consists of spongy cancellous bone made up of an anastomosing network of **trabeculae** (2) in the form of spicules of bone tissue. The outer portion consists of a dense layer of bone tissue known as **compact bone** (3). Its thickness varies in different parts of the bone. The shaft of this bone, the **diaphysis** (4), is also made up of **compact bone** (5) and, in its interior, **spongy bone** (6). Also within the shaft of the bone is **bone marrow** (7), which at this stage of life is in the form of hemopoietic tissue. Cartilage is also a component of the bone, present as an **articular surface** (8) and as a **growth plate** (9). The latter is described on Plate 21.

Bone, human, H&E, x178.
The region within the top right box on the orientation micrograph, containing compact bone in the epiphysis, is shown here at higher magnification. The lighter staining area is **cartilage** (1). It serves as the articular surface of the epiphysis. Note the presence of isogenous groups of **chondrocytes** (2), a characteristic feature of growing cartilage. Below the cartilage is **bone tissue** (3). It can be distinguished from the cartilage by the arrangement of its cells, the **osteocytes** (4). The osteocytes lie within the bone matrix, but are typically recognized only by their nuclei. Because bone matrix is laid down in layers (lamellae), bone characteristically shows linear or circular patterns that appear as striations. The irregular spaces seen within the bone tissue are **vascular channels** (5) that contain bone-forming tissue as well as vessels.

Bone, human, H&E, x135.
Bone from the diaphysis within the bottom right rectangle on the orientation micrograph is shown here at higher magnification. The outer surface of the bone is covered by dense connective tissue known as **periosteum** (6). The remaining tissue in the micrograph is compact bone. The **osteocytes** (7) are recognized by their nuclei within the bone matrix. Another feature worth noting in this growing bone is the presence of bone resorbing cells, known as **osteoclasts** (8). They are large, multinucleate cells found at sites in bone where remodeling is taking place.

Bone, human, H&E, x135.
The area in the top left rectangle on the orientation micrograph, containing spongy bone in the epiphysis, is shown here at higher magnification. Although the bone tissue at this site forms a three-dimensional structure consisting of branching trabeculae, its structural organization and components are the same as compact bone. Note the **osteocyte nuclei** (9). As bone matures, the bone tissue becomes reorganized and forms **osteons** (10), which consist of a central vascular channel and surrounding layers (lamellae) of bone matrix. The two circular spaces are sites in which bone tissue has been removed and will be replaced by new tissue in the form of osteons. The spaces surrounding the trabecular bone contain bone marrow, consisting mainly of adipocytes. Cells that have the capacity to form bone or hemopoietic tissue are also present.

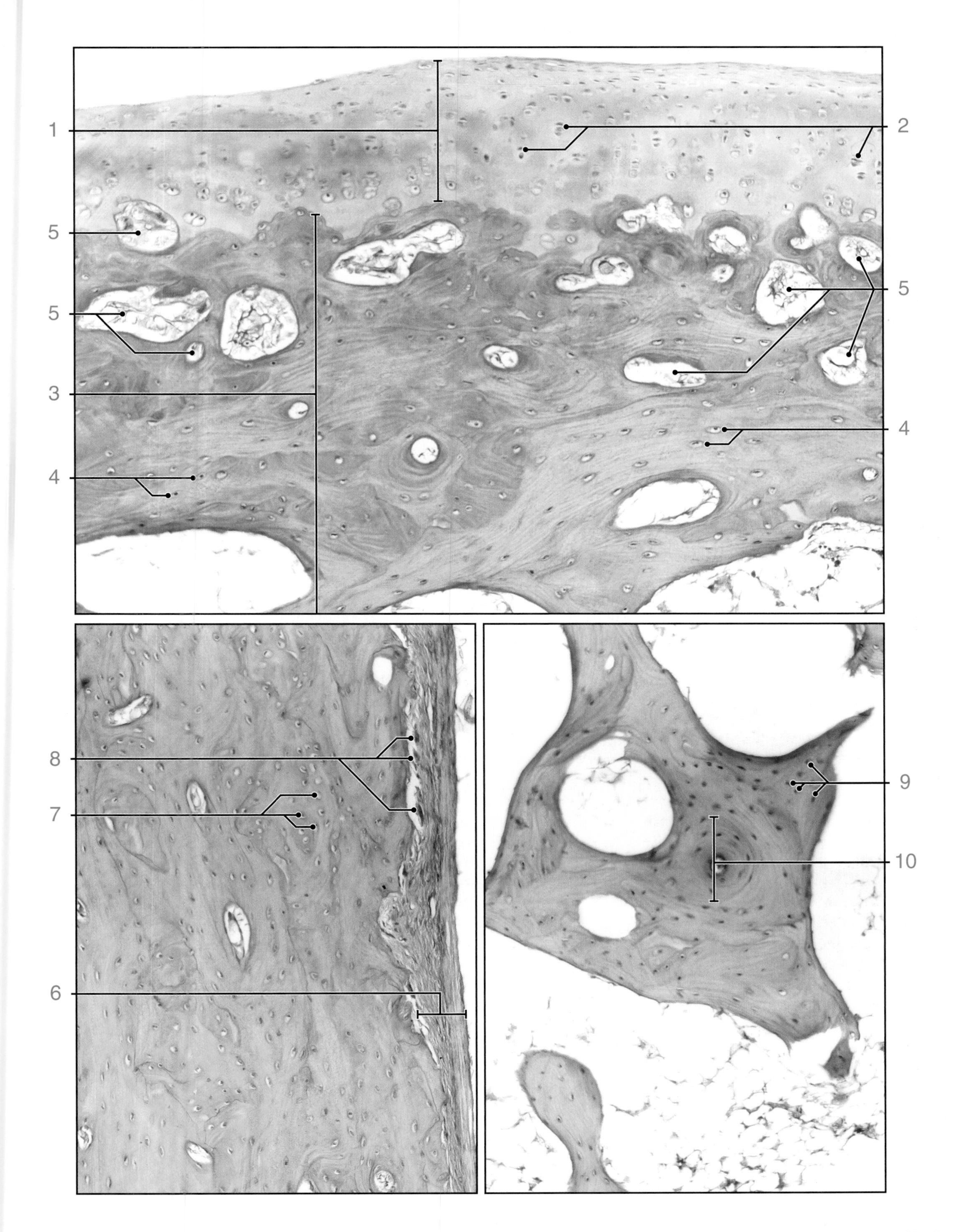

PLATE 17. BONE AND BONE TISSUE

PLATE 18. **BONE, GROUND SECTION**

Bone tissue can also be examined by a procedure in which the tissue is not decalcified. This method, referred to as ground bone, is accomplished by collecting a bone specimen, allowing it to dry out, then cutting thin slices using a fine saw. The dry, cut bone is then ground down, using a flat abrasive surface, to a thinness that can be mounted on a slide and viewed in the light microscope. This procedure results in loss of the cellular components of the bone, but allows better visualization of the structural organization of the bone matrix. When the specimen is placed on the slide and covered by a mounting medium and cover slip, air is trapped within the spaces previously occupied by osteocytes. These spaces then appear black, exposing the site of the osteocyte cell body and its extremely thin processes within the lamellar structure of the bone.

Another technique that permits visualization of the osteocyte is treating decalcified sections of bone with thionin and picric acid (Schmorl's technique). This method causes the lacunae, housing the osteocyte cell body, and the canaliculi, housing the cell processes, to appear red to dark brown. The bone matrix appears yellow, and the osteocyte nuclei appear red. Schmorl's technique is not a true stain, since the color occurring in the lacunae and canaliculi is a precipitate of the coloring material. This precipitation also occurs to a considerable extent in the vascular spaces of the bone, making the structural components difficult to differentiate.

Bone, ground, human, unstained, x90; inset x200.

This micrograph is from a dried long bone that was first cut across its diameter, then a small piece was cut from the outer part of the bone, and this piece was ground to appropriate thinness for observation in the microscope. The outer portion of the bone reveals a peripheral layer of concentrically oriented lamellae of bone matrix referred to as **circumferential lamellae** (1). The **inset** shows the circumferential lamellae at higher magnification. The black elongate structures are **lacunae** (2), spaces that were occupied by osteocytes. Note how the lacunae appear in linear arrays. The surrounding material is the bone matrix. The disposition of the cell layers is correlated with their age. As the bone grew in diameter, new layers of matrix were laid down on the pre-existing layer of matrix, thus the youngest cells are at the periphery of the bone. Beneath the circumferential lamellae, the bone matrix is organized into **osteons** (3) and remnants of pre-existing osteons. Osteons are cylindrical structures oriented in the direction of the long axis of the bone. They consist of concentric layers of mineralized matrix material. At the center of each osteon is an **osteonal (Haversian) canal** (4). Haversian canals contain small blood vessels, lymphatic capillaries, nerve fibers, and a variety of other cell types, including macrophages, osteoprogenitor cells, and osteoclasts. During the growth period of an individual as well as during adult life, there is constant internal remodeling of bone. This involves osteonal destruction and formation of new osteons. Typically, the destruction or breakdown of an osteon is not complete and part of an osteon may remain intact. The remnants of pre-existing osteons are referred to as **interstitial lamellae** (5). The vessels within the Haversian canals are supplied by blood vessels from the marrow via perforating (Volkmann) canals that extend between Haversian canals. A short segment of a **Volkmann's canal** (6) is seen in this specimen. Note that it is oriented perpendicular to the Haversian canals.

Bone, ground, human, unstained, x300.

The osteon within the rectangle of the above micrograph is shown here at higher magnification. The **Haversian canal** (7) appears to contain dried cellular and other precipitated material. Surrounding the Haversian canal are layers of bone matrix in which one can readily see the **lacunae** (8) that contained the osteocytes. Also note the fine black striations that emanate from the lacunae, extending from one layer to the next. These threadlike profiles are canaliculi, spaces within the bone matrix that contained the cytoplasmic processes of the osteocytes. The canaliculi of each lacuna communicate with canaliculi of neighboring lacunae, creating a three-dimensional channel system throughout the bone matrix. At the top of the micrograph is the **Volkmann canal** (9). Note that this vascular space is not surrounded by concentric rings of bone matrix, as is the Haversian canal. This feature identifies it as a Volkmann canal.

Bone, Schmorl's technique, x150; inset x1100.

The technique used for this specimen was designed to show lacunae and canaliculi with intact osteocytes. At this magnification, the **osteons** (10) and **interstitial lamellae** (11) are clearly defined. Like the previous micrograph, the osteons are seen in cross-section with a central **Haversian canal** (12). The material within the canal is very heavily colored, making it difficult to define the structures that are present. Also evident in this micrograph is a **Volkmann's canal** (13) that is coursing through interstitial lamellae and entering an osteon to join the Haversian canal within that osteon. The **inset** shows two **osteocytes** (14) in adjoining lamellae. Note the cytoplasmic processes that extend across the lamellae from each of these osteocytes.

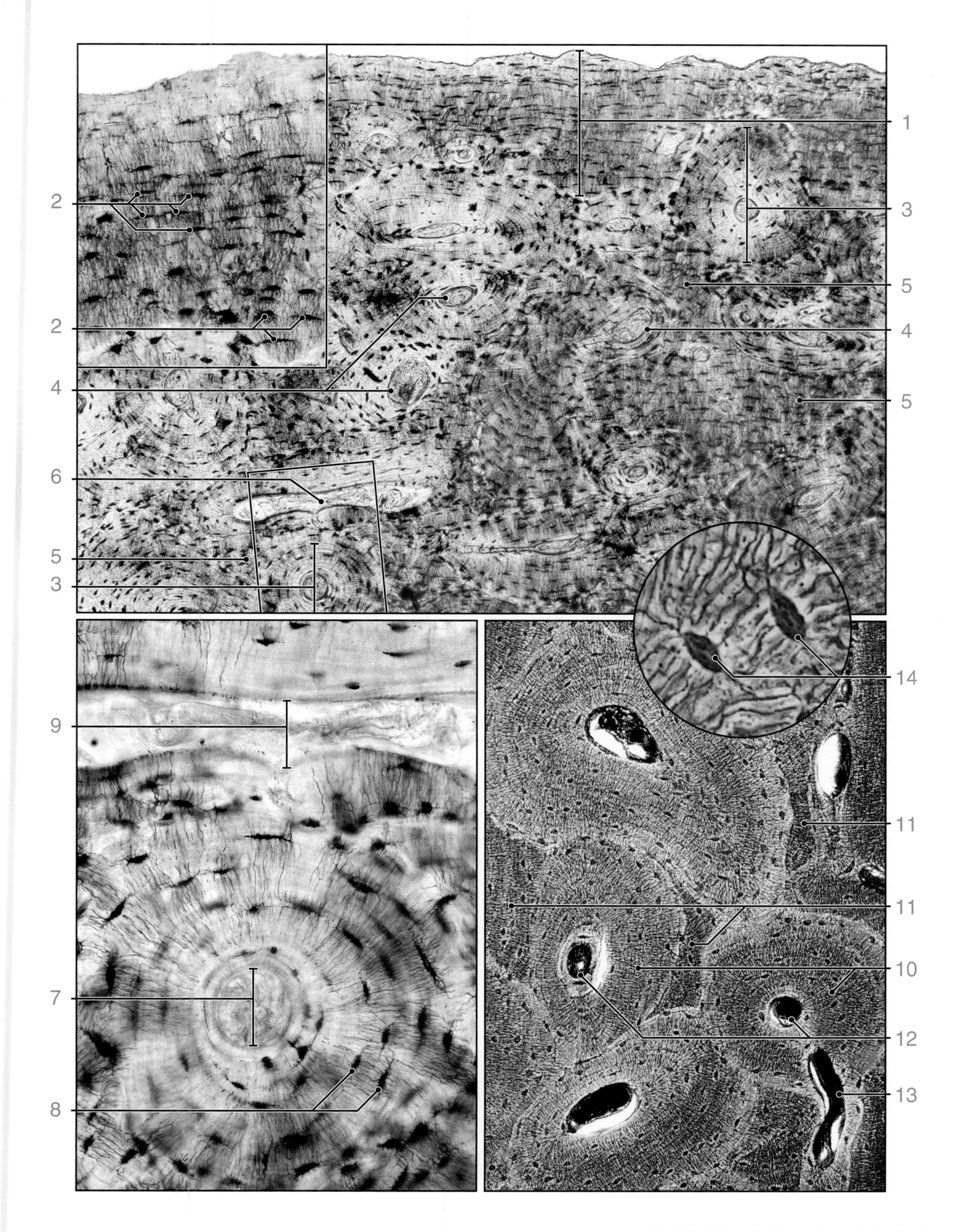

PLATE 18. BONE, GROUND SECTION

PLATE 19. BONE, ELECTRON MICROSCOPY

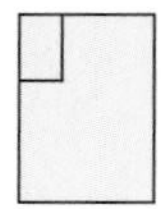

Cortical bone, rabbit, electron micrograph, x3000; inset x10,000.

This electron micrograph shows, at low magnification, an osteon from the femur of a young rabbit. The bone was decalcified, leaving the cells and extracellular soft tissue components (collagen) essentially intact. The particular osteon illustrated here consists of three complete lamellae, numbered (1) to (3) and, immediately surrounding the **Haversian canal** (4), an incomplete lamella (*asterisks*). The outer limit of the osteon is marked by a **cement line** (5), and beyond this is an **interstitial lamella** (6).

The most apparent structures in the Haversian canal are the blood vessels. Both are **capillaries** (7). The smaller vessel is surrounded by pericytes, whereas the larger one does not possess this additional cellular investment. It should be noted that the closely packed **osteoprogenitor cells** (8) around the capillaries can be mistaken easily for smooth muscle cells of larger blood vessels with the light microscope. However, the electron microscope shows that veins and arteries are not present in Haversian canals. The cells within the Haversian canal (other than those associated with the capillaries) are either endosteal or osteoprogenitor cells. **Endosteal cells** (9) line the Haversian canal. Osteoprogenitor cells differentiate into osteoblasts to replace osteoblasts that become incorporated into the bone matrix that they produce. In examining the osteocytes, it should be noted that one of them exhibits a process extending into a canaliculus (*double arrows*) of the incomplete lamella. Similarly, the **osteocyte** (10) in the lower portion of the figure also shows cytoplasmic processes entering a canaliculus (*arrows*). The osteocyte within the boxed area is shown at higher magnification in the **inset**. Note the **cytoplasmic process** (11) extending from the cell body where it is entering a canaliculus.

The **canaliculi** (12) within the lamellae, and those extending across the lamellae, appear less numerous in this micrograph than in a light micrograph (see previous plate). The difference is a reflection of the thickness of the section; more canaliculi are included in a relatively thick light microscope section than in the thin sections used in electron microscopy.

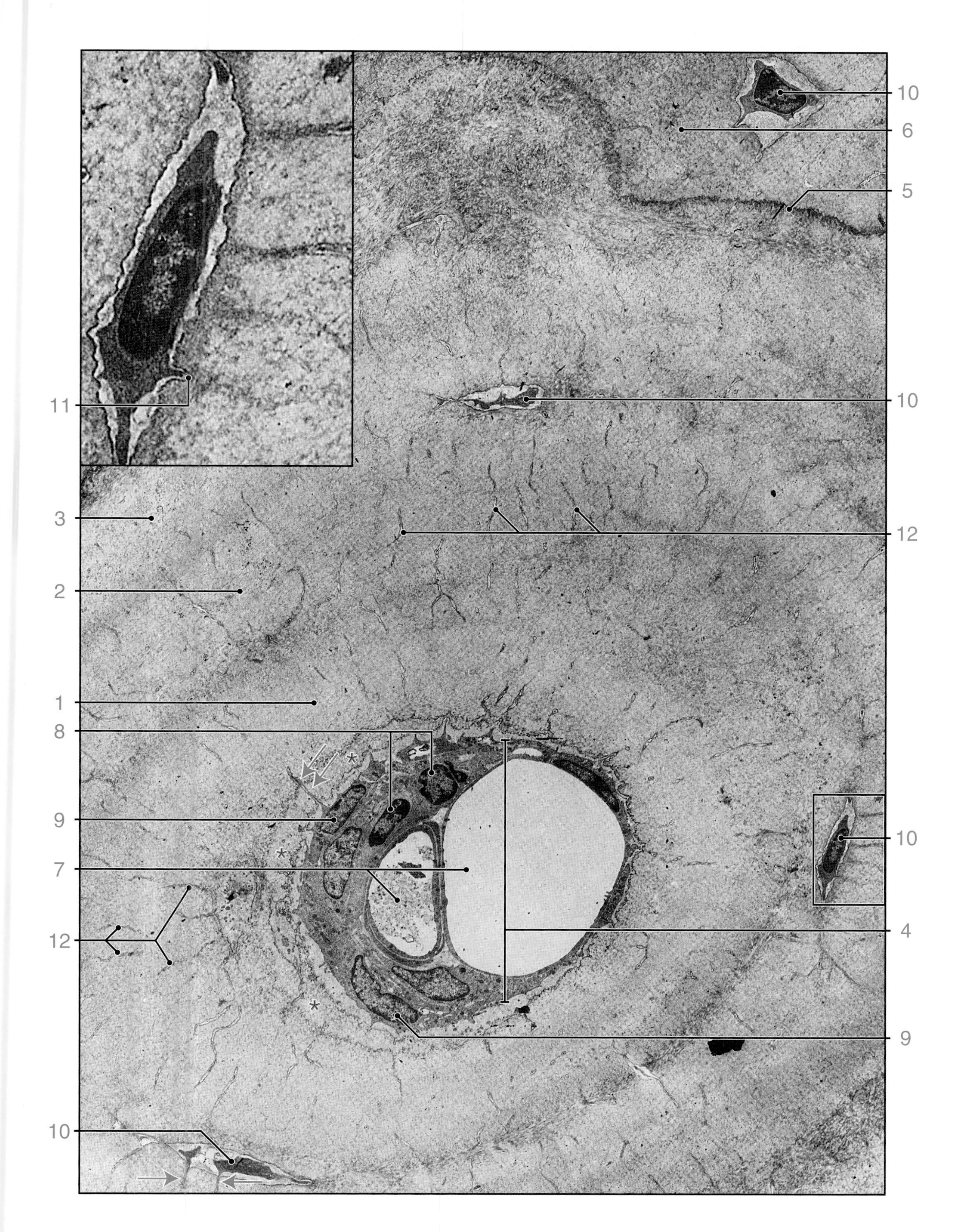

PLATE 19. BONE, ELECTRON MICROSCOPY

PLATE 20. ENDOCHONDRAL BONE FORMATION I

Endochondral bone formation involves a cartilage model, which is produced as a precursor to the bone that will be formed. The cartilage model is a miniature version of the future bone. Throughout the growth period of an individual, cartilage is present. Bones that arise through this method include the bones of the extremities and the vertebrae that bear weight. The first sign of bone formation is the appearance of bone forming cells around the diaphysis (shaft) of the cartilage model. The bone forming cells, known as osteoblasts, arise from the surrounding mesenchyme. Osteoblasts secrete the collagens, bone sialoproteins, osteocalcin, and other constituents of the bone matrix. The initial deposition of these components is referred to as osteoid; osteoid later becomes calcified. Growth of the bone in diameter is accomplished by continued deposition of bone and is referred to as appositional growth. With the initial establishment of the periosteal bony collar, the chondrocytes in the center of the cartilage model become hypertrophic (see top micrograph), leading to their death, and the cartilage matrix in this region becomes calcified. Concomitantly, blood vessels grow through the thin diaphyseal bony collar and vascularize the site, allowing the production of bone marrow and bone forming cells. Later, in the case of long bones, this process is repeated in the epiphyses of the cartilage model (see bottom micrograph). The process of bone deposition is described and illustrated in Plate 21.

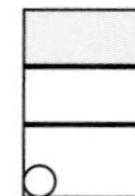

Developing bone, monkey, H&E, x240.
An early stage in the process of endochondral bone formation is shown in this micrograph. The mid-region of this long bone reveals **chondrocytes** (1) that have undergone marked hypertrophy. The cytoplasm of these chondrocytes appears very clear, or washed out. Their nuclei, when included in the plane of section, appear as small basophilic bodies that are not much different in size from the nuclei present in the chondrocytes at the distal regions of the cartilage model. Note how the cartilage matrix has been compressed into narrow bands surrounding the chondrocytes. At this stage of development, bone tissue has been produced to form the early **bony collar** (2) around the cartilage model. This bone tissue is produced by appositional growth from bone forming cells that were derived from the mesenchyme in the tissue surrounding the cartilage. This process represents intramembranous bone formation, which will be described later (see Plate 23).

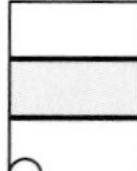

Developing bone, human, H&E, x60.
The bone shown in this micrograph represents a much later stage in development. Most of the diaphysis of the bone contains **marrow** (3), part of which is highly cellular and represents hemopoietic bone marrow. The non-staining areas consist of **adipose tissue** (4), which occupies much of the remainder of the marrow cavity. The thin bony collar seen earlier has now developed into a relatively thick mass of **diaphyseal bone** (5). The part of the bone in which bone tissue is being formed by **endochondral bone formation** (6) is seen at the ends of the marrow cavity. Note that its eosinophilic character is similar to the diaphyseal bone.

Developing bone, human, H&E, x60.
This specimen shows considerable developmental advancement beyond that of the bone seen in the above micrograph. A **secondary ossification center** (7) has been established in the proximal epiphysis of this long bone. At a slightly later time, a similar epiphyseal ossification center will form at the distal end of the bone. The process is identical to the process that occurs in the diaphysis, described in the introduction above. With time, these epiphyseal centers of ossification will increase in size to form much larger cavities. The consequence of this activity is that an epiphyseal plate is formed. The *dashed line* encompasses an area that, after enlargement of the secondary ossification center, will represent the **epiphyseal plate** (8). This plate, consisting of cartilage, separates the ossification centers at the distal ends of the bone from the central, or diaphyseal, ossification center in the shaft of the bone. This cartilaginous plate is essential for the longitudinal growth of the bone and will persist until bone growth ceases. The **inset** shows the secondary ossification center at higher magnification. Within this area, **new bone** (9) is already being produced. The new bone appears eosinophilic in contrast to the lighter staining of the surrounding **cartilage** (10). Note that its staining characteristic is identical to the more abundant **endochondral bone** (11) at the upper end of the diaphysis.

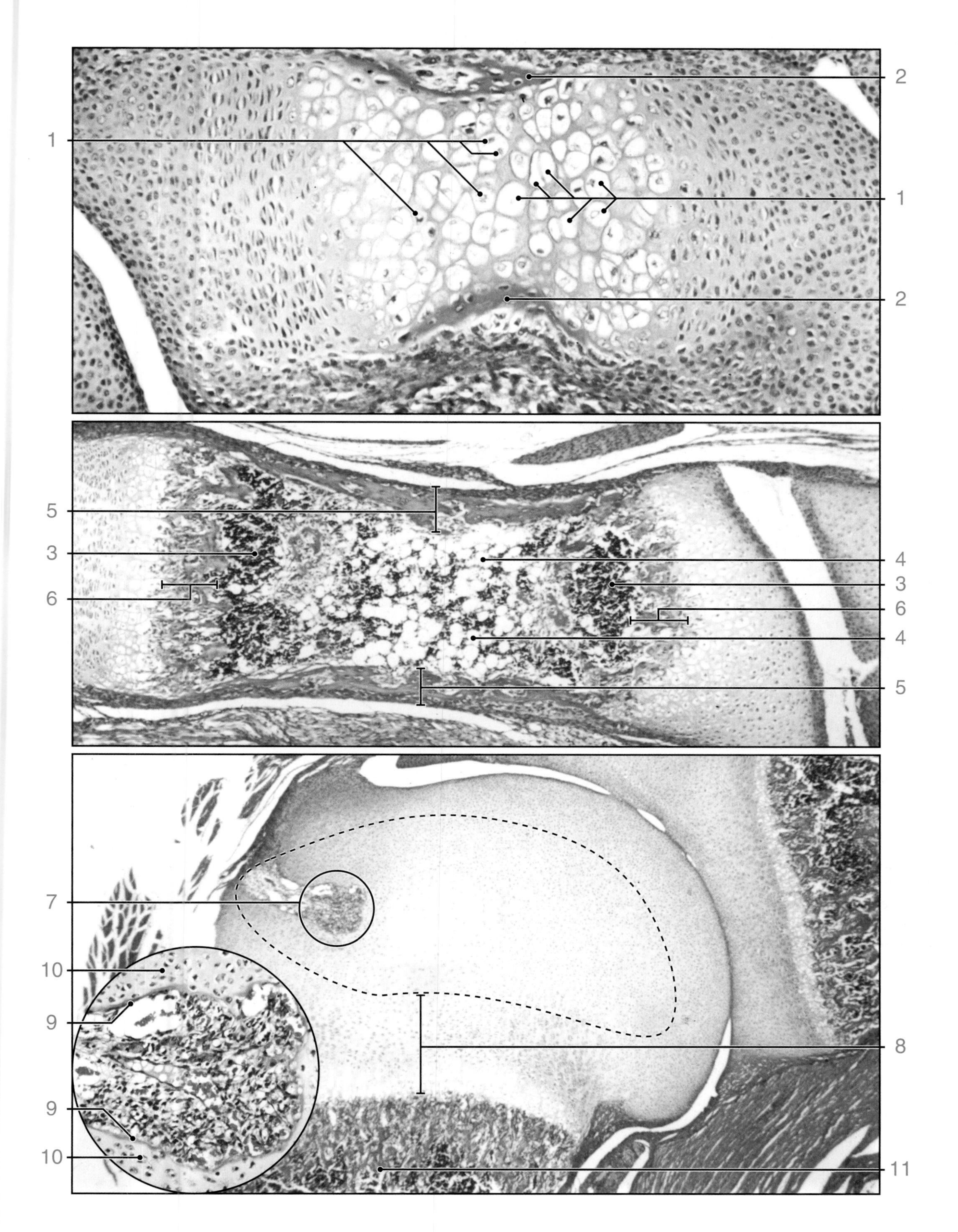

PLATE 20. ENDOCHONDRAL BONE FORMATION I

PLATE 21. ENDOCHONDRAL BONE FORMATION II

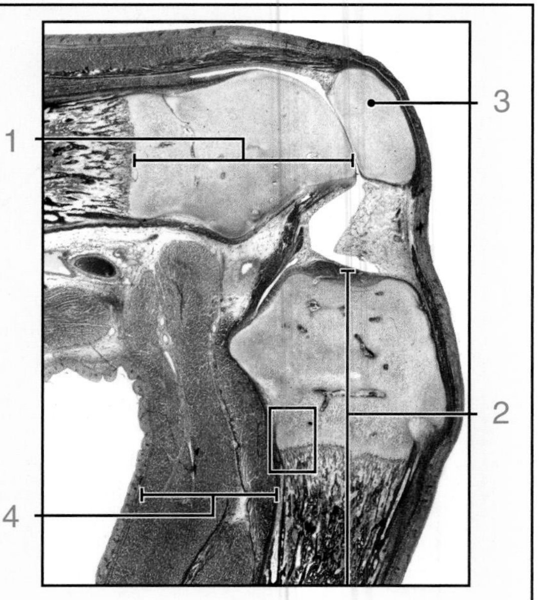

By the time bone is being produced in the process of endochondral bone formation, distinct zoning can be recognized in the epiphyseal cartilage at both ends of the early-formed marrow cavity. In the part of the cartilage that is furthest from the marrow cavity at both ends of the growing bone, individual chondrocytes, separated by cartilage matrix, are evident. Because of the nature of their distribution, this region is defined as the zone of reserve cartilage. Adjacent to this zone, towards the marrow cavity, the cartilage cells have undergone division and are organized into distinct columns. These cells are larger than those of the reserve zone and have produced new cartilage matrix. This area is known as the zone of proliferation. The production of cartilage matrix here causes lengthening of the bone. The next zone is referred to as the zone of hypertrophy. In this region, the chondrocytes undergo considerable enlargement. The cytoplasm of the chondrocyte is clear, a reflection of the abundant glycogen that these cells have accumulated. The cartilage matrix between the columns of these hypertrophied chondrocytes is compressed and forms linear bands. In the next zone, the zone of calcified cartilage, the bands of cartilage matrix have become calcified, and the hypertrophied cells begin to degenerate. The chondrocytes located in the more proximal part of this zone (closest to the marrow cavity) undergo apoptosis. The calcified cartilage located here will serve as an initial scaffold for the deposition of new bone. The last zone, the zone of resorption, is in direct contact with connective tissue of the marrow cavity. Blood vessels and accompanying connective tissue invade this region. The blood vessels here are the source of osteoprogenitor cells, which will differentiate into the bone producing cells that will come to reside on the calcified cartilage scaffold and produce bone matrix. In assessing these events, it should be recognized that new cells, as they are produced in the zone of proliferation, undergo a series of changes (i.e., hypertrophy and dying) without actually moving. Again, it is only the production of new cartilage cells and matrix material that causes the bone to lengthen. Just prior to birth, secondary ossification centers begin to form in the epiphyses, creating the epiphyseal plate and permitting subsequent epiphyseal bone growth.

ORIENTATION MICROGRAPH: This micrograph shows, at very low magnification, a portion of the **femur** (1), **tibia** (2), and **patella** (3) at the knee joint of a very late fetus (approximately 37 weeks). Also evident is **early developing muscle** (4).

Developing bone, human, Mallory, x90.
This micrograph shows the area within the rectangle of the orientation micrograph. It includes part of the **epiphyseal cartilage** (1) and the upper portion of the **marrow cavity** (2) where bone is being produced. Also evident is part of a **tendon** (3) and adjacent **developing muscle** (4). Within the epiphyseal cartilage, **vascular elements** (5) are evident. This is the initial stage in the formation of the secondary ossification center. The Mallory stain colors connective tissue elements blue. Thus, the cartilage appears light blue, and as it calcifies (6), it becomes a darker blue. The very dark blue within the marrow is **bone** (7). In contrast, the marrow cells and bone forming cells stain red.

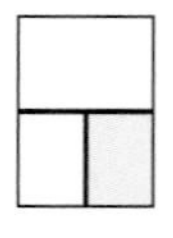

Developing bone, human, Mallory, x160.
The area in the rectangle on the left side of the above micrograph is shown here. It reveals the **zone of reserve cartilage** (8) at higher magnification. Beneath this is the **zone of proliferation** (9). Note how the cartilage cells are arranged in column-like arrays with their elongated nuclei appearing almost like stacks of discs. Again, it is this zone where new cartilage matrix has been produced by the proliferation of new chondrocytes.

Developing bone, human, Mallory, x160.
This micrograph shows the area within the rectangle on the right side of the above micrograph. In this view, the **zone of hypertrophy** (10) is readily recognized. The chondrocytes in this zone are in the process of degenerating. Those near the top of the micrograph show relatively intact nuclei, whereas those in the lower portion of the cartilage appear as swollen ghosts. In this region, the cartilage matrix has become calcified (11). Below this site, the calcified **cartilage spicules** (12) are surrounded by marrow, which has invaded the area previously occupied by degenerating chondrocytes.

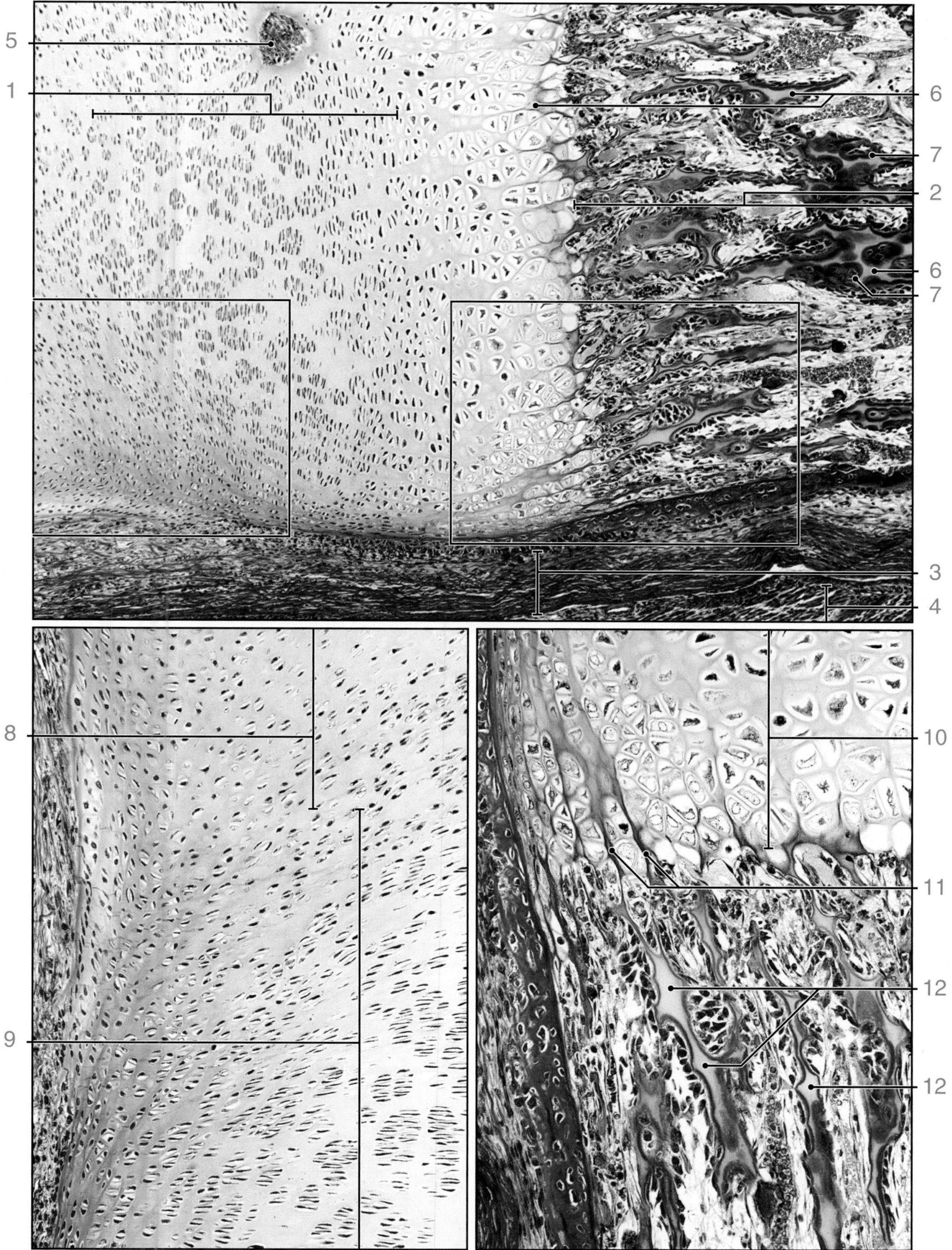

PLATE 21. ENDOCHONDRAL BONE FORMATION II

PLATE 22. **ENDOCHONDRAL BONE FORMATION III**

From the time that bone is first formed and throughout life, it undergoes constant change known as remodeling. During fetal development and subsequent growth periods, bone remodeling allows the progressive enlargement of the skeletal structure. The process of remodeling, once bone matrix has been produced, is dependent on osteoclasts. These are large, multinucleate cells that have the ability to resorb bone. They arise from the fusion of mononuclear progenitor cells in the marrow. Osteoclasts also give rise to the neutrophilic granulocyte and monocyte. The cooperative activities of osteoblasts and osteoclasts permit the remodeling of bone structure. As a bone grows in length, it must also grow in thickness. This occurs by osteoblast activity on the surface of the bone (appositional growth); however, older bone tissue must concomitantly be removed from the inner (marrow) surface to prevent excessive mass of the bone. Similarly, osteons produced during secondary bone formation are reorganized as new, replacement osteons are created to accommodate natural stresses placed on the bone. During adult life, remodeling continues, albeit at a slower pace, especially during the elderly years.

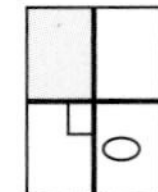

Developing bone, human, Mallory, x255.

This micrograph is a higher magnification of the specimen seen on the lower right of Plate 21. The major identifiable structural features are a **bone spicule** (1) and several **calcified cartilage spicules** (2). The bone spicule is readily identified by the **osteocytes** (3) within its matrix. Note that the calcified cartilage does not exhibit any cells within its matrix. A number of **osteoblasts** (4) are present on one side of the bone spicule, indicating growth of bone. The other side of the bone spicule is essentially free of osteoblast activity. Much of the surface of the cartilage spicules shows bone production. The pale blue area, which represents calcified cartilage, is covered in most areas by a dark blue **layer of bone** (5). The cells in apposition to this bone tissue are **osteoblasts** (6). Also of interest is that at several sites, a single **osteocyte** (7) is seen within the very thin bone matrix. As growth of the spicule continues, other osteoblasts will become incorporated into the bone matrix that they produced.

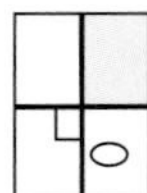

Developing bone, human, Mallory, x535.

This micrograph is a higher magnification of the bone spicule within the boxed area on the top left micrograph. Note that one side of the bone spicule is lined by numerous **osteoblasts** (8), indicating a growing surface. On the same side are numerous elongate cells that represent the developing **periosteum** (9). Further removed is a well-formed **dense connective tissue** (10), which will ultimately develop into a tendon. The vascularization of the growing bone is evidenced by a **blood capillary** (11).

Developing bone, human, Mallory, x500; inset x1300.

This micrograph, from the same specimen as the top left micrograph, shows several **osteoclasts** (12) removing bone from the spicules. Both spicules show a core of **calcified cartilage** (13) surrounded by bone matrix. The osteoclast within the rectangle is shown at higher magnification in the **inset**. The osteoclast has removed some of the bone matrix, forming a shallow resorption bay also known as Howship's lacuna. The **ruffled border** (14) of the osteoclast appears as a light band adjacent to the bone matrix at Howship's lacuna. The ruffled border consists of numerous infoldings of the plasma membrane, which create membrane bound folds of cytoplasm that are free of organelles. Also, note the multiple **nuclei** (15) in the osteoclast.

Decalcified bone, electron micrograph, x4800; inset x12,000.

This micrograph shows a region comparable to the boxed area on the inset of the bottom left micrograph. The light staining area represents **bone matrix** (16) after removal of the calcium component by the decalcification process. A thin layer of **calcified matrix** (17) remains at the site where the cytoplasm of the osteoclast is in apposition to the bone. The resorption front of the osteoclast is marked by numerous **infoldings of the plasma membrane** (18). These infoldings produce the ruffled-border appearance in the light microscope. The oval **inset** reveals the presence of calcium hydroxyapatite crystals, resorbed from the bone and located between the infolded membranes. Eventually, the calcium and phosphate components will enter the bloodstream for mobilization.

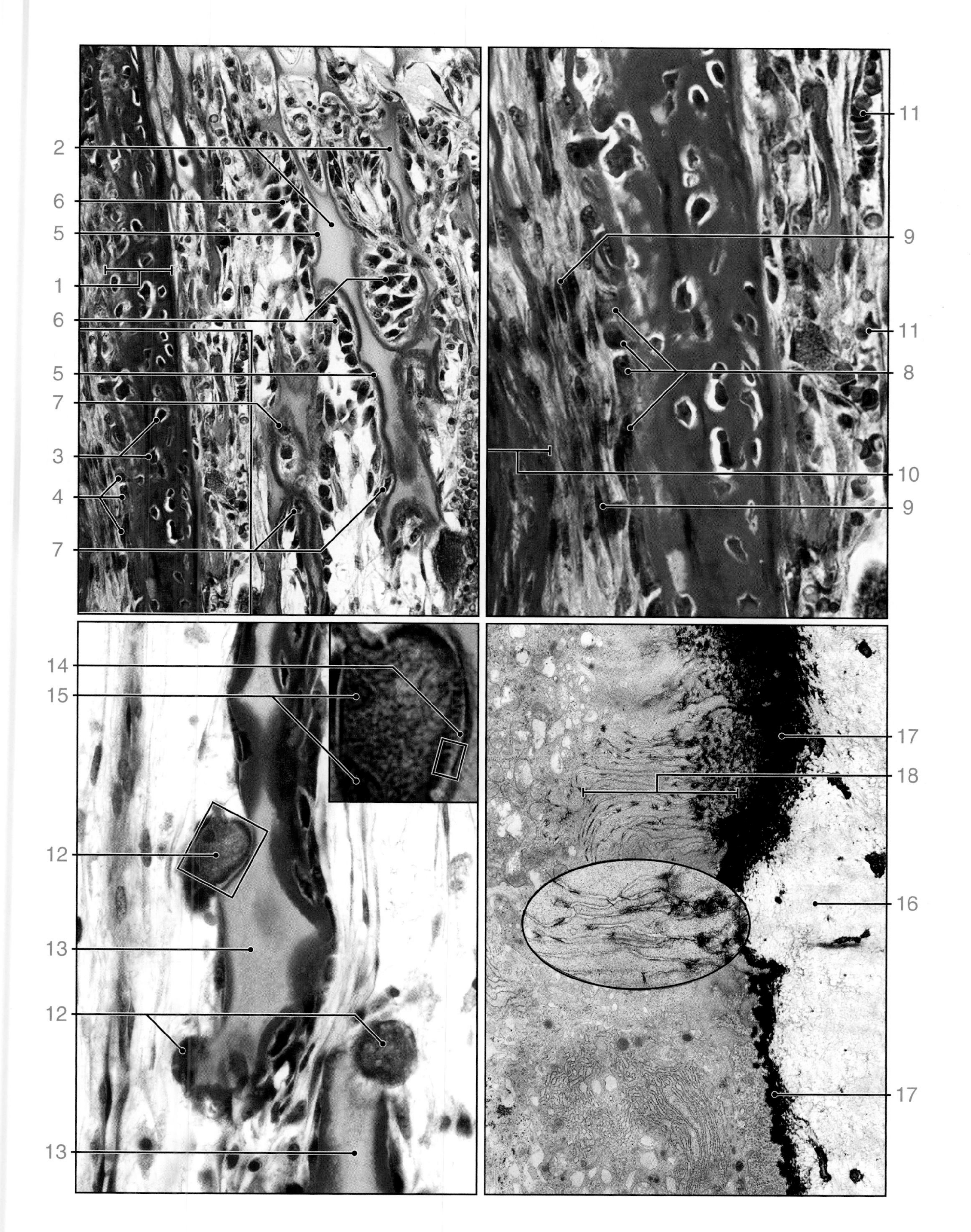

PLATE 22. ENDOCHONDRAL BONE FORMATION III

PLATE 23. INTRAMEMBRANOUS BONE FORMATION

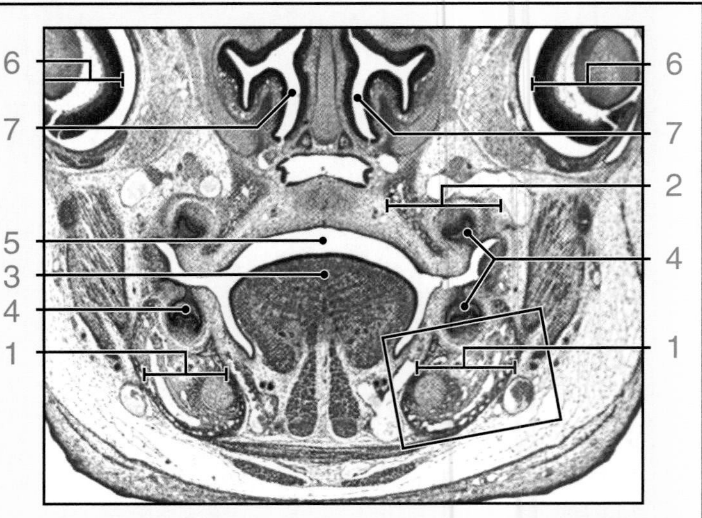

Intramembranous ossification is a process in which bone develops initially through the proliferation and differentiation of mesenchymal cells to become osteoblasts. No cartilage model exists as a precursor to the forming bone. Intramembranous ossification occurs in sites where bearing weight is not a required function of the developing bone, such as the flat bones of the skull and face, the mandible, and the clavicle. The process begins around the eighth week of gestation in humans. At this time, some of the mesenchymal cells migrate to the required site and multiply. The condensation or accumulation of mesenchymal cells initiates the ossification process by differentiating into osteoprogenitor cells. The site then becomes vascularized, and the osteoprogenitor cells become larger, rounded, and their cytoplasm changes from eosinophilic to basophilic. The cytological changes result in the differentiated osteoblast, which then secretes the components of bone matrix. As the matrix is produced, the osteoblasts become more distant from one another, but they remain attached by thin cytoplasmic processes. The matrix then becomes calcified, and the interconnecting processes of the bone forming cells, which now can be called osteocytes, are enclosed within canaliculi. As this occurs, surrounding mesenchymal cells proliferate and produce more osteoprogenitor cells. Some of the osteoprogenitor cells come into apposition with the initially formed calcified matrix, then become osteoblasts and create more matrix. By this process of appositional growth, the spicules become larger and form a trabecular network that will have the general shape of the mature bone. It should be pointed out that the continued growth of the precursor bone is the same in both endochondral bone formation and intramembranous bone formation—the only distinction is that bone production in endochondral bone formation uses a cartilage model, whereas in intramembranous bone formation, prior development of a cartilage model does not occur.

ORIENTATION MICROGRAPH: The specimen shown here is a frontal section through the head of a late fetus. Some of the structural features that can be identified are the **developing mandible** (1), **upper maxilla** (2), **tongue** (3), **developing teeth** (4), **oral cavity** (5), **developing eye** (6), and **nasal cavity** (7).

Intramembranous ossification, human, H&E, x90.

The area in the rectangle on the orientation micrograph is shown here at higher magnification. **Bone tissue** (1) has already been formed, giving an incomplete template of the mandible. (The *dashed line* encloses the area that will become the definitive mandible.) With the exception of **Meckel's cartilage** (2), which serves as a temporary structural form and will later be removed, there is no cartilage associated with the developing mandible. **Committed mesenchymal cells** (3) proliferate and give rise to osteoprogenitor cells at the site where bone is continuing to form. They will differentiate into **osteoblasts** (4) and lay down bone matrix.

Intramembranous ossification, human, H&E, x180.

This is a higher magnification of the area in the rectangle of the above micrograph. At this magnification, **bone tissue** (5), with its entrapped osteocytes, can be readily recognized. The site where new bone is being produced is seen in the upper right of the micrograph. In the encircled area (6), cells are in the process of becoming osteoblasts. Adjacent to this area is a population of cells that can already be identified as **osteoblasts** (7). They are producing new bone matrix at this site. Other points of interest are the nearby **mesenchyme** (8) and a site of **developing muscle** (9).

Intramembranous ossification, human, H&E, x180.

This micrograph is from an area on the opposite side of the mandible and shows a very slim, newly formed **bone spicule** (10). At one side of the spicule are **blood vessels** (11). On the opposite side of the bone spicule are **osteoblasts** (12) producing more bone matrix. One osteoblast has surrounded itself to become an **osteocyte** (13). At other sites along the spicule, no matrix is being produced. The cells in apposition to the spicule may be regarded as **osteoprogenitor cells** (14). Note their flattened nuclei compared to the osteoblast nuclei, which are cuboidal in shape. Later, the osteoprogenitor cells may be signaled to produce bone matrix, and thus assume the features characteristic of an osteoblast. Other notable features include **early periosteal formation** (15) and **mesenchyme** (16) with its elongate, fibroblast-like cells.

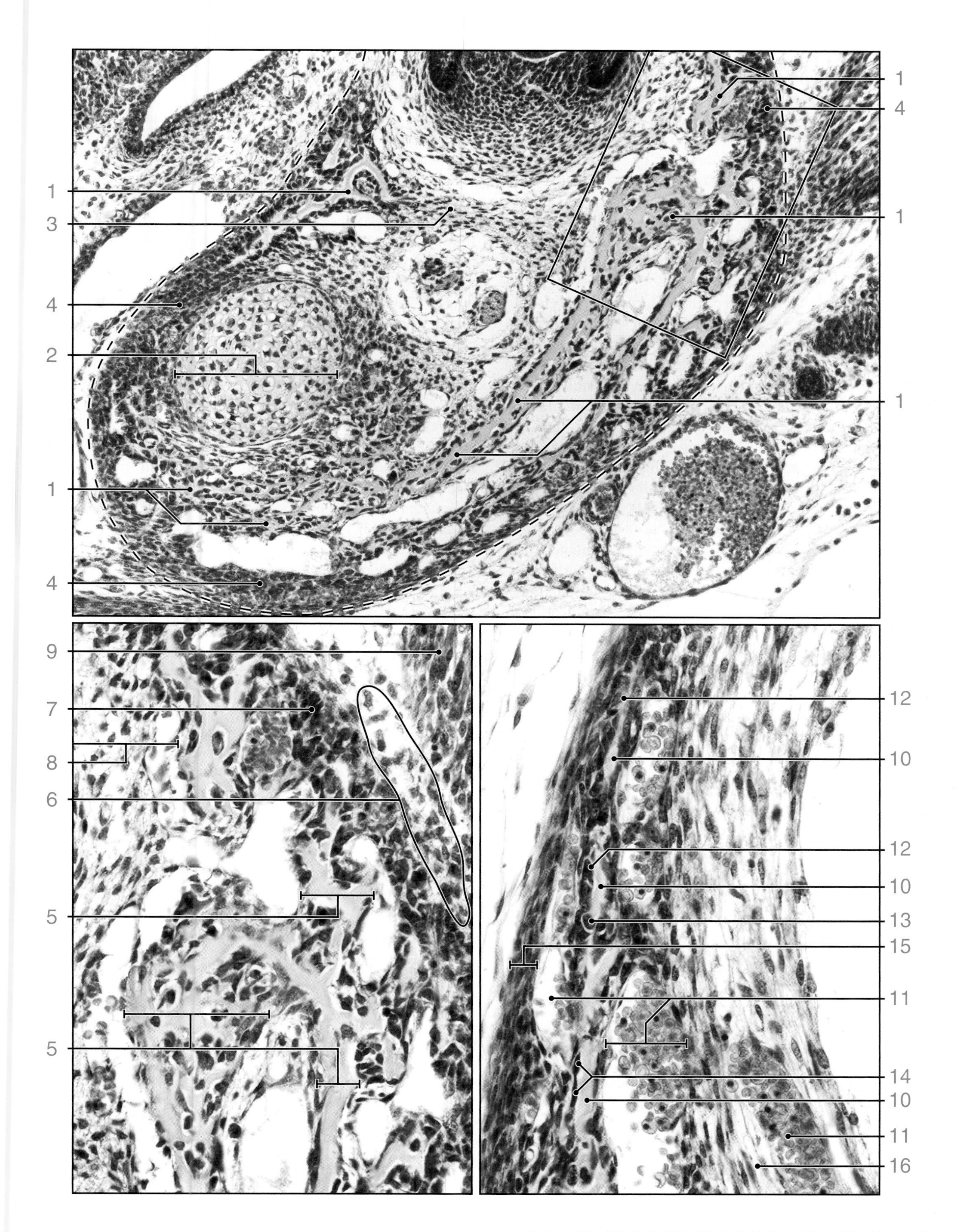

PLATE 23. INTRAMEMBRANOUS BONE FORMATION

PLATE 25. DEVELOPING BONE II, ELECTRON MICROSCOPY

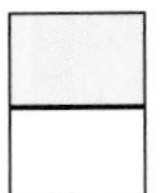

Developing bone, electron micrograph, x14,000.

This micrograph shows several osteoblasts. A single cuboidal-shaped osteoblast occupies the major portion of the micrograph. On the far left is a small portion of an adjoining osteoblast and on the far right is another osteoblast in which the **nucleus** (1) is visible. At the top of the micrograph is part of a **osteoprogenitor cell** (2) and at the bottom of the micrograph some of the **osteoid** (3) that has been secreted by the osteoblast is visible. The osteoblasts shown in this electron micrograph are comparable to those seen in the Plate 24. The higher magnification of this micrograph reveals the cytological detail of the cells.

In characterizing the osteoblast as well as understanding the nature of its product, it is important to realize that osteoblasts are derived from connective tissue cells that are indistinguishable from fibroblasts. They retain much of the cytological morphology of fibroblasts, evidenced by this micrograph. In active fibroblasts, there is a large amount of rough-surfaced endoplasmic reticulum and an extensive Golgi apparatus. The osteoblasts pictured here also contain a large amount of **rough endoplasmic reticulum** (4) and a prominent **Golgi apparatus** (5), indicative of their role in the production of osteoid. The Golgi apparatus visible in the osteoblast contains typical flattened sacs and transport vesicles. It also shows enlarged vesicles that contain material of various densities. Two of the large, elongated Golgi vesicles (*arrows*) contain a filamentous component. Based on special staining techniques for electron microscopy, these filamentous components are considered to be collagen precursors.

These osteoblasts (and those in the Plate 24) are cuboidal and, because of their close apposition, they resemble cuboidal cells in an epithelial sheet. Although the osteoblasts are indeed arranged on the surface of the developing bone in a sheetlike manner, they exhibit properties of connective tissue cells, not epithelial cells. It should be noted that no basal lamina is associated with the osteoblasts. Collagen fibrils are occasionally observed between osteoblasts. The osteoblast is highly polar; the surface of the osteoblast that faces the osteoid is the secretory face, or secretory pole, of the cell.

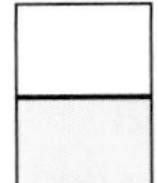

Developing bone, electron micrograph, x30,000.

The secretory face of the osteoblast is shown adjacent to the osteoid at higher magnification in this micrograph. Both the round and the somewhat larger, irregularly shaped profiles of the osteoid are **collagen fibrils** (6). Ground substance occupies the space between the collagen fibrils. Remember that the osteoblast moves away from the bone, leaving its product behind. The collagen fibrils closest to the cell are of small diameter and represent the most recently formed fibrils. Over time, as the cell recedes and as the mineralization front approaches, the collagen fibrils increase in diameter. Once the mineralization front passes the fibrils, they no longer increase in diameter. Mineralization results in an impregnation of both the collagen fibrils as well as the ground substance with calcium hydroxyapatite.

The view also includes processes of the bone-forming cells (*arrow*). These processes will become contained in canaliculi as the osteoblast becomes an osteocyte.

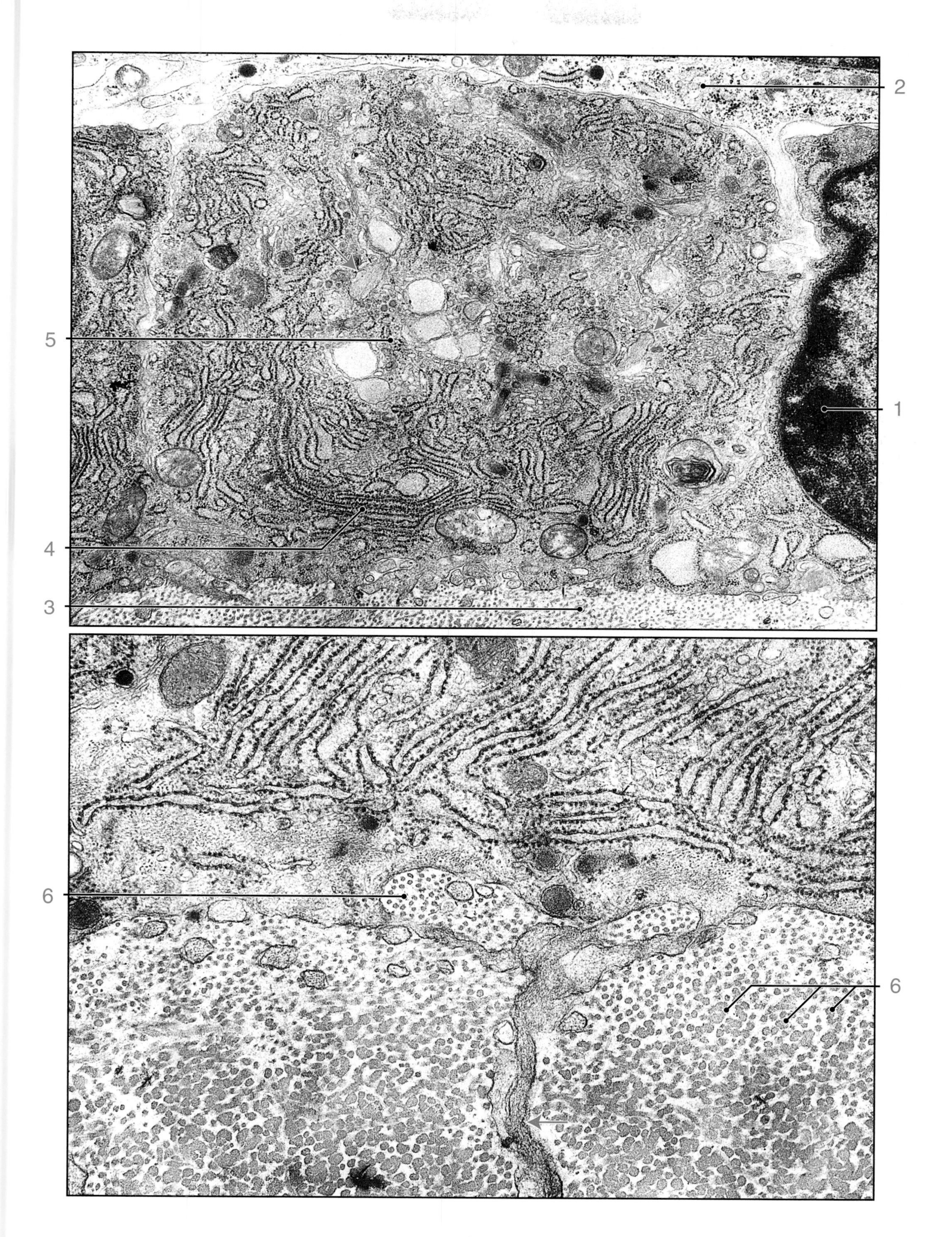

PLATE 25. DEVELOPING BONE II, ELECTRON MICROSCOPY

CHAPTER 6

Blood and Bone Marrow

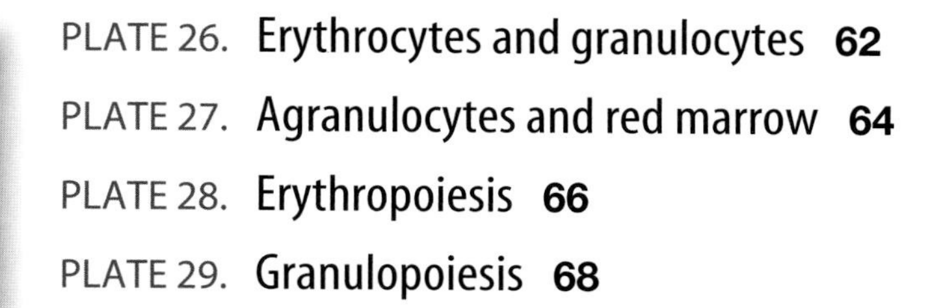

PLATE 26. ERYTHROCYTES AND GRANULOCYTES

Blood is regarded as a connective tissue. It is fluid in character and consists of formed elements and plasma. Red blood cells (erythrocytes), white blood cells (leukocytes), and blood platelets constitute the formed elements. Collectively, they make up 45% of the blood volume. Erythrocytes transport and exchange oxygen and carbon dioxide, and they constitute 99% of the total blood cell count. Leukocytes are categorized as agranulocytes or granulocytes. The agranulocytes are further classified as lymphocytes or monocytes. The granulocytes, named for the granules visible in their cytoplasm, consist of neutrophils, eosinophils, and basophils. Each type of leukocyte has a specific role in immune and protective responses in the body. They typically leave the circulation and enter the connective tissue to perform their specific role. In contrast, erythrocytes function only within the vascular system. Blood platelets are responsible for blood clotting and, consequently, have an essential role in incidents of small vessel damage.

Blood smears are used for microscope examination and identification of relative numbers of leukocytes in circulating blood. The blood smear is prepared by placing a small drop of blood on a microscope slide and then smearing it across the slide with the edge of another slide. When properly executed, this method provides a uniform, single layer of blood cells, which is allowed to air dry before it is stained. Wright's stain, a modified Romanovsky stain, is generally used. When examining smears under the microscope, it is useful to use a low magnification to find areas where the blood cells are uniformly distributed, like those seen in the smear on the micrograph at the top of this Plate. Switching to a higher magnification, one can identify the various types of white blood cells and, in fact, determine the relative number of each cell type. A normal cell count is as follows: neutrophils, 48.6–66.7%; eosinophils, 1.4–4.8%; basophils, 0–0.3%; lymphocytes, 25.7–27.6%; monocytes, 8.6–9.0%.

Blood smear, human, Wright's stain, x200. This low magnification micrograph shows part of a blood smear in which the blood cells are uniformly distributed. The majority of cells are erythrocytes. Because of their bi-concave shape, most of the erythrocytes appear donut-shaped. Two leukocytes, both granulocytes, are visible. One granulocyte is a **neutrophil** (1); the other is an **eosinophil** (2). At this magnification, however, the major distinction is in the staining of their cytoplasm. Higher magnification, as in the images below, allows for a more precise characterization of cell type.

Blood smear, neutrophils, human, Wright's stain, x2200. Neutrophils exhibit variation in size and nuclear morphology that is associated with the age of the cell. The left micrograph shows the nucleus of a neutrophil that has just passed the band stage and has recently entered the blood stream. The cell is relatively small, and its cytoplasm exhibits distinctive, fine granules. The neutrophil in the middle micrograph is considerably larger, and its cytoplasm contains more fine granules. The nucleus still exhibits a U-shape, but **lobulation** (*arrows*) by the constriction of the nucleus at several points is becoming apparent. The neutrophil shown in the right micrograph exhibits greater maturity by its very distinctive lobulation. Here, the lobules are connected by a very thin nuclear "bridge." A very distinctive feature, associated with the nucleus of this cell, is the presence of a **"drumstick"** (*arrowhead*), indicative of blood that has been drawn from a female.

Blood smear, eosinophils, human, Wright's stain, x2200. The eosinophils seen in these micrographs similarly represent different stages of maturity. The eosinophil in the left micrograph is relatively small and is just beginning to show lobulation. The cytoplasm is almost entirely filled with eosinophilic granules that characterize this cell type. The lighter stained area, devoid of granules, probably represents the site of the **Golgi apparatus** (*arrow*). The eosinophil shown in the middle micrograph is larger, and its nucleus is now distinctively bilobed. At one site, three distinct **granules** (*arrowhead*) are evident. Note their spherical shape and their relatively uniform size. The eosinophil in the right micrograph is more mature, displaying at least three lobes. When adjusting through the focal range of these cells, the eosinophil granules often appear to light up due to their crystalline structure.

Blood smear, basophils, human, Wright's stain, x2200. The cells shown here are basophils and also represent different stages of maturation. The basophil in the left micrograph is relatively young and small. The granules are variable in size and tend to obscure the morphology of the nucleus. Also, they are less plentiful than the granules seen in the eosinophil. The nucleus of the basophil in the middle micrograph appears to be bilobed, but the granules that lie over the nucleus, again, obscure its true shape. The basophil in the right micrograph is probably more mature. The nuclear shape is almost entirely obscured by the granules.

Blood platelets (*arrowheads*) are present in several of the micrographs. Typically they appear as small, irregularly shaped bodies.

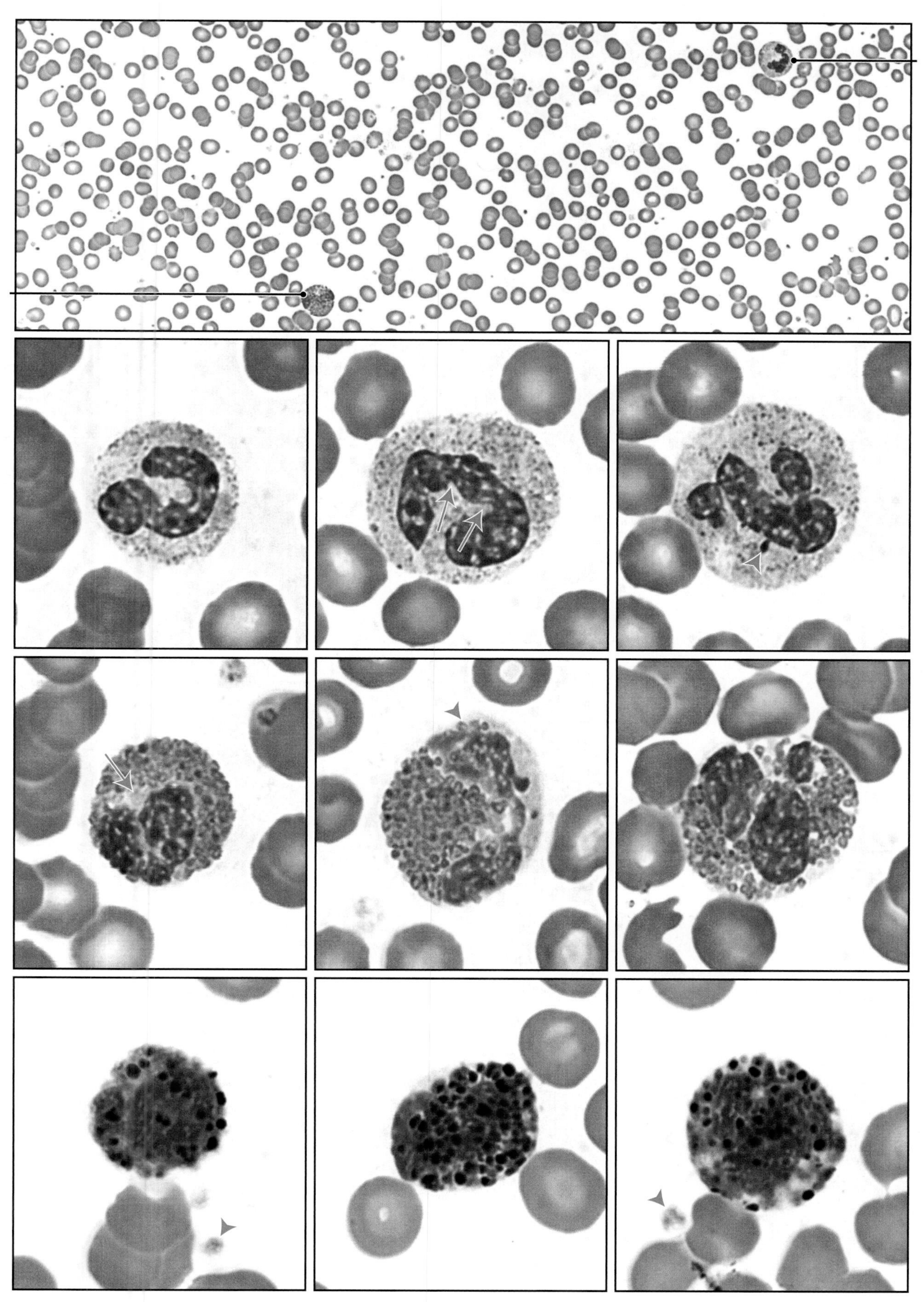

PLATE 26. ERYTHROCYTES AND GRANULOCYTES

PLATE 27. AGRANULOCYTES AND RED MARROW

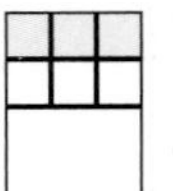

Blood smear, lymphocytes, human, Wright's stain, x2150.

The lymphocytes shown here vary in size, but each represents a mature cell. Circulating lymphocytes are usually described as small, medium or large. A small lymphocyte is shown in the left panel. Lymphocytes in this category range in size from 7 to 9 μm. The lymphocyte in the middle panel is medium sized. A large lymphocyte is seen in the right panel. These cells may be as large as 16 μm. Differences in lymphocyte size are attributable mostly to the amount of cytoplasm present. The nucleus also contributes to the size of the cell, but to a lesser degree. In differential counts, lymphocyte size is disregarded. **Platelets** (*arrows*) are evident in the left panel.

Blood smear, monocytes, human, Wright's stain, x2150.

The white cells in these panels are mature monocytes. Monocytes range in size from approximately 13 to 20 μm, with the majority falling in the upper size range. The nucleus exhibits the most characteristic feature of the monocyte, namely, an indentation that is sometimes so prominent that it exhibits a U-shape, as seen in the right micrograph. The cytoplasm is very weakly basophilic. Small, azurophilic granules (lysosomes) are also characteristic of the cytoplasm and are similar to those seen in neutrophils. **Platelets** (*arrows*) are present in the left and middle micrographs.

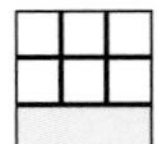

Bone marrow smear, human, Giemsa, x180.

This micrograph shows a bone marrow smear at low magnification. This type of preparation allows the examination of developing red and white cells. A marrow smear is prepared similarly to a peripheral blood smear. A sample of bone marrow is aspirated from a bone, placed on a slide, and spread into a thin monolayer of cells. A wide variety of cell types are present in a marrow smear.

Most of the cells are developing granulocytes and developing erythrocytes. **Mature erythrocytes** (1) are also present in large numbers. They are readily identified by their lack of a nucleus and eosinophilic staining. Often intermixed with these cells are small groups of reticulocytes. Reticulocytes are very young erythrocytes that contain residual ribosomes in their cytoplasm. The presence of the ribosomes gives the reticulocyte a just-perceptible blue coloration compared to the mature, eosinophilic erythrocyte. The reticulocytes are best distinguished at higher magnifications. **Adipocytes** (2) are present in variable numbers. In specimens such as this, the lipid content is lost during preparation, and recognition of the cell is based on the presence of a clear or unstained, round space. Another large cell that is typically present is the **megakaryocyte** (3). The megakaryocyte is a polyploid cell that exhibits a large, irregular nuclear profile. It is the platelet-producing cell.

At this low magnification, it is difficult to distinguish the earlier stages of the developing cell types. However, examples of each stage of development in both cell lines are presented in Plates 28 and 29. In contrast, many cells in their late stage of development, particularly in granulocytes, can be identified with some degree of certainty at low magnification. For example, some **band neutrophils** (4) and young **eosinophils** (5) can be identified by their morphology and staining characteristics.

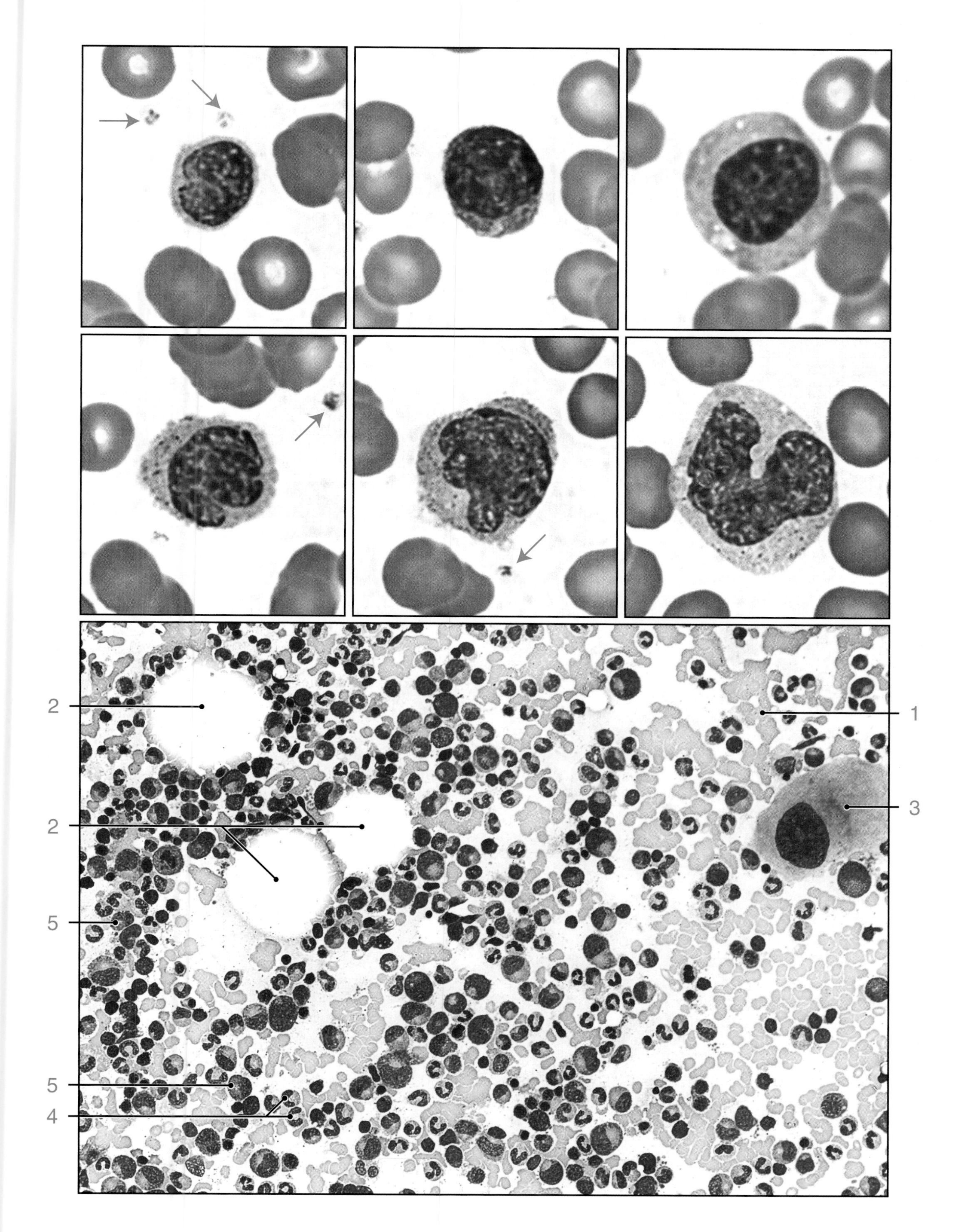

PLATE 27. AGRANULOCYTES AND RED MARROW

PLATE 28. **ERYTHROPOIESIS**

Erythropoiesis is the process by which the concentration of erythrocytes in the peripheral blood stream is maintained under normal conditions in a steady state. Stimulation of erythroid stem cells by hormonal action results in a proliferation of precursor cells that undergo differentiation and maturation in the bone marrow. The earliest recognizable precursor of the red blood cell is the proerythroblast, also called pronormoblast or rubriblast. These cells lack hemoglobin. Their cytoplasm is basophilic, and the nucleus exhibits a dense chromatin structure and several nucleoli. The Golgi apparatus, when evident, appears as a light staining area. The basophilic erythroblast is smaller than the proerythroblast, from which it arises by mitotic division. Its nucleus is smaller. The cytoplasm shows strong basophilia due to the increasing number of ribosomes involved in hemoglobin synthesis. The accumulation of hemoglobin in the cell gradually changes the staining reaction of the cytoplasm so that it begins to stain with eosin. The presence of hemoglobin in the cell, recognizable by its staining, signifies the cell's transition to the polychromatophilic erythroblast. The cytoplasm in the earlier part of this stage may exhibit a blue-grey color. As increasing amounts of hemoglobin are synthesized, decreasing numbers of ribosomes are present. The nucleus of the polychromatophilic erythroblast is smaller than that of the basophilic erythroblast and the heterochromatin is much coarser. At the end of this stage, the nucleus has become much smaller and the cytoplasm more eosinophilic. This is the last stage in which mitosis occurs. The next definable stage is the orthochromatophilic erythroblast, also called normoblast. Its nucleus is smaller than earlier stages and is extremely condensed. The cytoplasm is considerably less blue, leaning more to a pink, or eosinophilic, color. It is slightly larger than a mature erythrocyte. At this stage, it is no longer capable of division. In the next stage, the polychromatophilic erythrocyte, more commonly called a reticulocyte, loses its nucleus and is ready to pass into the blood sinusoids of the red bone marrow. Some ribosomes that can still synthesize hemoglobin are present in the cell. These ribosomes give the cell a very slight basophilia. Comparison of this cell to typical mature erythrocytes in the marrow smear reveals a slight difference in coloration.

Bone marrow smear, proerythroblast, human, Giemsa, x2200.
The proerythroblast shown here is a large cell, larger than the cells that follow in the developmental process. Note the very large size of the nucleus, occupying most of the cell volume. Several **nucleoli** (1) are evident. The cytoplasm is basophilic. Division of this cell results in the basophilic erythroblast.

Bone marrow smear, basophilic erythroblast, human, Giemsa, x2200.
The basophilic erythroblast shown here is smaller than its predecessor. The nuclear–cytoplasmic ratio is decreased. The greater abundance of cytoplasm is deeply basophilic compared to that of the proerythroblast. Typically, nucleoli are absent. As maturation continues, the cell decreases in size.

Bone marrow smear, polychromatophilic erythroblast, human, Giemsa, x2200.
Two polychromatophilic erythroblasts are seen in this micrograph. The larger and less mature cell exhibits pronounced clumping of its chromatin. The cytoplasm is basophilic, but is considerably lighter in color than that of the basophilic erythroblast. The cytoplasm also exhibits some eosinophilia, which is indicative of hemoglobin production. The smaller cell represents a later stage of a polychromatophilic erythroblast. Note how much more dense the chromatin appears as well as how much smaller the nucleus has become. Also, the cytoplasm now favors eosinophilia; however, some basophilia is still evident.

Bone marrow smear, orthochromatophilic erythrocyte, human, Giemsa, x2200.
Two orthochromatophilic erythrocytes are seen in this micrograph. Their nuclei have become even smaller and the nucleus exhibits a compact, dense staining. The cytoplasm is predominantly eosinophilic, but still possesses a degree of basophilia. Overall, the cell is only slightly larger than a mature erythrocyte. At this stage, the cell is no longer capable of division.

Bone marrow smear, polychromatophilic erythrocyte, human, Giemsa, x2200.
A **polychromatophilic erythrocyte** (2) is seen in this micrograph. Its nucleus has been extruded and the cytoplasm exhibits a slight basophilia. In proximity are several mature **erythrocytes** (3). Compare the coloration of the polychromatophilic erythrocyte with that of the mature erythrocytes. Polychromatophilic erythrocytes can also be readily demonstrated with special stains that cause the remaining ribosomes in the cytoplasm to clump and form a visible reticular network; hence the polychromatophilic erythrocyte is also commonly called a reticulocyte.

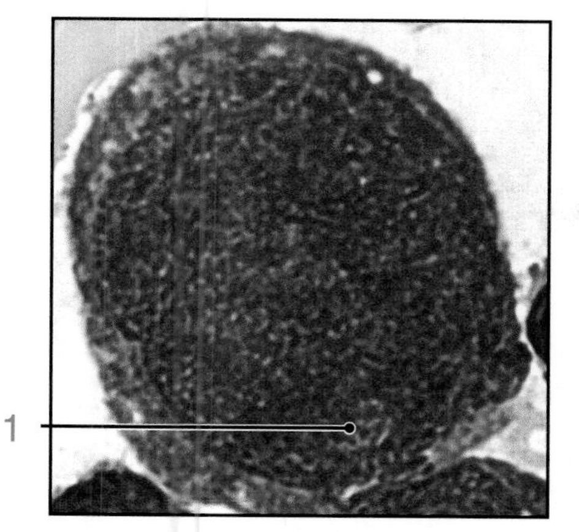

proerythroblast
(pronormoblast)

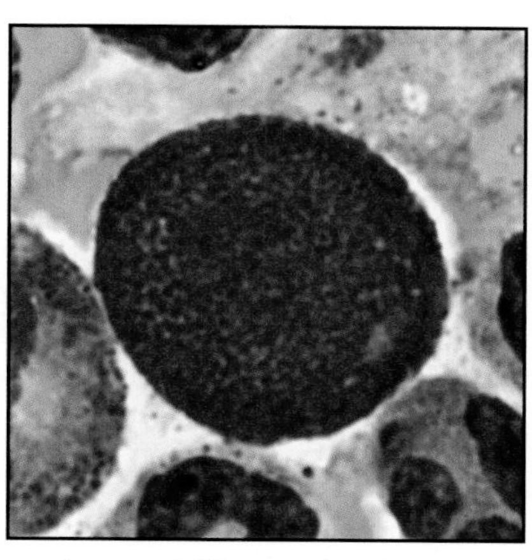

basophilic erythroblast
(basophilic normoblast)

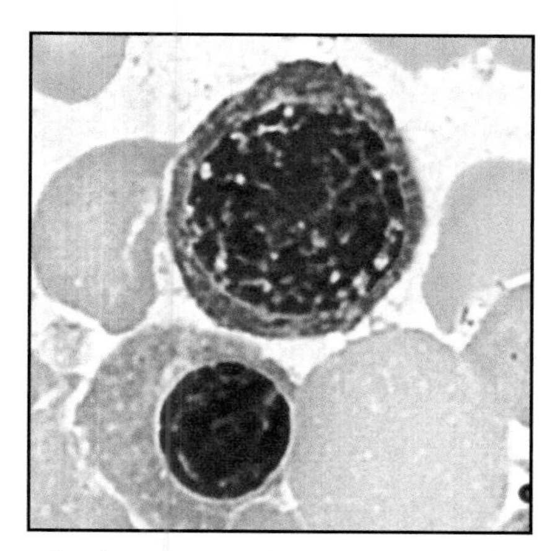

polychromatophilic erythroblast
(polychromatophilic normoblast)

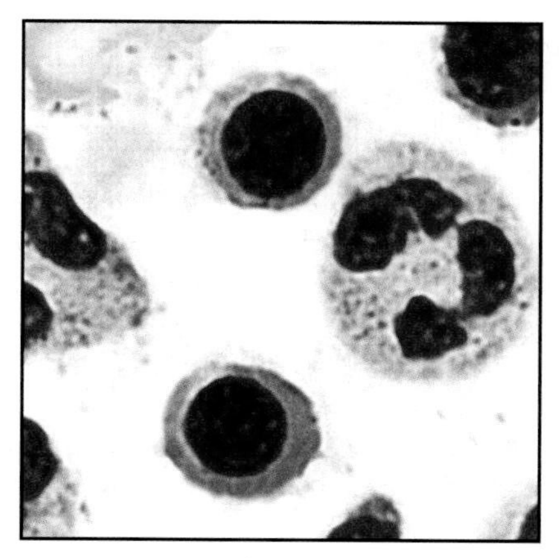

orthochromatic erythroblast
(normoblast)

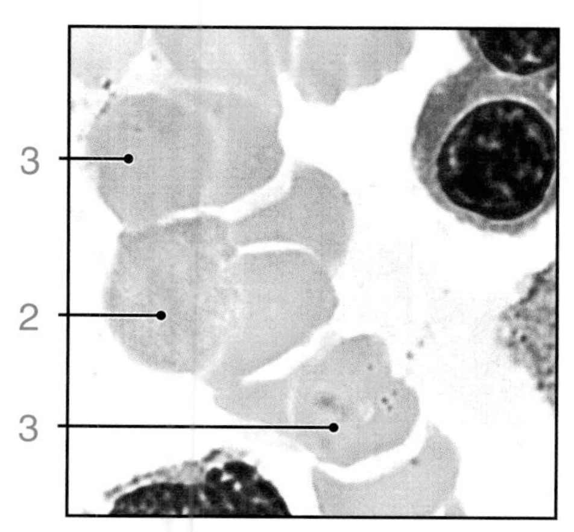

polychromatophilic erythrocyte
(reticulocyte)

PLATE 29. GRANULOPOIESIS

Granulopoiesis is the process by which the granulocyte blood cells (neutrophils, eosinophils, and basophils) differentiate and mature in the bone marrow. The earliest recognizable stage is the myeloblast, which is followed consecutively by the promyelocyte, myelocyte, metamyelocyte, band cell, and finally, the mature granulocyte. It is not possible to differentiate eosinophil, basophil, or neutrophil precursors morphologically until the myelocyte stage is reached—when specific granules characteristic of each cell type appear. The cells of the basophil lineage are extremely difficult to locate in a marrow smear because of the minimal number of these cells in the marrow.

The myeloblast is characterized by a large, euchromatic, spherical nucleus with three to five nucleoli. The cell measures 14 to 20 μm in diameter. The cytoplasm stains deeply basophilic. The presence of a light or poorly staining area indicates a Golgi apparatus. The promyelocyte exhibits a similar size range, 15 to 21 μm; nucleoli are present. Promyelocyte cytoplasm stains similarly to that of the myoblast but it is distinguished by the presence of large, blue/black, primary azurophilic granules, also called nonspecific granules. The myelocyte ranges from 16 to 24 μm. Its chromatin is more condensed than its precursor, and nucleoli are absent. The cytoplasm of the neutrophilic myelocyte is characterized by small, pink-to-red specific granules with some azurophilic granules present. The eosinophilic lineage has a similar appearing nucleus, but its specific granules are large. The metamyelocyte ranges from 12 to 18 μm. The nuclear cytoplasmic ratio is further decreased, and the nucleus assumes a kidney shape. There are few azurophilic granules at this stage in cells, and there is a predominance of small, pink-to-red specific granules. The eosinophilic metamyelocyte shows an increased number of specific granules compared to the neutrophilic metamyelocyte. The band cells are further reduced in size, 9 to 15 μm. The chromatin of the nucleus exhibits further condensation and has a horseshoe shape. In the neutrophilic band cell, the small, pink-to-red specific granules are the only granule type present. The eosinophilic band cell shows little or no change relative to the specific granules, but the nucleus exhibits a kidney shape. Mature granulocytes are shown on Plate 26.

Bone marrow smear, myeloblast, human, Giemsa, x2200.
The myeloblast shown here exhibits a deep blue cytoplasm with a lighter region that represents the **Golgi area** (1). The nucleus is round. Several **nucleoli** (2) are evident.

Bone marrow smear, promyelocyte, human, Giemsa, x2200.
The promyelocyte exhibits a round nucleus with one or more **nucleoli** (3) present. The cytoplasm is basophilic and exhibits relatively large blue-black **azurophilic granules** (4).

Bone marrow smear, eosinophilic myelocyte, human, Giemsa, x2200.
The eosinophilic myelocyte exhibits a nucleus the same as that described for the neutrophilic myelocyte. The cytoplasm, however, contains the large specific granules characteristic of eosinophils, but they are fewer in number than in the mature eosinophil.

Bone marrow smear, neutrophilic myelocyte, human, Giemsa, x2200.
The neutrophilic myelocyte retains the round nucleus, but nucleoli are now absent. The cytoplasm exhibits small, pink-to-red specific granules.

Bone marrow smear, eosinophilic metamyelocyte, human, Giemsa, x2200.
The eosinophilic metamyelocyte exhibits a kidney- or bean-shaped nucleus. The cytoplasm exhibits numerous, characteristic eosinophilic granules throughout the cytoplasm.

Bone marrow smear, neutrophilic metamyelocyte, human, Giemsa, x2200.
The neutrophilic metamyelocyte differs from its precursor by the presence of a kidney- or bean-shaped nucleus. The small, pink-to-red specific granules are now seen in the cytoplasm, and few or no azurophilic granules are present.

Bone marrow smear, eosinophilic band cell, human, Giemsa, x2200.
The eosinophilic band cell exhibits a horseshoe-shaped nucleus. Its cytoplasm is filled with eosinophilic granules.

Bone marrow smear, neutrophilic band cell, human, Giemsa, x2200.
The band or non-segmented neutrophil exhibits a horseshoe-shaped nucleus with abundant small, pink-to-red specific granules in the cytoplasm.

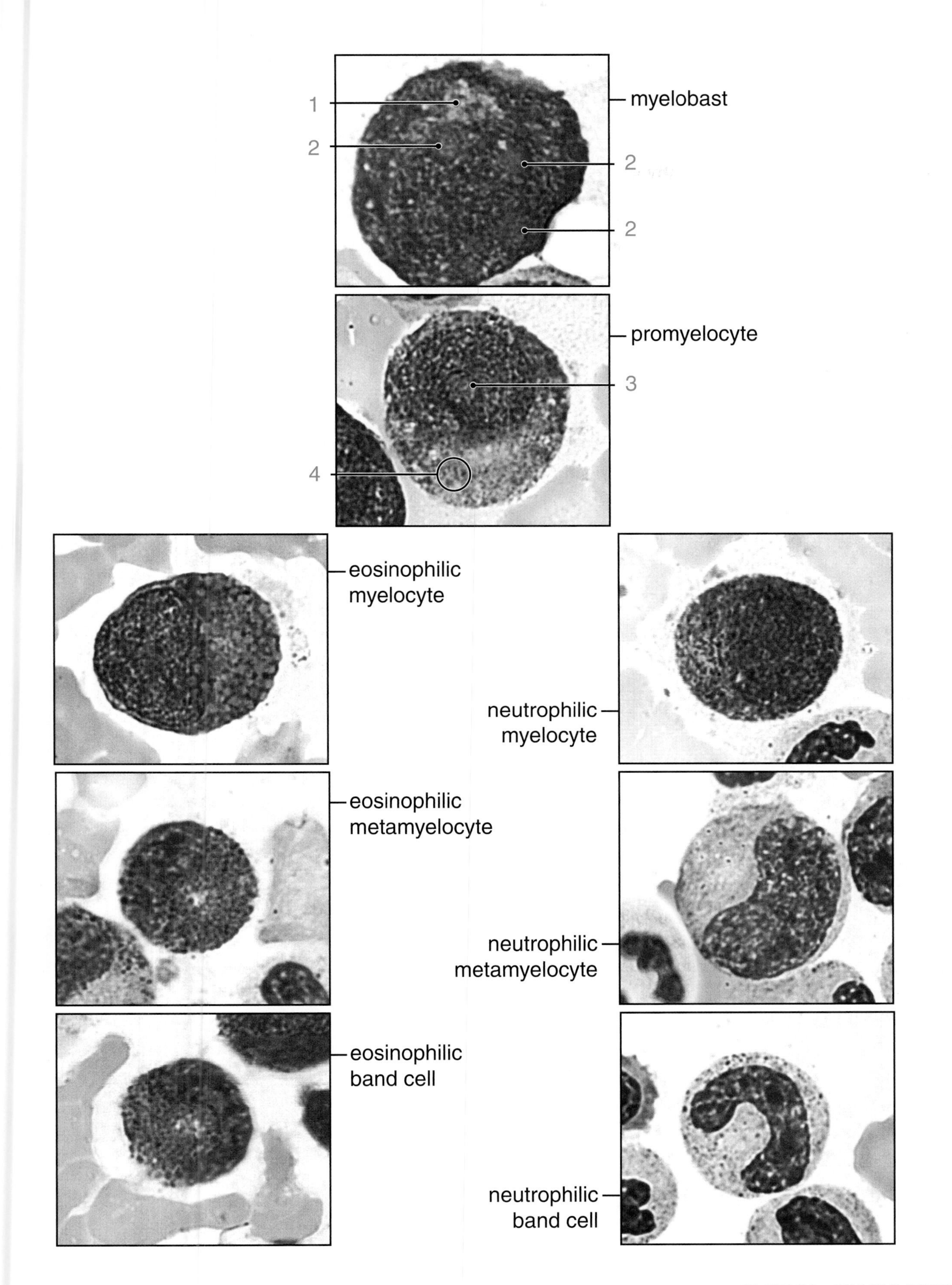

PLATE 29. GRANULOPOIESIS

CHAPTER 7
Muscle Tissue

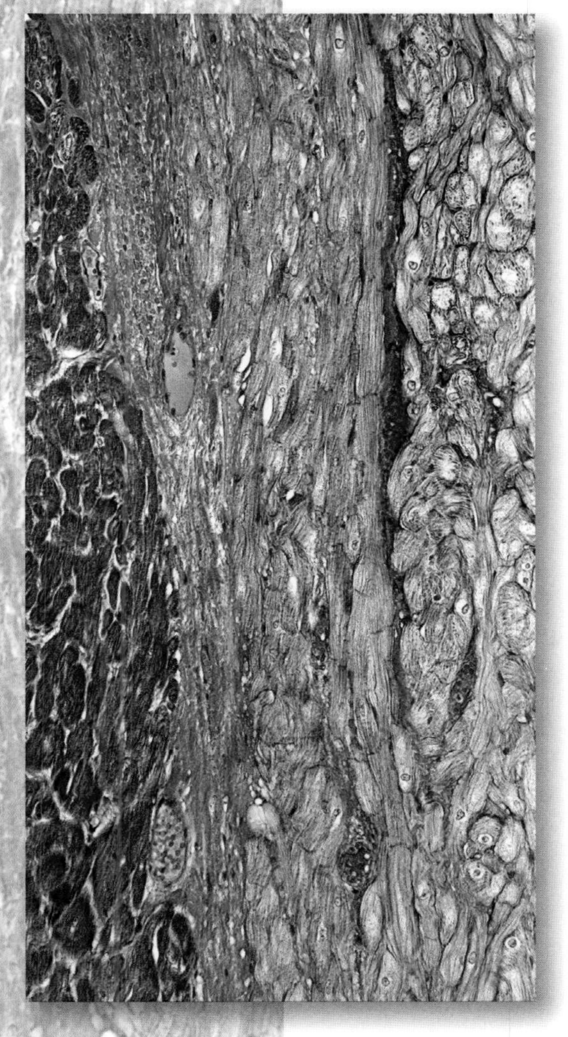

PLATE 30. SKELETAL MUSCLE I

Muscle tissue is classified based on the appearance of its contractile cells. Two major types are recognized: striated muscle, in which the cells exhibit a cross striation pattern when observed at the light microscope level; and smooth muscle, in which the cells lack striations.

Striated muscle is subclassified based on location, namely, skeletal muscle, visceral striated muscle, and cardiac muscle. Skeletal muscle is attached to bone and is responsible for movement of the axial and appendicular skeleton and maintenance of body position and posture. Visceral striated muscle is morphologically identical to skeletal muscle but is restricted to soft tissues, such as the tongue, pharynx, upper part of the esophagus, and diaphragm. Cardiac muscle is found in the heart and the base of the large veins that empty into the heart.

The cross striations in striated muscle are due to the organization of the contractile elements present in the muscle cell, namely, thin filaments, composed largely of the protein actin, and thick filaments, composed of the protein myosin II. These two types of myofilaments occupy the bulk of the cytoplasm. The skeletal and visceral striated muscle cell, more commonly called a fiber, is a multinucleated syncytium, formed by the fusion of individual, small muscle cells called myoblasts.

For individual muscle cells to work together collectively to create transduction of force, they are bundled by collagenous fibers. Surrounding each fiber is a delicate mesh of collagen fibrils referred to as endomysium. Bundles of muscle fibers that form functional units within a muscle are surrounded by a thicker connective tissue layer called perimysium. Lastly, the sheath of dense connective tissue that surrounds the muscle is referred to as epimysium. The force generated by individual muscle fibers is transferred to the collagenous elements of each of these connective tissue elements, which end in a tendon.

Skeletal muscle, H&E, x33.
This low power micrograph shows a longitudinal section of striated muscle. The muscle tissue within the muscle is arranged in a series of **fascicles** (1). The individual muscle fibers within a fascicle are in close proximity to one another and are not individually discernable. The small, blue, dot-like structures are nuclei of the fibers. Between the fascicles, though difficult to see at this magnification, is connective tissue, the **perimysium** (2). Also evident in this micrograph is a **nerve** (3).

Skeletal muscle, H&E, x33.
This micrograph reveals part of a muscle that has been cut in cross-section. Again, bundles of muscle fibers, or **fascicles** (4), can be readily identified. In contrast to the previous micrograph, individual **muscle fibers** (5) can be identified in many of the fascicles. The general cross-sectional shape of the fascicle is also readily visible. Each is bounded by connective tissue, which constitutes the **perimysium** (6). Also identifiable in this micrograph is a dense connective tissue surrounding the muscle, namely, **epimysium** (7).

Skeletal muscle, H&E, x256; inset x700.
This higher magnification of a longitudinal section of a muscle reveals two **muscle fibers** (8). At this magnification, the cross banding pattern is just perceptible. With few exceptions, the **nuclei** (9), which tend to run in linear arrays, belong to the muscle fibers. They are larger than the nuclei of the fibroblasts present in the endomysium, which are relatively few in number in a given section. Also evident in this micrograph is a small **blood vessel** (10). The **inset**, taken from a glutaraldehyde fixed plastic embedded specimen, is a much higher magnification of a portion of two muscle fibers. The major bands are readily identifiable at this magnification and degree of specimen preservation. The thick, darkly stained bands are the A bands. Between A bands are lightly stained areas, the I bands, which are longitudinally bisected by a Z line. The two elongate **nuclei** (11) belong to the muscle fibers. The muscle cell nuclei exhibit more euchromatin, with a speckling of heterochromatin, giving them a lighter staining appearance. Below them are a **capillary** (12) and a portion of an **endothelial cell nucleus** (13). At this higher magnification, the endothelial nuclei, as well as the nuclei of the fibroblasts, can be distinguished from the muscle cell nuclei by their smaller size and heterochromatin, giving them a darker stain.

Skeletal muscle, H&E, x256.
The muscle shown here has been cut in cross-section. In this plane, individual **muscle fibers** (14) are readily discernable, unlike in longitudinal sections. For example, when viewing a longitudinal section across a number of cells (see *dashed line*), the close proximity of the muscle cells can mask the boundary between individual cells within a fascicle. The **connective tissue** (15) that is readily apparent here belongs to the perimysium, which separates fascicles. The nuclei of the individual fibers are located at the periphery of the cell. At this magnification, it is difficult to distinguish between the nuclei of occasional fibroblasts, belonging to the endomysium, and the nuclei of the muscle cells.

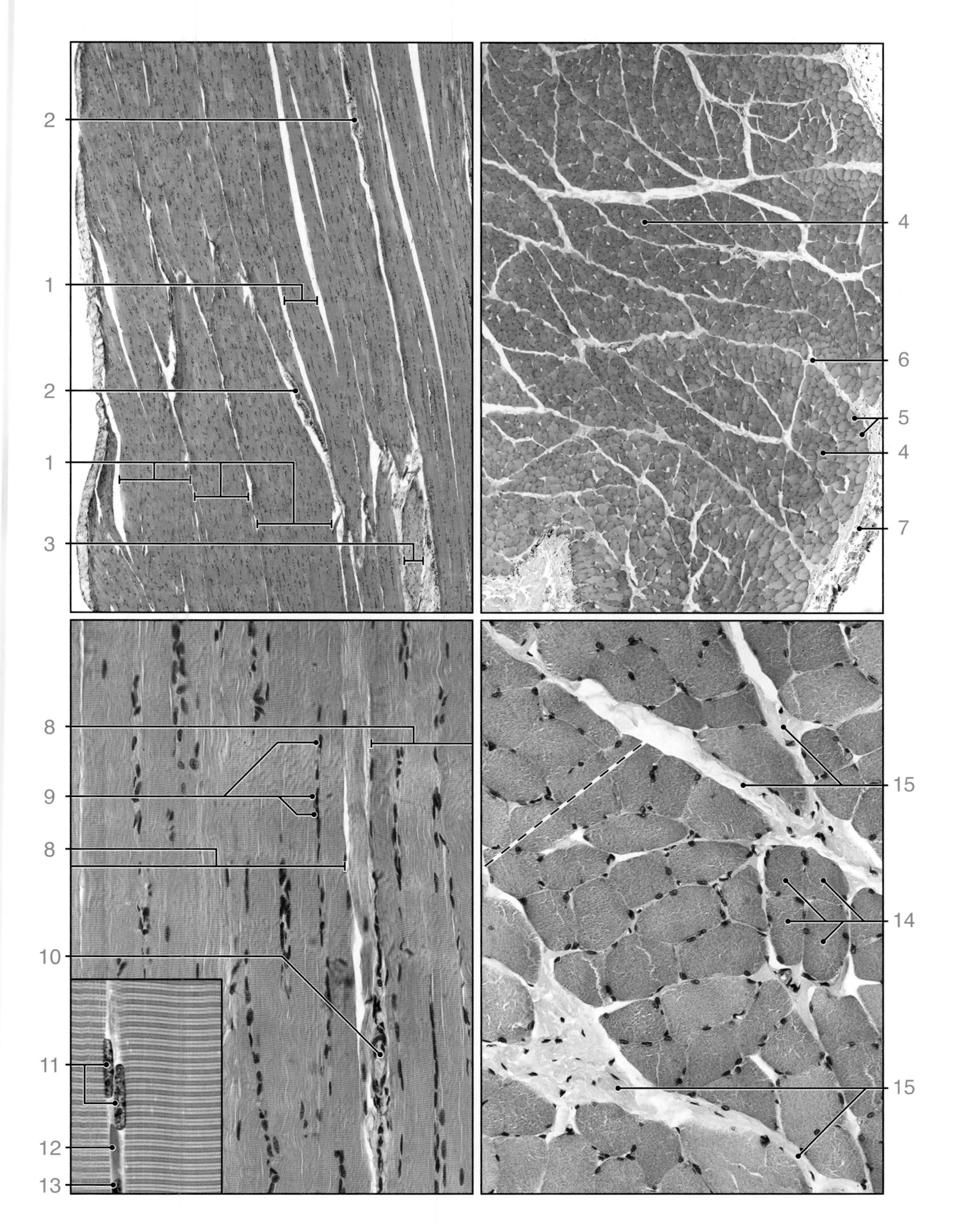
2
1
2
1
3
4
6
5
4
7
8
9
8
10
11
12
13
15
14
15

PLATE 31. SKELETAL MUSCLE II, ELECTRON MICROSCOPY

The myofibril is the structural and functional subunit of a muscle fiber. Myofibrils are best seen at higher magnification in the light microscope, in cross sections of the cells, where they appear as dot-like structures. Their presence gives the cytoplasm a stippled appearance. Each myofibril is composed of a bundle of two types of myofilaments. One type is the myosin II thick filament. The other is the thin filament, made up of actin and its associated proteins. The arrangement of the thick and thin filaments produces density differences that result in the cross striations of the myofibril when viewed in longitudinal section. The overlapping of thin and thick filaments produces the dark A band. The light-appearing I band contains only thin filaments. Careful examination of the A band in the light microscope reveals a lightly staining area in the middle of the A band. This is referred to as the H band—an area occupied by thin filaments and devoid of thick filaments. At the middle of each I band is the thin, dense Z line, to which the thin filaments are attached.

The distance between Z lines is referred to as a sarcomere. When a muscle contracts, the sarcomeres and I bands shorten. The filaments, however, maintain a constant length. Thus, contraction is produced by an increase in the overlap between the two filament types.

Skeletal muscle, H&E, x512; inset x985.
This micrograph reveals a cross section of a muscle fascicle. The individual **muscle fibers** (1) exhibit a polygonal shape, but vary only slightly in width. Of the many nuclei that can be seen in this plane of section, only some belong to the muscle fibers. The **nuclei** (2) that belong to the muscle fibers appear to be embedded within the extreme periphery of the fiber. In contrast, **fibroblast nuclei** (3), belonging to the endomysium, lie clearly outside of the muscle fiber. They are typically smaller and exhibit greater density than the nuclei of the muscle fibers. Also present between the muscle fibers are cross-sectioned **capillaries** (4). The **endothelial cell nuclei** (5) are also relatively dense. Other nuclei that may be present, but are very difficult to identify, belong to satellite cells. The **inset**, which shows the boxed area at higher magnification, reveals several nuclei; two belong to the muscle fibers (2). The small, very dense nucleus (3) probably belongs to a fibroblast of the endomysium. The striking feature at this magnification is the muscle cells' myofibrils, which appear as dot-like structures.

Skeletal muscle, H&E, x512; inset x985.
This micrograph, a longitudinal section of a glutaraldehyde fixed, plastic-embedded specimen, reveals four **myofibers** (6). Although they appear to be markedly different in width, the difference is due mainly to the plane of section through each of the fibers. Because the nuclei of the myofibers are located at the periphery of the cells, their position seems to vary when observed in longitudinal section. For example, three **nuclei** (7) are seen in what appears to be the center of a fiber. This is because the cut grazed the periphery of this fiber. The clear space at either end of two of these nuclei represents the cytoplasmic portion of the cell, which contains organelles and is devoid of myofibrils. Other **myofiber nuclei** (8) can be seen at the periphery of the myofibers. Note that they exhibit a similar chromatin pattern as the three nuclei previously described. Also present in this micrograph is a **capillary** (9), coursing along the center of the micrograph. In this plane of section, it is difficult to clearly distinguish between the endothelial cell nuclei and the nuclei of fibroblasts in the endomysium. Perhaps the most significant feature of a longitudinal section of a muscle fiber is the striations that they exhibit. The **inset** shows, at higher magnification, the banding pattern of the myofiber. The darkly staining lines represent the A band. The light-staining area is the I band, which is bisected by the darkly staining Z line.

Skeletal muscle, electron micrograph, x5000.
Compare this low power electron micrograph to the inset of the longitudinally sectioned myofibers in the top right micrograph. The micrograph reveals portions of three **myofibers** (10), two of which exhibit a **nucleus** (11). Between cells, various amounts of collagen fibers are present, representing the **endomysium** (12). This micrograph clearly illustrates the banding pattern of the myofibrils. In contrast to the longitudinally sectioned muscle in the top right inset, individual **myofibrils** (13) can be identified in this micrograph. They correspond to the dot-like structures seen in the top left inset, showing cross-sectioned myofibers. Note that adjacent myofibrils are aligned with one another with respect to their banding pattern and also that they exhibit different widths. Each myofibril is essentially a cylindrical structure, much like a dowel; thus, when sectioned in a longitudinal plane, the width of each myofibril will vary depending on what portion of the cylindrical structure has been cut. The sarcoplasmic reticulum, a membrane system that is present between myofibrils and the nature of the bands in a myofibril are shown to advantage in the next plate.

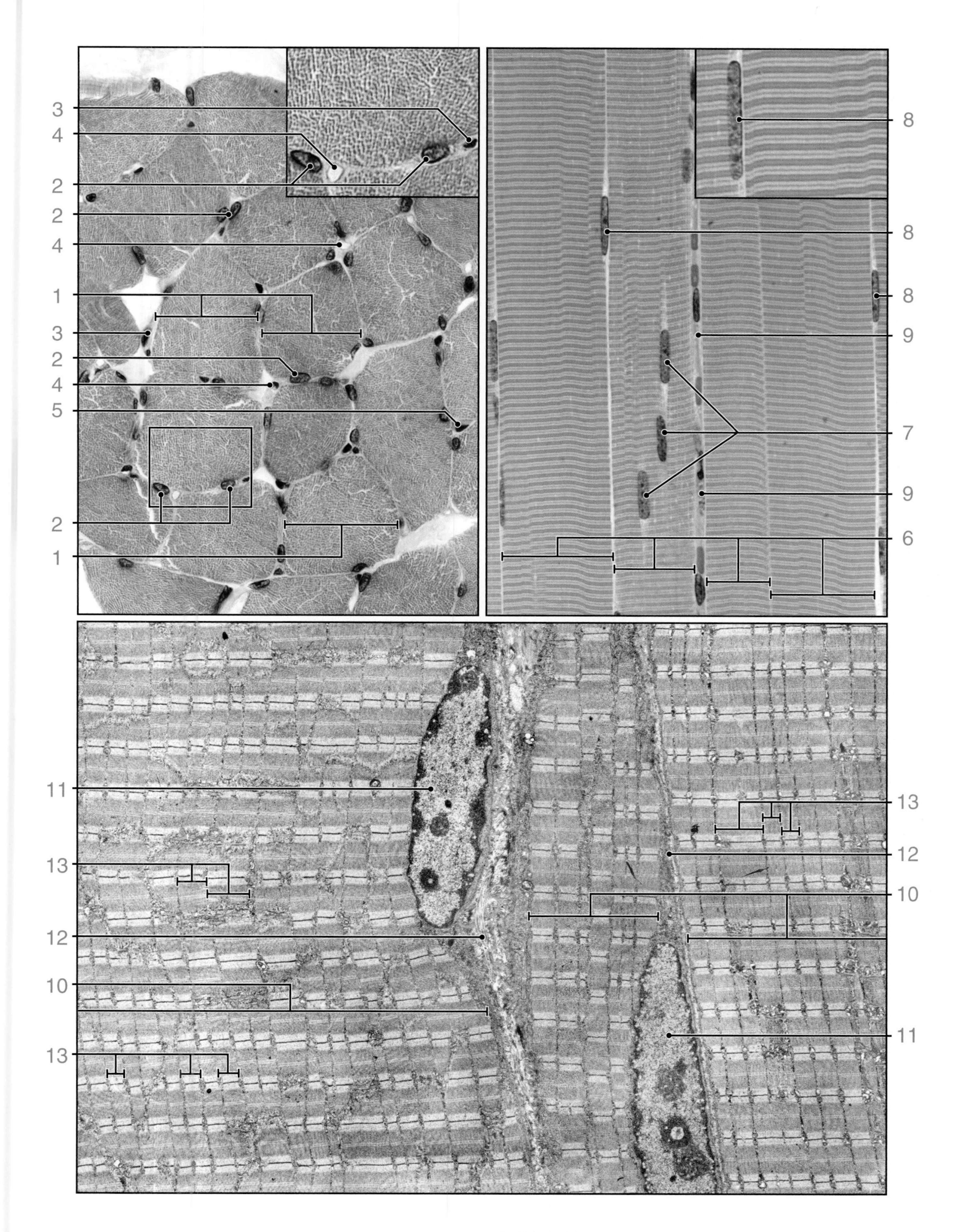

PLATE 31. SKELETAL MUSCLE II, ELECTRON MICROSCOPY

PLATE 32. SKELETAL MUSCLE III, ELECTRON MICROSCOPY

Skeletal muscle, electron micrograph, x45,000; insets x52,000.

The electron micrograph shown here illustrates the nature of the sarcoplasm, especially the membrane system that pervades it. It also demonstrates the filamentous components that make up the myofibril. The muscle fiber is longitudinally sectioned but has been turned so that the direction of the fiber has a vertical orientation. The various bands and lines of a **sarcomere** (1) are labeled in the myofibril on the left of the illustration. The thin actin filaments are especially well resolved in the **I band** (2) in the lower left corner and can easily be compared with the thicker myosin filaments of the **A band** (3). The point of juncture or insertion of the actin filaments into the **Z line** (4) is also readily apparent.

In contrast to the relatively nondescript character of the sarcoplasm as seen in the light microscope, the electron microscope reveals a well-developed membrane system called the sarcoplasmic reticulum. The sarcoplasmic reticulum consists of segments of anastomosing tubules that form a network around each myofibril. This micrograph fortuitously depicts a relatively wide area of **sarcoplasm** (5) in a plane between two myofibrils. Prominent in this area are numerous glycogen particles, which appear as dense, dot-like structures, and **mitochondria** (6). Somewhat less apparent but nevertheless evident are the anastomosing tubules of the **sarcoplasmic reticulum** (7). Near the junction of the A and I bands, the tubules of the reticulum become confluent, forming flattened, saclike structures known as **terminal cisternae** (8). These cisternae come in proximity to another membrane system, the **T (transverse tubular) system** (9).

The T system consists of membranous tubular structures, with each tubule originating from an invagination of the sarcolemma. The tubules course transversely through the muscle fiber. Because of the uniform register of the myofibrils, each T tubule comes to surround the myofibrils near the juncture of the A and I bands. In effect, the T system is not simply a straight tubule, rather it is a grid-like system that surrounds each myofibril at the A–I junction. Thus, in a section passing tangentially to a myofibril, as is seen in the center of the micrograph and in the **upper inset**, the lumen of the T tubules may appear as elongate channels bounded by a pair of membranes. The inner facing pair of membranes belongs to the T tubule. The outer membranes, on either side, belong to the terminal cisternae of the sarcoplasmic reticulum. The communication between a terminal cisterna and the tubular portion of the sarcoplasmic reticulum is marked by an *arrowhead* in the **upper inset**. In contrast, a longitudinal section passing through two adjacent myofibrils (**lower inset**) reveals the T tubule to be flattened with the terminal elements of the sarcoplasmic reticulum on either side. The combination of the T tubule and the adjoining dilated terminal cisternae of the sarcoplasmic reticulum on either side is referred to as a **triad** (10).

The nature and geometric configuration of the triad helps explain the rapid and uniform contraction of a muscle fiber. The depolarization of the sarcolemma continues along the membranes of the T tubules and, thereby, results in an inward spread of excitation to reach each myofibril at the A–I junction. This initiates the first stage in the contraction process (i.e., the release of calcium ions from the immediately adjacent terminal cisternae). Relaxation occurs through the recapture of calcium ions by the sarcoplasmic reticulum. In terms of energetics, it is also of interest that the mitochondria occupy a preferential site in the sarcoplasm, being oriented in a circular fashion around the myofibrils in the region of the I band.

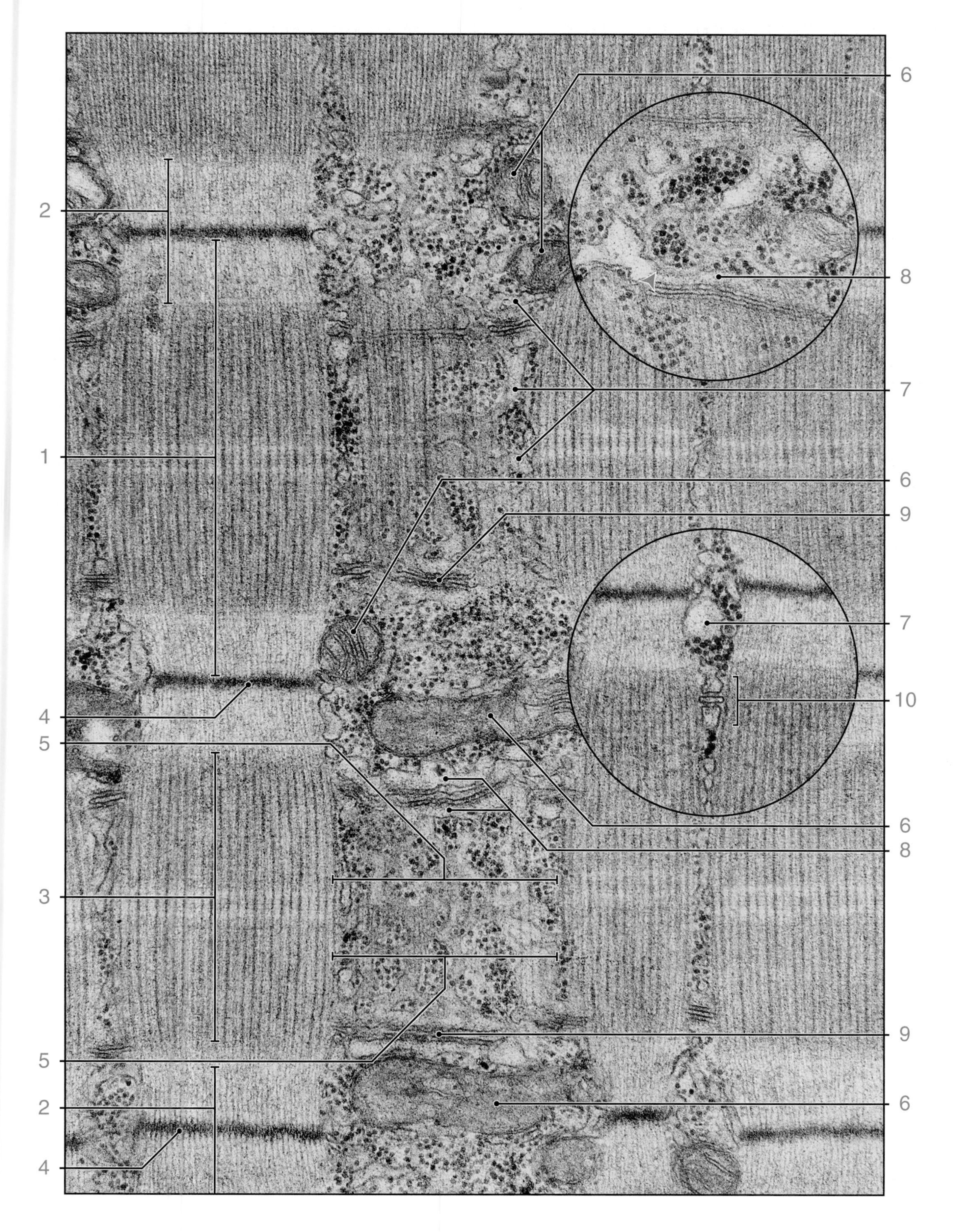

PLATE 32. SKELETAL MUSCLE III, ELECTRON MICROSCOPY

PLATE 33. **MUSCULOTENDINOUS JUNCTION**

The work to allow body movement is effected by skeletal muscle and the tendons to which the muscle fibers are attached. The site of attachment between a muscle fiber and the collagen of a tendon is referred to as the musculotendinous junction. The muscle fibers at the junctional site end in numerous fingerlike cytoplasmic projections. At the end of each projection and between projections, the collagen fibrils of the tendon attach to the basal lamina of the muscle cell (see the electron micrograph at the bottom of this Plate). In the light microscope, these fingerlike projections appear to merge with the tendon. The detail of this relationship is visible at the electron microscope level. The last sarcomeres in the muscle fiber end where the fingerlike projections begin. At this point, the ending sarcomere lacks its Z line, and the actin filaments from the A band continue into the cytoplasmic fingers, ending at the sarcolemma.

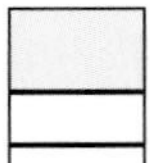

Musculotendinous junction, monkey, H&E, x365.

This light micrograph shows a **tendon** (1) and several **muscle fibers** (2). The tendon contains scattered tendinocytes whose **nuclei** (3) are compressed between the collagenous bundles of the tendon. Several of the **muscle fibers** (4) are seen at the point where they terminate and attach to the tendon fibers. The boxed area is shown at higher magnification in the micrograph below.

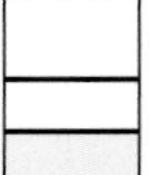

Musculotendinous junction, monkey, H&E, x1560.

The **muscle fiber** (5) in this micrograph is seen at its musculotendinous junction. Note the banding pattern of the muscle fiber. At this magnification, the fingerlike projections (*arrows*) at the end of the muscle fiber are clearly visible. Between the fingerlike structures are the collagen fibers of the tendon. The nuclei of the **tendinocytes** (6) are evident in the tendon.

Musculotendinous junction, electron micrograph, x24,000.

This micrograph also shows the end of a muscle. Note that the last **sarcomere** (7) lacks a Z line. The actin filaments appear to extend from the A band, continuing along the length of the finger projections, and then seem to attach to the sarcolemma. Between the finger projections are the collagen fibrils (*arrows*) that make up the tendon. (Micrograph courtesy of Douglas E. Kelly.)

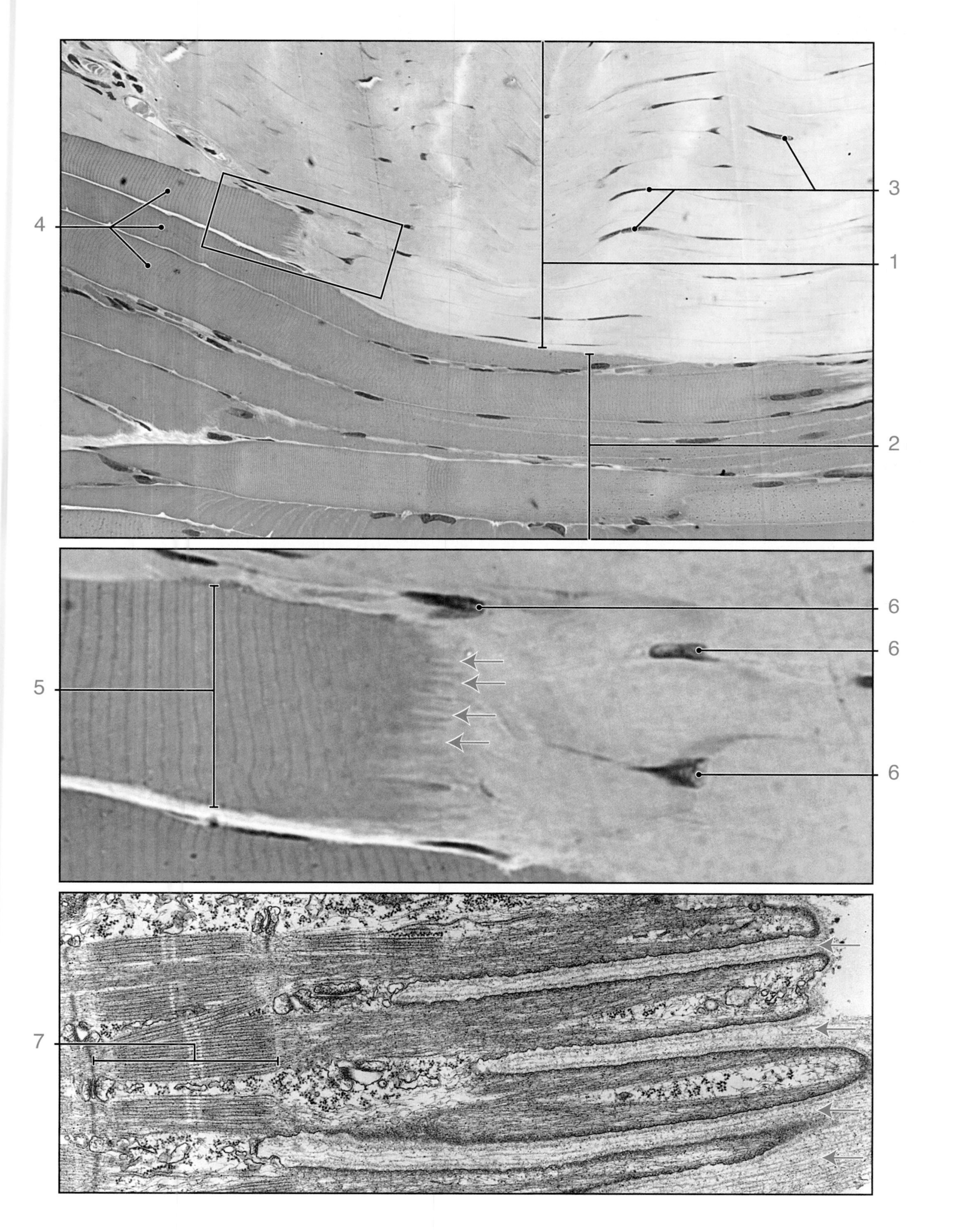

PLATE 33. MUSCULOTENDINOUS JUNCTION

PLATE 34. MUSCLE SPINDLES AND DEVELOPING SKELETAL MUSCLE

Muscle spindles are narrow, elongate (0.75 to 7 mm and sometimes longer) structures that are wider in their mid-region. They are arranged parallel to the bundles of muscle fibers and tend to be embedded within the muscle bundle. Their functions are both motor and sensory. Each spindle is surrounded by a connective tissue capsule, enclosing a variable number of muscle fibers and individual nerve fibers, as well as blood vessels and nerve terminals. These elements are contained within a fluid medium bound by the capsule. The muscle fibers of the spindle are narrow fibers, much smaller in diameter than the muscle fibers surrounding the spindle. They are referred to as intrafusal fibers. Each spindle is innervated by motor nerves that terminate at the intrafusal fibers, and each has typical motor plates. In addition, the spindle has one or more thicker, sensory nerve fibers whose axons are covered by a thin layer of Schwann cell cytoplasm. Identifying muscle spindles in histologic sections is accomplished best by examining cross-sectioned striated muscle.

Striated muscle arises from mesoderm, providing a self-renewing population of multipotential myogenic stem cells. These cells differentiate into myoblasts. The early myoblasts fuse end-to-end, forming early, or primary, myotubes, which are chain-like structures. The myotubes exhibit multiple central nuclei with the peripheral portion of the myotube exhibiting myofilaments. Subsequently, the myotubes become innervated through direct contact with nerve terminals, thereby defining the myotubes as late, or secondary, myotubes. The secondary myotubes continue to be formed by sequential fusion of late myoblasts into the already formed secondary myotubes at random positions along their length thus elongating the myotube. The mature, multinucleated muscle fiber arises when the nuclei occupy a position in the peripheral sarcoplasm just beneath the plasma membrane.

Skeletal muscle, H&E, x128.
This relatively low magnification of striated muscle shows cross-sectioned profiles of two **muscle spindles** (1). They tend to be embedded within a muscle bundle. The cross-sectioned profiles are more readily encountered than specimens wherein the muscle is longitudinally sectioned.

Skeletal muscle, H&E, x512.
The muscle spindle in the lower left corner of the previous micrograph is shown here at higher magnification. The extremely thin cytoplasm of the **capsule** (2) is clearly evident. The small circular structures are the **intrafusal fibers** (3). The intrafusal fibers are bound together by **connective tissue** (4). Also evident is a **capillary** (5). The amorphous, eosin-stained material is **precipitated protein** (6) from the fluid within the spindle.

Developing muscle fibers, human fetus, Mallory-Azan, x128.
This low power micrograph shows several **muscle fascicles** (7). The connective tissue fibers of the developing **epimysium** (8) and the **perimysium** (9) are stained light blue. Also evident in the micrograph is a developing **tendon** (10). The more abundant presence of collagen fibers in the tendon gives it a dark blue color. At this stage of development, the muscle fascicles are made up of thin cylindrical myotubes.

Developing muscle fibers, human fetus, Mallory-Azan, x512; inset x1000.
This micrograph shows the myotubes of a muscle fascicle at higher magnification. Their **nuclei** (11) occupy a central location in the myotube. In other sites, the **banding pattern** (12) of the myofibrils is clearly evident. Between the myotubes, the lightly staining **endomysium** (13) is evident. The elongate nuclei in the endomysium belong to **fibroblasts** (14). The **inset** shows, at a higher magnification, a centrally located nucleus and the peripherally organized myofibrils in one of the myotubes.

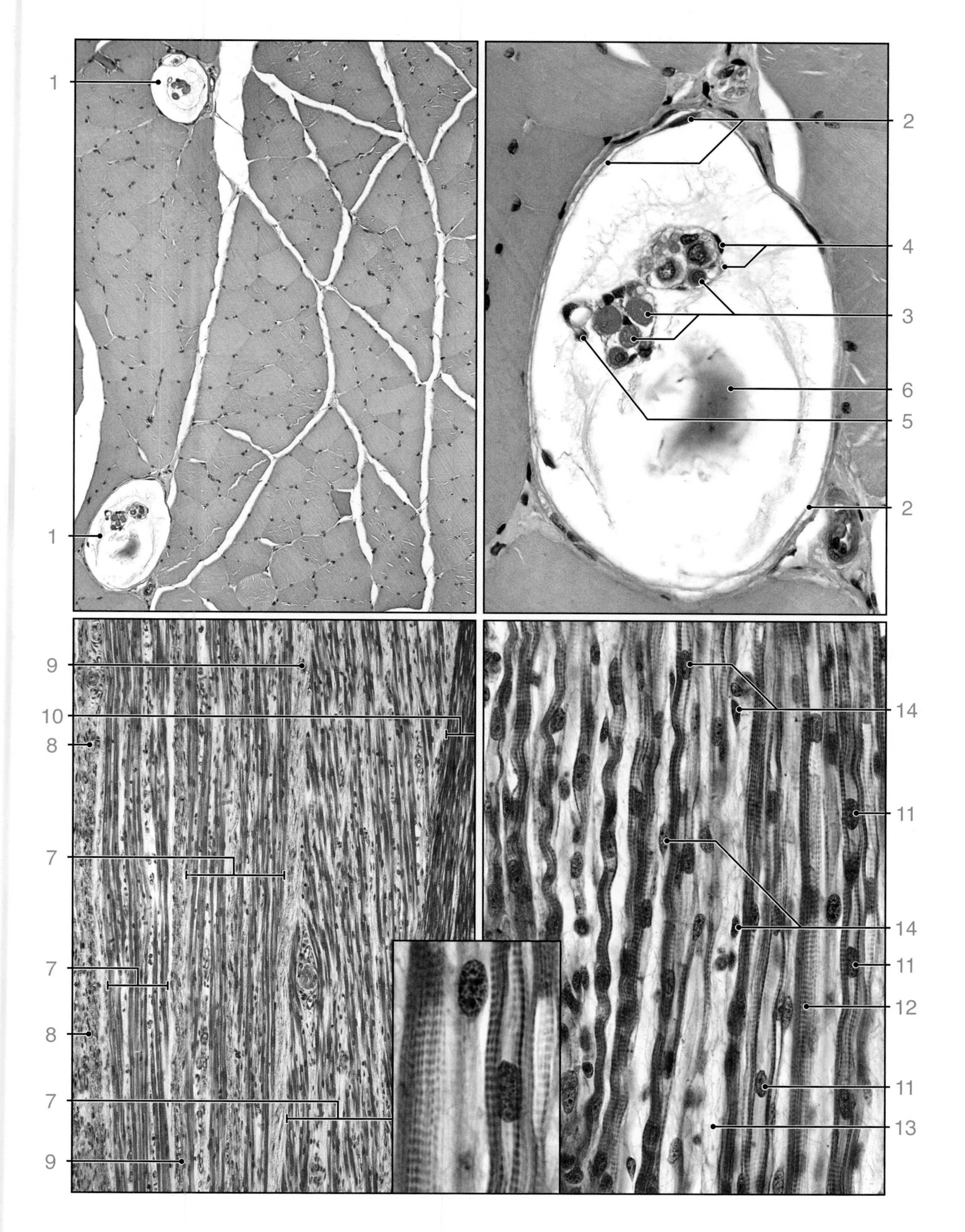

PLATE 34. MUSCLE SPINDLES AND DEVELOPING SKELETAL MUSCLE

PLATE 35. NEUROMUSCULAR JUNCTION

Nerve supply to skeletal muscle is from myelinated cerebrospinal afferent and efferent nerves. Each muscle receives one or more nerves that contain motor fibers, sensory fibers to muscle spindles, and neurotendinous sensory endings as well as autonomic nerve endings that supply its blood vessels. Functionally, a muscle is composed of motor units, which consist of a single nerve fiber and the muscle fiber or fibers it innervates. In the case of very delicate movement, as in the muscles of the eye, a single nerve fiber supplies each muscle fiber. In larger muscles, such as those of the body trunk, a single nerve may supply many muscle fibers. As a nerve fiber, or axon, approaches a muscle fiber, it loses its myelin sheath and is covered only by an exceedingly thin layer of Schwann cell cytoplasm and its external lamina. The end of the axon then divides into several end branches. Each end branch comes to lie in a shallow depression on the muscle fiber and provides a number of receptor sites, or neuromuscular junctions. The plasma membrane of the muscle fiber at the junctions exhibits many deep folds, referred to as junctional folds, or subneural folds. Between this area of the muscle cell and the neuromuscular junctions, there is a narrow space, referred to as the synaptic cleft. Within the synaptic cleft is the external lamina, which extends from the Schwann cell and the muscle cell. The nerve ending, when examined by electron microscopy, is largely filled with mitochondria and small vesicles. The vesicles contain acetylcholine, whose release into the synaptic cleft initiates the polarization of the plasma membrane of the muscle cell, resulting in its contraction.

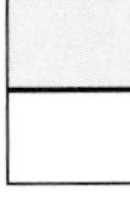

Neuromuscular junction, human, silver stain, x1000.

This light micrograph shows a **nerve** (1) overlying several muscle fibers. The specimen was prepared by teasing out some muscle fibers from a muscle and prepared for light microscopy without sectioning. In certain areas along the edges of the muscle fibers, the characteristic **striations** (2) are evident. The nerve consists of a number of **axons** (3), several of which can be seen leaving the nerve bundle and terminating on the muscle fibers. At its end, the axon branches and each branch (*arrows*) continues along the surface of the muscle fiber, giving off terminal sites of neuromuscular junctions. The **neuromuscular junctions** (4) appear as dense, black spots.

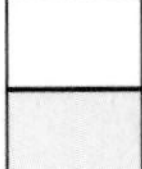

Neuromuscular junction, human, electron micrograph, x45,000.

This electron micrograph of a neuromuscular junction reveals a **terminal component of an axon** (5), which contains numerous **mitochondria** (6) and **acetylcholine-containing synaptic vesicles** (7). The part of the motor axon ending that is not in apposition with the muscle fiber is covered by **Schwann cell cytoplasm** (8), consisting of a very thin layer of cytoplasm, and is devoid of a myelin covering. The muscle fiber exhibits a number of **junctional folds** (9). The space around the junctional folds and the axon terminal is referred to as the **synaptic cleft** (10). The external lamina of the muscle fiber and the Schwann cell is barely evident in this micrograph, but it occupies the space of the synaptic cleft. (Micrograph courtesy of Dr. George Pappas.)

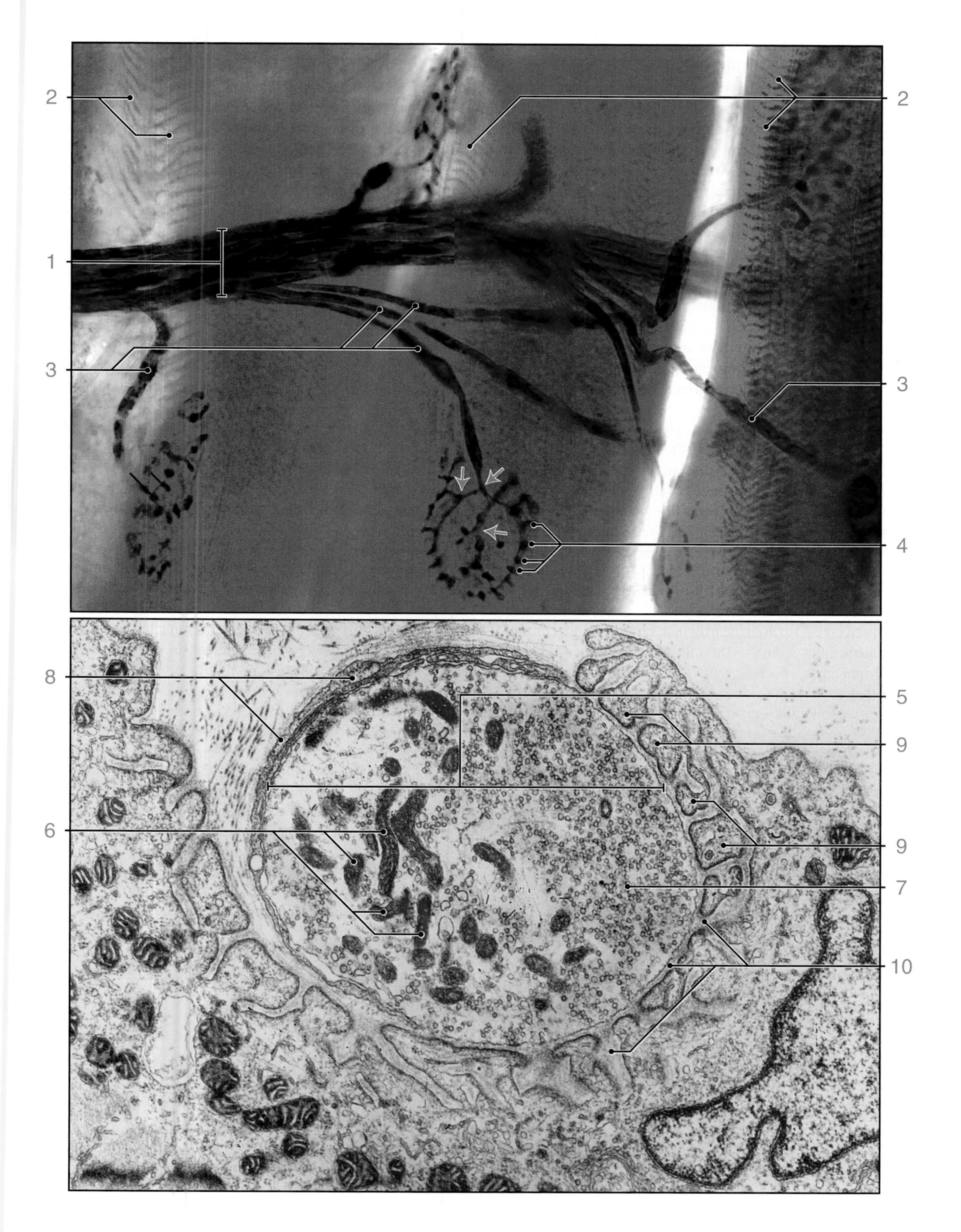

PLATE 35. NEUROMUSCULAR JUNCTION

PLATE 37. CARDIAC INTERCALATED DISC, ELECTRON MICROSCOPY

Cardiac muscle, monkey, electron microscopy, x40,000.

This micrograph shows a small portion of two cardiac muscle cells and the intercalated disc that joins them. Also evident in the micrograph is an area of **unattached lateral surfaces** (1) of each cell. The **myofibrils** (2) are oriented diagonally across the micrograph. Due to the contracted state of the muscle cells, the **Z lines** (3) are readily recognizable, but the adjacent I bands have become almost obscured. Other features that are readily recognized include **mitochondria** (4), **T tubules** (5) with adjacent profiles of sarcoplasmic reticulum (diads), and a few **lipid droplets** (6). The intercalated disc, which marks the line of junction between the two cells, takes an irregular, steplike course, making a number of right angle turns. This pattern provides a more extensive surface area between the two cells, compared to two flat adjoining surfaces. Viewing the junctional complex, starting on the left, the **macula adherens** (7) is shown enlarged in the **upper inset**. Note the dense **intracellular attachment plaques** (8) of the adjoining cells and the **intermediate line** (9) in the extracellular space. The **fascia adherens** (10), shown at higher magnification in the **lower inset**, is considerably more extensive relative to surface area between the adjoining cells than is the desmosome. Where the cells adjoin one another perpendicular to the aforementioned surface (lateral component of the intercalated disc), the two cells form a **gap junction** (11). Of significance here is that this surface is not exposed to the contractile forces of the adjoining cells and therefore would not represent a weak point in the end-to-end attachment between the two cells. The **middle inset** shows this junction at higher magnification. Three dense **matrix granules** (12) in the mitochondrion in the cell on the left are visible.

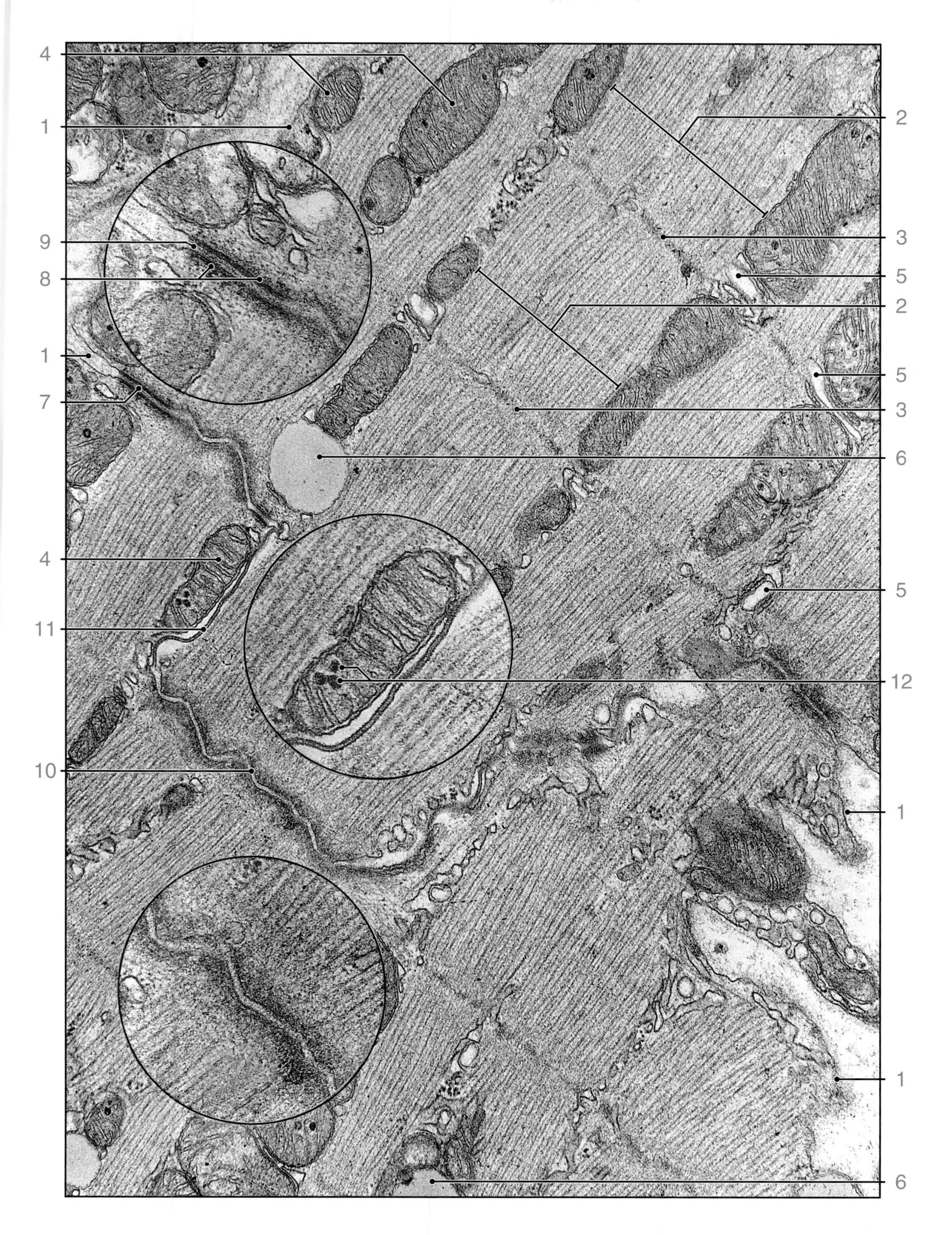

PLATE 37. CARDIAC INTERCALATED DISC, ELECTRON MICROSCOPY

PLATE 38. **CARDIAC MUSCLE, PURKINJE FIBERS**

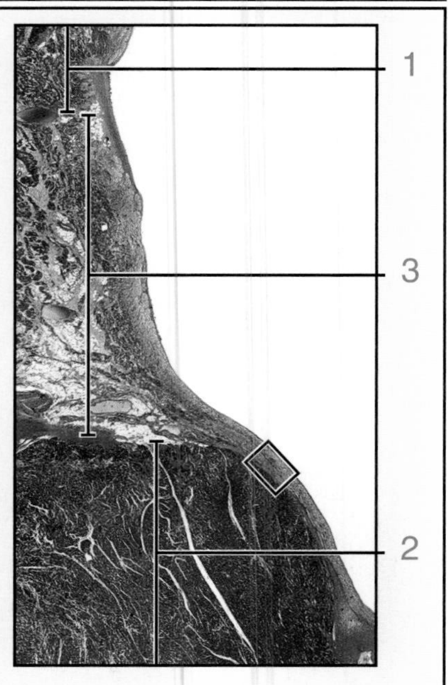

Cardiac muscle cells possess the ability for spontaneous rhythmic contractions. The contraction, or beat, of the heart is regulated and coordinated by specialized and modified cardiac muscle cells that are found in nodes and muscle bundles. The beat of the heart is initiated at the sinoatrial (SA) node, which consists of a group of specialized cardiac muscle cells located at the junction of the superior vena cava in the right atrium. The impulse spreads from this node along the cardiac muscle fibers of the atria. The impulse is then received at the atrioventricular (AV) node, which is located in the lower part of the right atrium on the interatrial septum adjacent to the tricuspid valve. Specialized cells that leave the AV node are grouped into a bundle, the AV bundle (of His). It conducts impulses from the AV node to the ventricles. This bundle then divides into two main branches that descend along the ventricular septum and into the ventricular walls. The left bundle branch pierces the interventricular septum and goes to the left ventricle and the right bundle branch goes to the right ventricle. The specialized conducting fibers carry the impulse approximately four times faster than regular cardiac muscle fibers. They are responsible for the final distribution of the electrical stimulus to the myocardium. While the SA node exhibits a constant or inherent rhythm on its own, it is modulated by the autonomic nervous system. Thus, the rate of the heartbeat can be decreased by parasympathetic fibers from the vagus nerve or increased by fibers from sympathetic ganglia. The specialized conducting cells within the ventricles are referred to as Purkinje fibers. The cells that make up the Purkinje fibers differ from cardiac muscle cells in that they are larger and have their myofibrils located mostly at the periphery of the cell. Their nuclei are also larger. The cytoplasm between the nucleus and the peripherally located myofibrils stains poorly, a reflection, in part, of the large amount of glycogen present in this part of the cell.

ORIENTATION MICROGRAPH: The specimen shown here is a sagittal section revealing part of the **atrial wall** (1) and the **ventricular wall** (2). Between these two portions of the heart is the **atrioventricular septum** (3). The clear space is the interior of the atrium.

Purkinje fibers, human, Masson, x180.

This micrograph shows the boxed area from the orientation micrograph. At this site, the **endocardium** (1) has been divided by bundles of **Purkinje fibers (bundle of His)** (2) coursing along the ventricle wall. Normally, the endocardium consists of three layers. The **endothelium** (3), lining the ventricle, is the most superficial but is barely detectable at this magnification. Beneath the endothelium is a middle layer consisting of **dense irregular connective tissue** (4), in which elastic fibers are present as well as some smooth muscle cells. The third layer, the **deepest part of the endocardium** (5), consists of more irregularly arranged connective tissue with blood vessels and occasional fat cells. At the bottom of the micrograph is **myocardium** (6), containing cardiac muscle fibers. Note how darkly stained the cardiac muscle fibers are compared to those of the Purkinje fibers.

Purkinje fibers, human, Masson, x365; inset x600.

This is a higher magnification of the boxed area on the top micrograph. It reveals the endothelial cells of the **endocardium** (7) and the underlying connective tissue, containing **smooth muscle cells** (8). Where the Purkinje fibers are cross-sectioned or obliquely sectioned, the **myofibrils** (9) are seen at the periphery of the cell. The cytoplasm in the inner portion of the cell appears unstained. Where the nuclei are included in the section of the cell, they are surrounded by the clear cytoplasm. In the lower portion of the figure, several longitudinally sectioned Purkinje fibers can be seen. Note the **intercalated discs** (10) when seen in this profile. The **inset** better reveals the intercalated discs and the myofibrils with their cross-banding. Note the clear, unstained cytoplasm surrounding the nuclei.

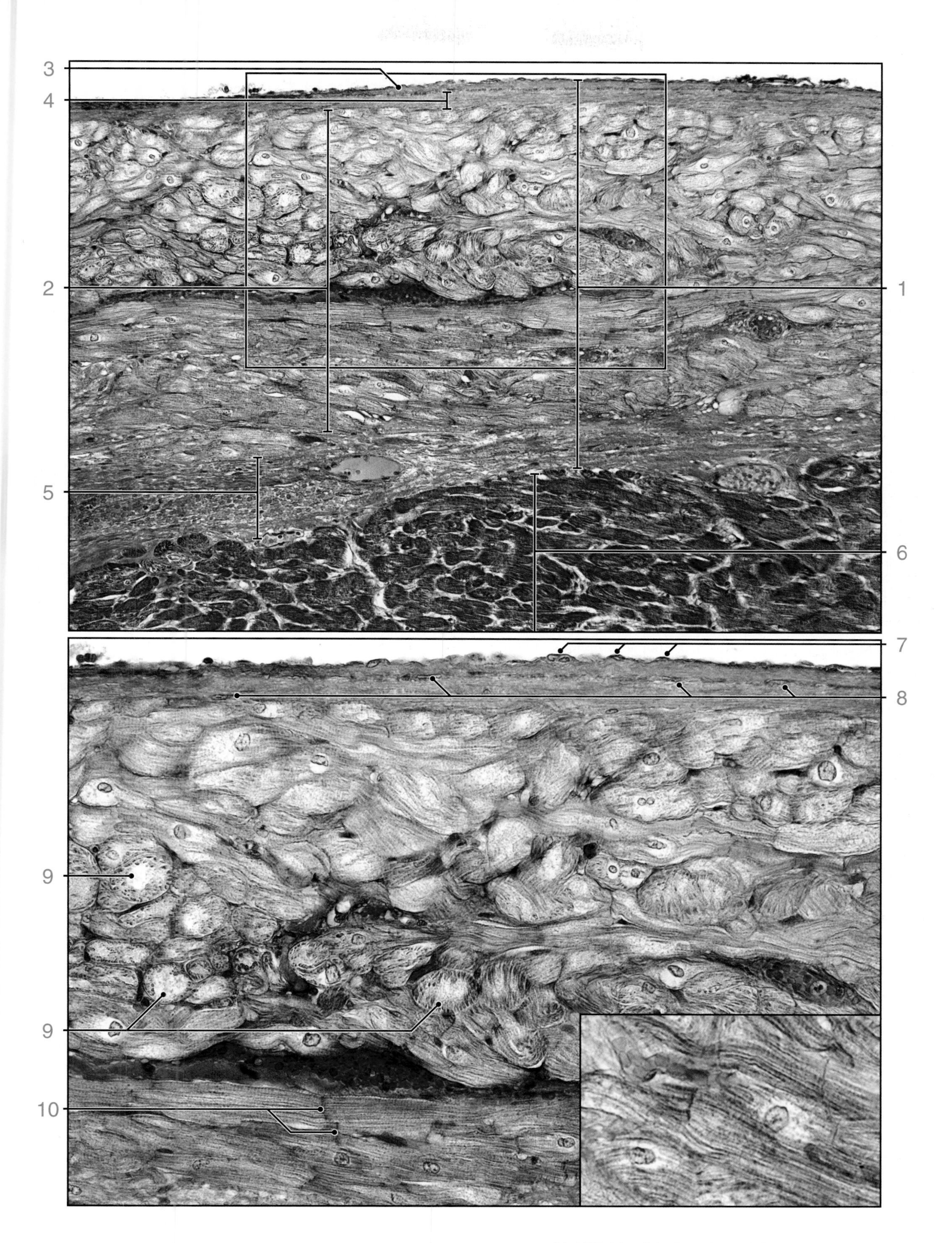

PLATE 38. CARDIAC MUSCLE, PURKINJE FIBERS

PLATE 39. SMOOTH MUSCLE I

Smooth muscle forms the intrinsic muscle layer of the alimentary canal, blood vessels, the genitourinary and respiratory tracts, and other hollow and tubular organs. It is also a component of the nipple, scrotum, skin (arrector pili muscle), and eye (iris). In most locations, smooth muscle consists of bundles, or layers, of elongate, fusiform cells. They range in length from 20 μm, in the walls of small blood vessels, to about 200 μm, in the intestinal wall. In the uterus, they may become as large as 500 μm during pregnancy. The smooth muscle cells are joined by gap junctions, which allow small molecules and ions to pass from cell to cell, and allow regulation of contraction of the entire bundle or sheet of smooth muscle. In routine H&E preparations, the cytoplasm of smooth muscle cells stains uniformly with eosin because of the concentration of actin and myosin in these cells. The nucleus is located in the center of the cell and is elongate with tapering ends, matching the shape of the cell. When the cell is maximally contracted, the nucleus exhibits a corkscrew shape. During lesser degrees of contraction, the nucleus may appear to have a slight spiral shape. Often in H&E preparations, smooth muscle stains much the same as dense connective tissue. A distinguishing feature of smooth muscle is that its nuclei are considerably more numerous than connective tissue, and they tend to look uniform, appearing elongate when longitudinally sectioned and circular when cross-sectioned. In contrast, the nuclei of dense connective tissue are fewer in number per unit area and appear in varying shapes in a given section.

Smooth muscle, small intestine, human, H&E, x256.

This low power micrograph shows part of the wall of the small intestine, the muscularis externa. The left side of the micrograph reveals two **longitudinally sectioned bundles** (1); whereas on the right side, smooth muscle bundles are seen in **cross-section** (2). Note that the nuclei of the smooth muscle cells in the longitudinally sectioned bundles are all elongate; in contrast, the nuclei in the cross-sectioned smooth muscle bundles appear circular. Intermixed between the bundles is **dense irregular connective tissue** (3). While the smooth muscle cells and the dense connective tissue both stain with eosin, the dense connective tissue exhibits a paucity of nuclei compared to the smooth muscle cell bundles.

Smooth muscle, small intestine, human, H&E, x512.

This higher magnification micrograph shows a bundle of **smooth muscle cells** (4). Note how the nuclei exhibit an undulating, or wavy, form indicating that the cells are partially contracted. The nuclei seen in the **dense connective tissue** (5), in contrast, exhibit a variety of shapes. The collagen fibers, in this and the top left micrograph, have a brighter red coloration than the cytoplasm of the smooth muscle cells, providing further distinction between the two types of tissue. This, however, is not always the case, and the two may appear similarly stained.

Smooth muscle, small intestine, human, H&E, x256.

This micrograph shows several bundles of cross-sectioned **smooth muscle** (6) at low magnification. Note how the smooth muscle bundles are separated from one another by **dense connective tissue** (7). Also note the numerous circular profiles of the smooth muscle cell nuclei.

Smooth muscle, small intestine, human, H&E, x512; inset x1185.

At higher magnification, the smooth muscle is again seen in cross-section. As is typical, the distribution of the smooth muscle cell nuclei is not uniform. Thus, in some areas there appears to be **crowding of nuclei** (8), whereas in other areas, there appears to be a **paucity of nuclei** (9). This reflects the side-by-side orientation of the smooth muscle cells—in this area, the cells are aligned such that their nuclei have not been included in the plane of this section. The **inset** is a higher magnification of this area of few nuclei and reveals the variably-sized, circular profiles of the cross-sectioned smooth muscle cells.

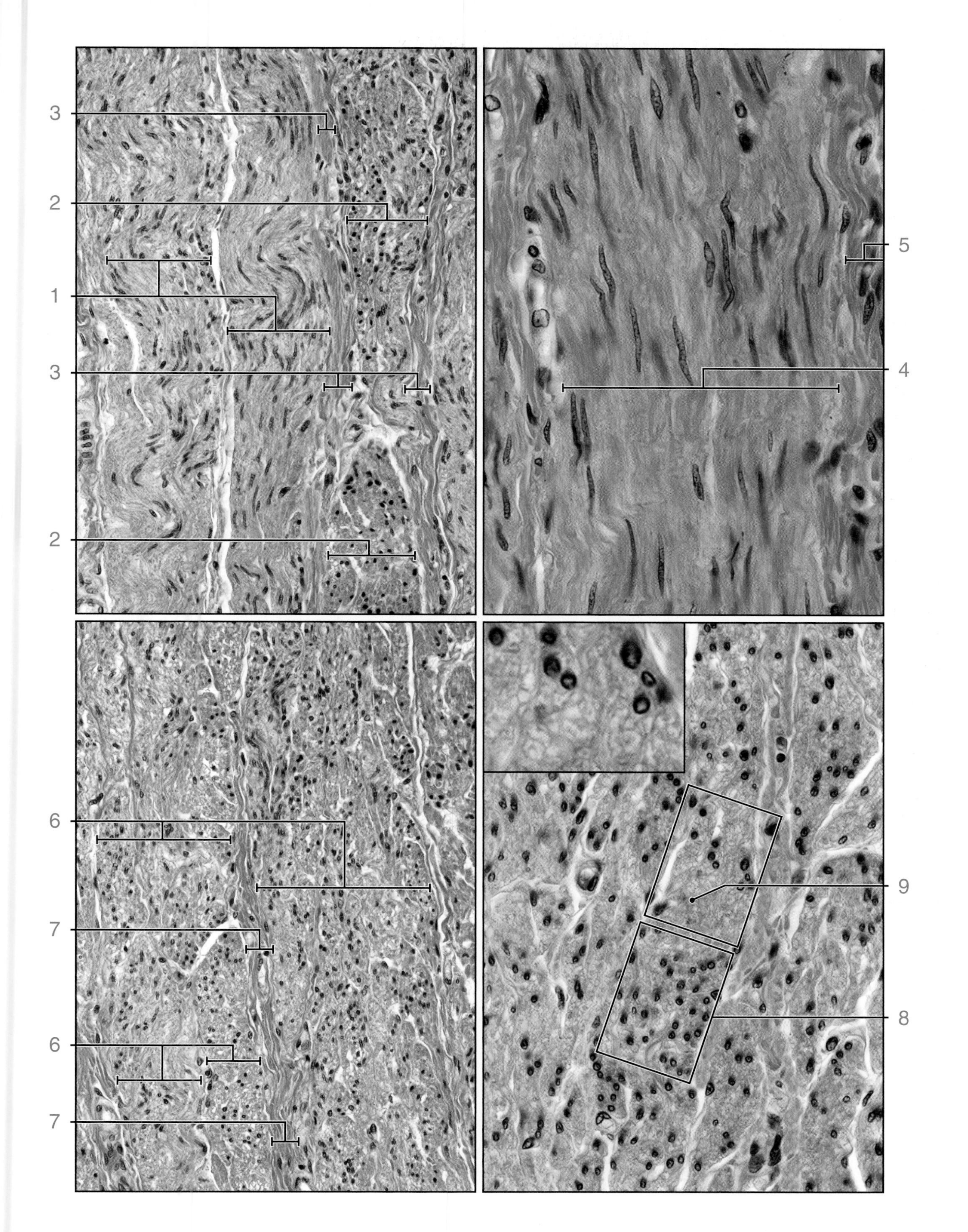
3
2
1
3
2
5
4
6
7
6
7
9
8

PLATE 40. **SMOOTH MUSCLE II**

Smooth muscle, cervix, human, H&E, x256.
The myometrium of the cervix is the thickest component and is arranged in three relatively distinct layers, namely, external, middle, and internal layers. The external and internal layers consist mostly of longitudinal fibers, with the internal layer also containing prominent circular fibers. The intermediate layer, shown here, presents no regularity in arrangement of the fibers. Bundles of smooth muscle cells are seen in **longitudinal profiles** (1), **cross-sectional profiles** (2), and **oblique profiles** (3). Varying amounts of **dense irregular connective tissue** (4) separate these bundles. The dense connective tissue is readily distinguished from the bundles of smooth muscle by the paucity of nuclei.

Smooth muscle, cervix, human, H&E, x512.
At this magnification, the nuclei in the longitudinally sectioned smooth muscle bundles exhibit the characteristic **corkscrew shape** (5). In contrast, where the muscle fibers have been cross-sectioned, the smooth muscle cell nuclei exhibit **circular profiles** (6). Typically, nuclei cut in this plane of section exhibit variable diameters. The smaller-appearing nuclei represent a section through the narrower ends of a nucleus.

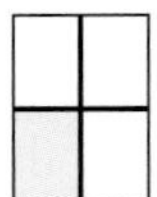

Smooth muscle, scrotum, human, H&E, x256.
The scrotum is a cutaneous and fibromuscular sac containing the testes. It consists of skin and an underlying layer of dartos muscle. The latter is a layer of smooth muscle fibers arranged in varying planes. Thus, in any given section, bundles of **longitudinally arranged fibers** (7) as well as **cross-sectioned fiber bundles** (8) and **obliquely sectioned bundles** (9) are evident. Between the muscle bundles is varying amounts of dense irregular connective tissue. Although the collagen fiber bundles in this tissue appear similar to the smooth muscle, in that they are cut in varying planes, they differ in their relative paucity of nuclei. In this specimen, several **lymphatic vessels** (10) are present.

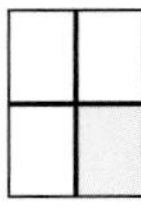

Smooth muscle, scrotum, human, H&E, x256.
This higher magnification micrograph shows an area of the scrotum in which the bulk of the tissue consists of dense irregular connective tissue. The **smooth muscle bundles** (11) are recognized by the concentration of their nuclei compared to the surrounding dense connective tissue rather than their staining characteristics.

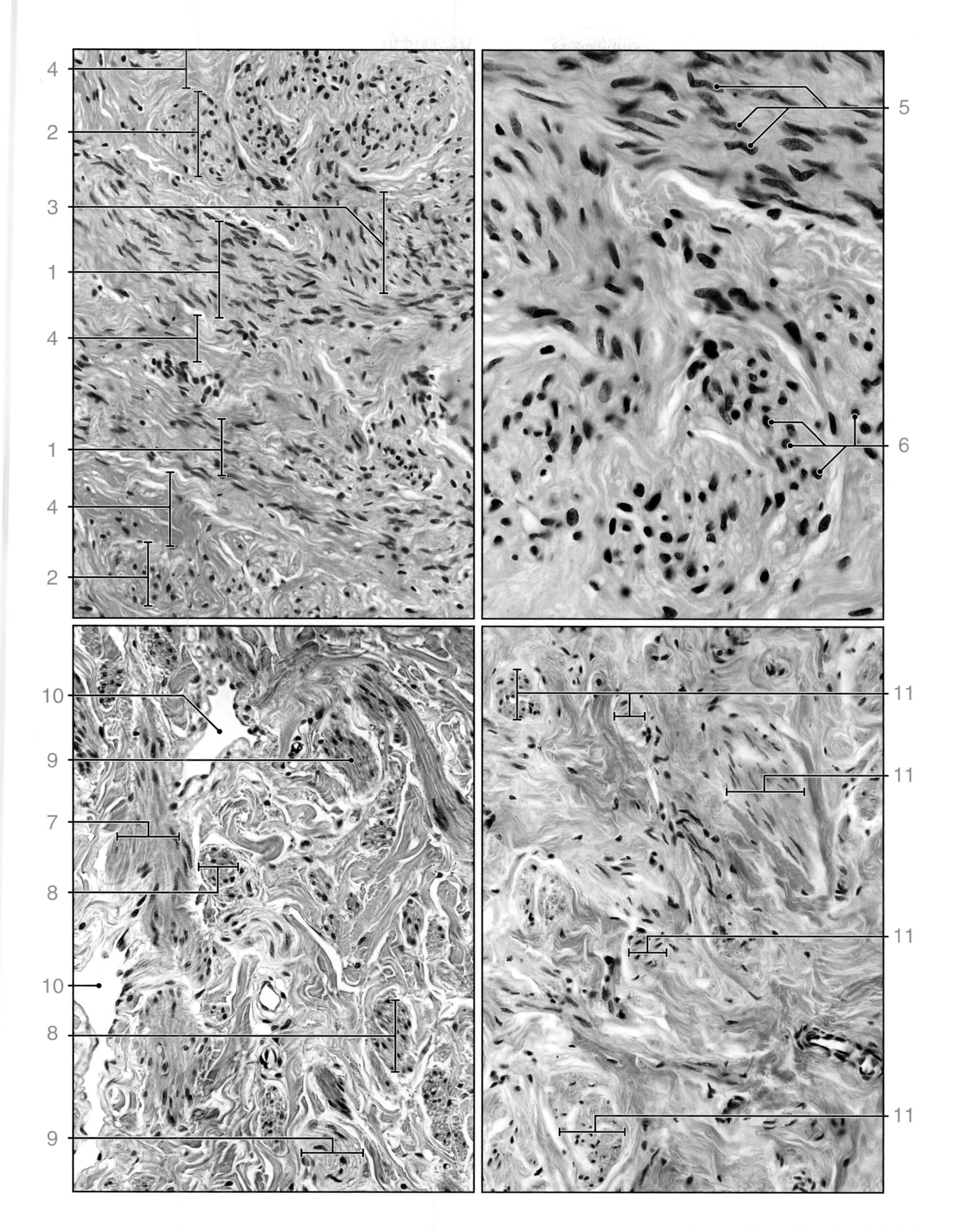

PLATE 40. SMOOTH MUSCLE II

PLATE 41. SMOOTH MUSCLE III, ELECTRON MICROSCOPY

Uterine tube, monkey, electron micrograph, x5500; insets x20,000.

This electron micrograph shows smooth muscle comparable with that shown in the light micrograph in the upper right micrograph on Plate 39. The muscle cells are longitudinally oriented and are seen in a relatively relaxed state, as evidenced by the smooth contour of their nuclei. The intercellular space is occupied by **collagen fibrils** (1), which course in varying planes between the cells.

At the low magnification used here, much of the cytoplasmic mass of the muscle cells has a homogeneous appearance. This homogeneous appearance is due to the contractile components of the cell, namely, the thin actin filaments. (The thicker, myosin filaments of mammalian smooth muscle cells are extremely labile and tend to be lost during tissue preparation.) The nonhomogeneous-appearing portions of the cytoplasm contain cytoplasmic densities (sites of actin filament attachment), mitochondria, and other cell organelles. To illustrate these differences, the area within the *lower left small circle* is shown at higher magnification in the **lower inset** on this micrograph. The particular site selected reveals a **filamentous area** (2), a region containing **mitochondria** (3), and a few profiles of rough endoplasmic reticulum (rER) (*arrows*). Although the distinction between the filament-containing regions of cytoplasm and those areas containing the remaining organelles are clearly evident in the electron microscope, routine, light microscopy, H&E preparation reveals only a homogeneous, eosinophilic cytoplasm.

In electron micrographs, fibroblasts and other connective tissue cells, when present, are readily discernible among the smooth muscle cells. In this micrograph, several **fibroblasts** (4) are evident. In contrast to the smooth muscle cell, their cytoplasm exhibits numerous profiles of rER as well as other organelles throughout all but the very attenuated cytoplasmic processes. For comparison, the area of the fibroblast in the *upper right small circle* is shown at higher magnification in the **upper inset**. Note the more numerous, dilated profiles of endoplasmic reticulum (*arrows*). The presence of rER in both the fibroblast and the smooth muscle cell is consistent with the finding that, in addition to their contractile role, smooth muscle cells have the ability to produce and maintain collagen and elastic fibers.

The tissue seen in this micrograph is from a young animal and contains fibroblasts in a relatively active state, hence, the well-developed endoplasmic reticulum and abundant cytoplasm. With the light microscope, it is unlikely that one would be able to distinguish readily between the fibroblasts shown here and the smooth muscle cells. Less active fibroblasts, as in a more mature tissue or in an older individual, have less extensive cytoplasm and, accordingly, are more easily distinguished from the smooth muscle cells.

Uterine tube, monkey, electron micrograph, x5500; inset x30,000.

This micrograph of cross-sectioned smooth muscle cells was obtained from the same specimen as the upper micrograph. It shows, at higher resolution, many smooth muscle cell features already seen with the light microscope. For example, the muscle cells can be seen arranged in bundles, comparable with those seen in the lower part of Plate 39. Even in electron micrographs, the bundle arrangement is not readily apparent when the muscle cells are longitudinally sectioned.

Generally, **cross-sectioned smooth muscle cells** (5) show an irregular contour when seen at the electron microscopic level. This is partially due to the better resolution obtained with the electron microscope but, even more so, reflects the extreme thinness of the section. The cross-sectioned profiles with the greatest diameter represent sections through the midportion of muscle cells, thus showing their nuclei; the profiles of smaller diameter represent sections through the tapered ends of cells and show only cytoplasm. The smooth muscle cells exhibit cytoplasmic areas that appear homogeneous due to the presence of myofilaments. The homogeneity is evident, but the individual myofilaments are not. Also seen within the cytoplasm are mitochondria. Other cytoplasmic constituents cannot be resolved at this relatively low magnification.

At this magnification, numerous sites are seen where a cell appears to contact a neighboring cell. Most of these sites do not reflect true cell-to-cell contacts. There are, however, places where the smooth muscle cells do make contact by means of a nexus, or gap junction. They are the structural sites that facilitate ion movement from cell to cell and permit the spread of excitation from one cell to another. In routinely prepared electron micrographs of gap junctions (*arrowhead*, **inset**), it appears as though the adjacent plasma membranes make contact (see Plate 37).

This micrograph also shows connective tissue cells associated with the bundles of smooth muscle cells. Fibroblasts are the most numerous of the connective tissue cells. The processes of the fibroblasts tend to delineate and define the limits of the smooth muscle bundles. Note the very thin **fibroblast processes** (6) coursing along the periphery of the two muscle bundles (one bundle occupying the upper half of the micrograph, the other occupying the lower half). Although the nuclei of the fibroblasts would be evident with the light microscope, their attenuated processes would not. In addition to fibroblasts, cells of another type are also seen between the bundles of smooth muscle cells. These are **macrophages** (7), readily identified with the electron microscope by the presence of lysosomal bodies within the cytoplasm. With the light microscope, these cells would be difficult to identify because the lysosomes are not evident without the use of special histochemical staining procedures. If, however, the macrophages contain phagocytosed particles of large size or cellular inclusions (e.g., hemosiderin), their identification with the light microscope, using immunocytochemical techniques, can precisely establish their nature.

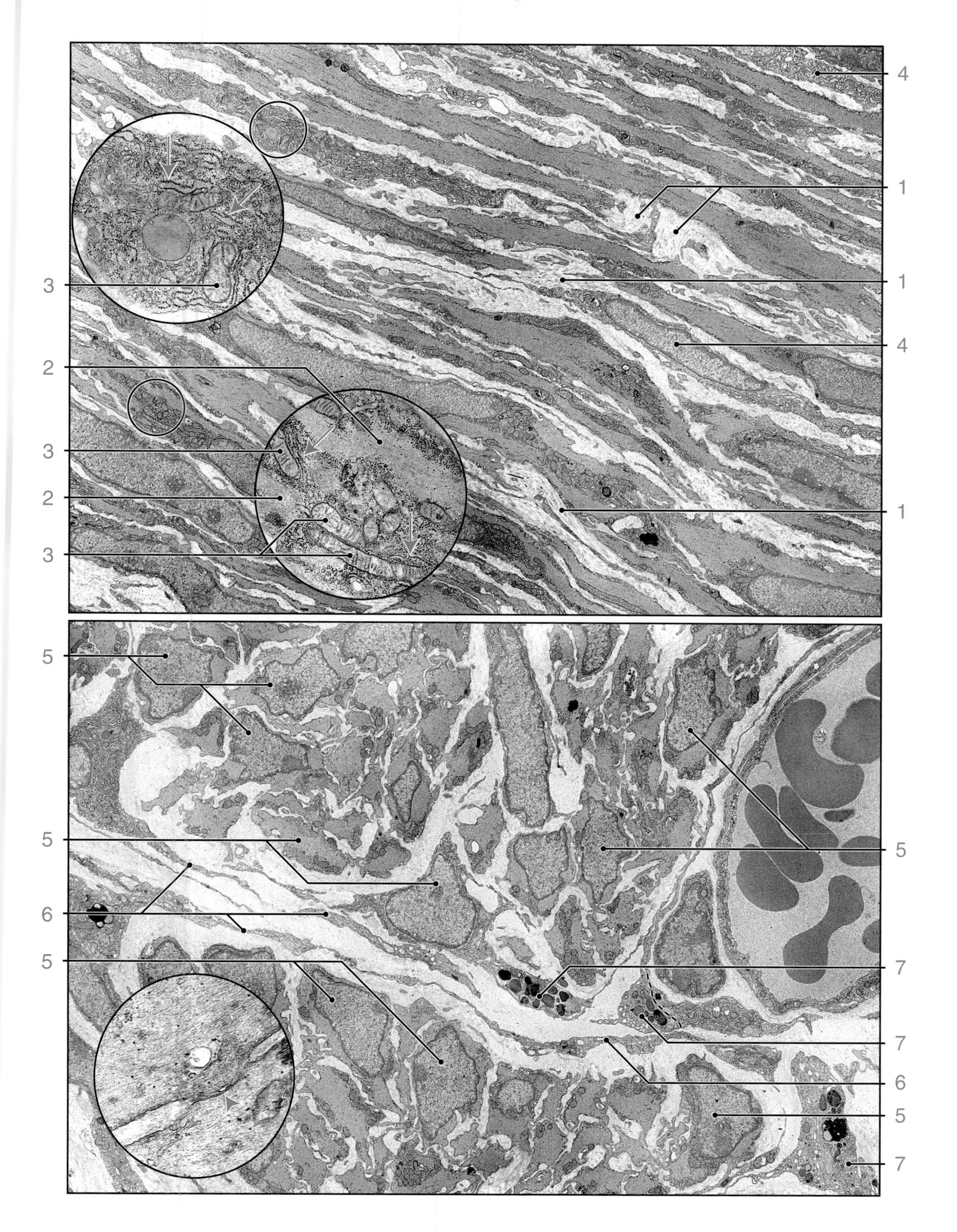

PLATE 41. SMOOTH MUSCLE III, ELECTRON MICROSCOPY

CHAPTER 8
Nerve Tissue

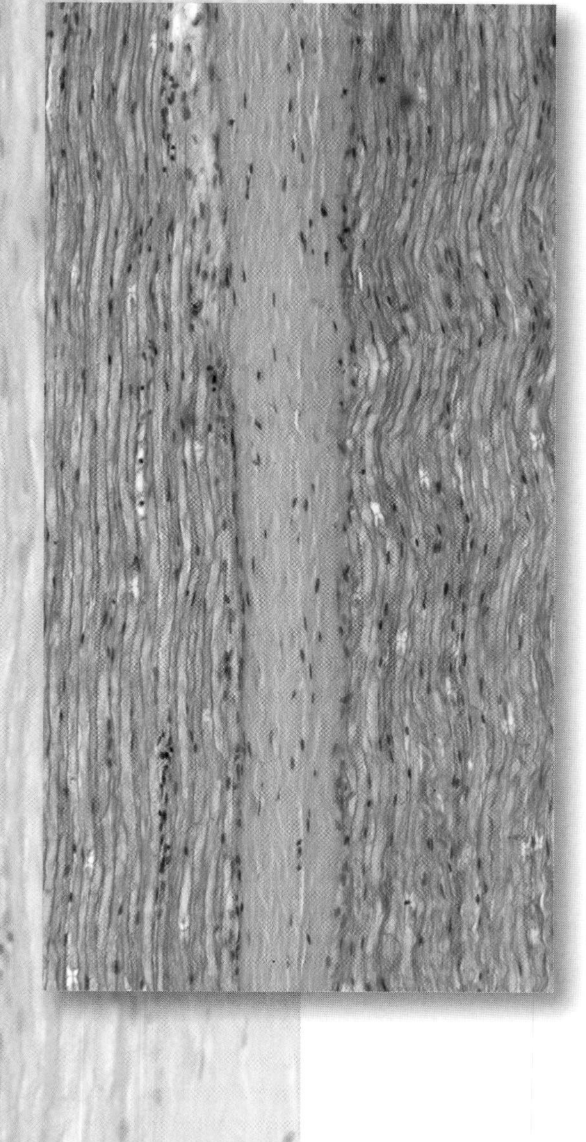

PLATE 42. **PERIPHERAL NERVE**

Peripheral nerves consist mainly of bundles of nerve fibers held together by connective tissue. The nerve bundles are generally referred to as fascicles. Each nerve fiber within a bundle, or fascicle, is invested by Schwann cell cytoplasm. They, in turn, are surrounded by a delicate connective tissue, referred to as endoneurium, that binds the nerve fibers of the fascicle together. Within the fascicle and associated with the connective tissue are fibroblasts, macrophages, and occasional mast cells. Surrounding each fascicle is a specialized connective tissue consisting of a variable number of cell layers; the number of cell layers increases with increasing thickness of the fascicle. This specialized tissue is the perineurium. The cells that make up the perineurium are contractile. Moreover, in each layer, they form a sheetlike epithelial layer of cells joined by tight junctions. In effect, the perineurial cells form a blood–nerve barrier. Lastly, each fascicle, with its perineurial sheath, is held together by dense connective tissue, the epineurium. It forms the outermost tissue of the nerve. The blood vessels that supply the nerve travel in the epineurium; their smaller branches penetrate into the nerve to travel between the nerve fibers, but are ensheathed with at least a single layer of perineurial cells. The cell bodies of peripheral nerves are located within the central nervous system or in peripheral ganglia. The ganglia contain clusters of neuronal cell bodies (see Plate 45) along with their nerve fibers, which lead to and from the cell body.

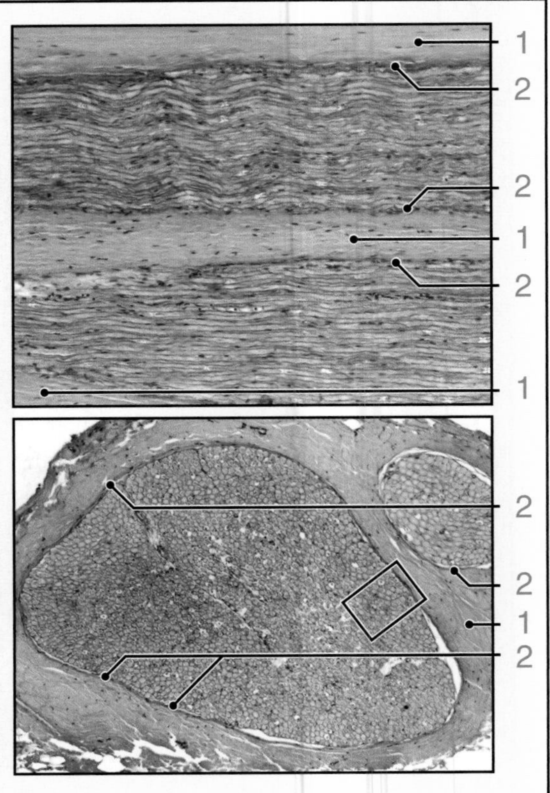

ORIENTATION MICROGRAPH: The *upper micrograph*, a longitudinal section of a peripheral nerve, reveals two nerve bundles, each surrounded by dense connective tissue, or **epineurium** (1). Between the epineurium and nerve fibers, there is a thin, dark-staining layer, the **perineurium** (2). A common feature seen in longitudinally sectioned nerves is the appearance of the nerve fibers in a wavy pattern. This is particularly evident in the lower nerve bundle and reflects contraction of the perineurial cells during preparation of the tissue. The *lower micrograph* is from a cross-sectioned peripheral nerve. It reveals two nerves; the smaller nerve is a branch of the larger nerve. Each is surrounded by dense connective tissue, the **epineurium** (1). At this very low magnification, the dark-staining layer, representing the **perineurium** (2), can be seen surrounding the nerve fiber bundles.

Nerve, longitudinal section, human, H&E, x145. At this magnification, the **epineurium** (1) is readily recognized as dense irregular connective tissue. Note the elongate nuclei, which belong to fibroblasts. The **perineurium** (2) stains more darkly than the epineurium and has more numerous nuclei. At this magnification, the nerve fibers are difficult to differentiate from one another, however, it is possible to distinguish numerous elongate nuclei associated with the nerve fibers. Most of these are Schwann cell nuclei. The Schwann cells and the nerve fibers are more clearly resolved in the bottom right micrograph.

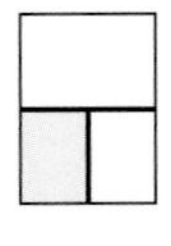

Nerve, longitudinal section, human, H&E, x725. This micrograph is a higher magnification of the boxed area of the micrograph above. The top of this micrograph includes a very small portion of the **perineurium** (3). Below this are the individual nerve fibers. Two of the nerve fibers reveal a well-defined **Schmidt-Lanterman cleft** (4). Note the narrowing of the nerve fiber at these sites. In this preparation, the axon is recognizable by the pale blue staining of its **cytoplasm** (5). The **myelin** (6) of these two nerve fibers resembles railroad track ties. This is due to the lipid extraction of the myelin leaving a protein framework that stains with eosin. The longer elongate nuclei belong to the **Schwann cells** (7). In addition, several **fibroblast nuclei** (8) of the endoneurium can be identified. They are smaller than the Schwann cell nuclei, but cannot always be identified with certainty.

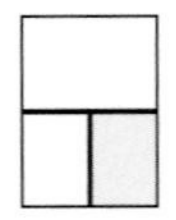

Nerve, cross section, human, H&E, x725. The boxed area on the lower orientation micrograph is shown here at higher magnification. **Epineurium** (9) is visible at the top of the micrograph. Beneath this is the **perineurium** (10). There are at least three cell layers present and a **capillary** (11) courses between the cell layers. Compare the number and appearance of the nuclei in the perineurial cells to the **fibroblast nucleus** (12) in the epineurium. The **axons** (13) appear as circular profiles surrounded by their **myelin sheaths** (14). Most of the nuclei present among the nerve fibers represent **fibroblast nuclei of the endoneurium** (15). The eosinophilic strands seen between the nerve fibers are **collagen fibers of the endoneurium** (16). One nucleus that is intimately related to the myelin of one of the fibers belongs to a **Schwann cell** (17). Note how it curves around the fiber.

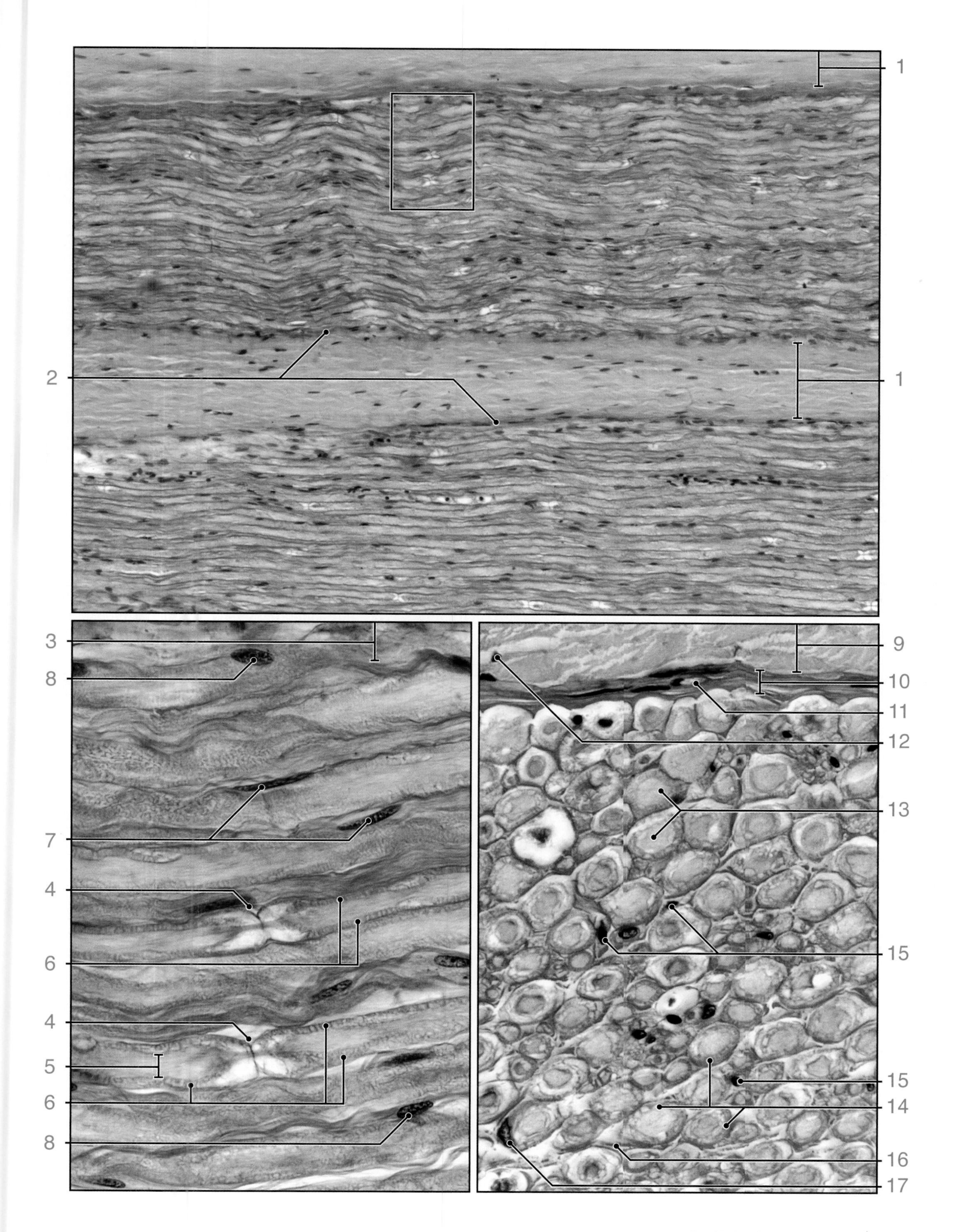

PLATE 42. PERIPHERAL NERVE

PLATE 43. PERIPHERAL NERVE AND STAINS

The structure of nerves can be better visualized by the use of certain stains that allow for differentiation of cellular and other integral components. For example, osmium tetroxide, which is a fixative that stabilizes and stains lipids, provides a good means of examining the myelin sheath. Trichrome stains, such as Mallory's trichrome stain, offer better visualization of cytoplasmic components and perineurial cells, which can be readily distinguished from fibroblasts in the adjacent epineurium. It also distinguishes the endoneurium, its cellular components, and its collagenous fibers better than H&E stain.

ORIENTATION MICROGRAPH: The *upper micrograph* reveals a **small nerve** (1) and a **larger nerve** (2), fixed and stained with osmium tetroxide. Both nerves are cut in cross-section. Between and surrounding each nerve is dense connective tissue forming the **epineurium** (3). The smaller nerve is presumably a branch of the larger nerve. The dense, black material is the preserved **lipid** (4) within adipocytes located in the connective tissue surrounding the epineurium. The *lower micrograph* reveals portions of three cross-sectioned nerves stained with Mallory's trichrome. The larger nerve appears to be surrounded by a relatively thick layer of **connective tissue** (5). When observed at higher magnification, it will be evident that this layer is composed partly of epineurium, with an underlying layer of perineurium.

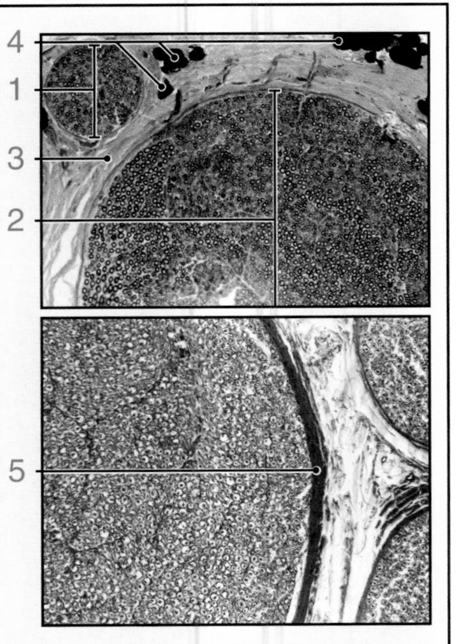

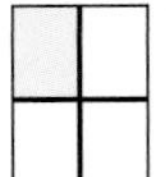

Nerve, cross section, human, osmium tetroxide, x1000.

The osmium staining in this specimen highlights the **myelin** (1) surrounding the axons. In contrast, the **axons** (2), surrounded by their myelin sheaths, appear lightly stained and, in some cases, unstained. The myelin appears as a black ring around the axonal cytoplasm. Note the marked differences in size of the various axons and myelin sheaths. Also note that the largest axons have the thickest myelin sheaths. In some areas within the nerve bundle, there is an absence of myelinated fibers. These sites, when observed at higher magnification, would reveal the presence of very small **unmyelinated axons** (3). These axons are covered by Schwann cell cytoplasm, but there is no myelin associated with these axons. (Nerves containing only unmyelinated axons are depicted in Plate 91.) The **epineurium** (4) appears as an almost amorphous structure in this preparation; however, the **perineurium** (5) appears as a layered structure, more darkly stained than the epineurium.

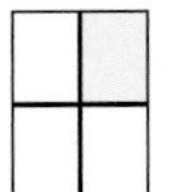

Nerve, longitudinal section, human, osmium tetroxide, x1000.

This longitudinal section provides a better perspective of the nature of the myelin sheath in relation to its axon. For example, the **node of Ranvier** (6) is clearly evident where included in the section. Note how the myelin is absent at the node. Likewise, **Schmidt-Lanterman clefts** (7) are readily apparent when the axon is longitudinally cut at its greatest diameter. Note how the clefts appear as unstained, angular, paired notches, or incisures. The clear area represents the cytoplasm of the Schwann cell, which is produced during the formation of the myelin as the Schwann cell wraps around the axon. Between the myelinated fibers are elements of the connective tissue, forming the **endoneurium** (8). At the very top of the micrograph is part of the **epineurium** (9) and, below that, the darker-staining **perineurium** (10).

Nerve, cross section, human, trichrome, x320.

This section of nerve includes the **epineurium** (11) and, below, the **perineurium** (12). The feature that distinguishes perineurium from epineurium is the presence of more numerous red-stained cells in the perineurium. The remainder of the micrograph reveals cross-sectioned nerve fibers. The axons of the nerve fibers, which have a similar appearance in an H&E-stained section, appear as dots surrounded by a clear area, representing the myelin sheath from which the lipid component was lost during slide preparation.

Nerve, cross section, human, trichrome, x640.

At this higher magnification, the distinction between epineurium and perineurium can be better visualized. Note the abundance of **perineurial cells** (13) in the perineurium compared to the paucity of cells in the dense irregular connective tissue that represents the **epineurium** (14). Examining the nerve fibers, the axons are the most prominent feature. As already mentioned above, the **myelin** (15) was lost during preparation, leaving a clear space around the axon; however, the protein component of the **myelin sheath** (16) precipitated, leaving a skeletal framework, which appears similarly to those seen in H&E preparations. The blue-staining material surrounding, and coursing between, the nerve fibers is connective tissue that represents the endoneurium. Lastly, several well-defined **capillaries** (17) are also evident.

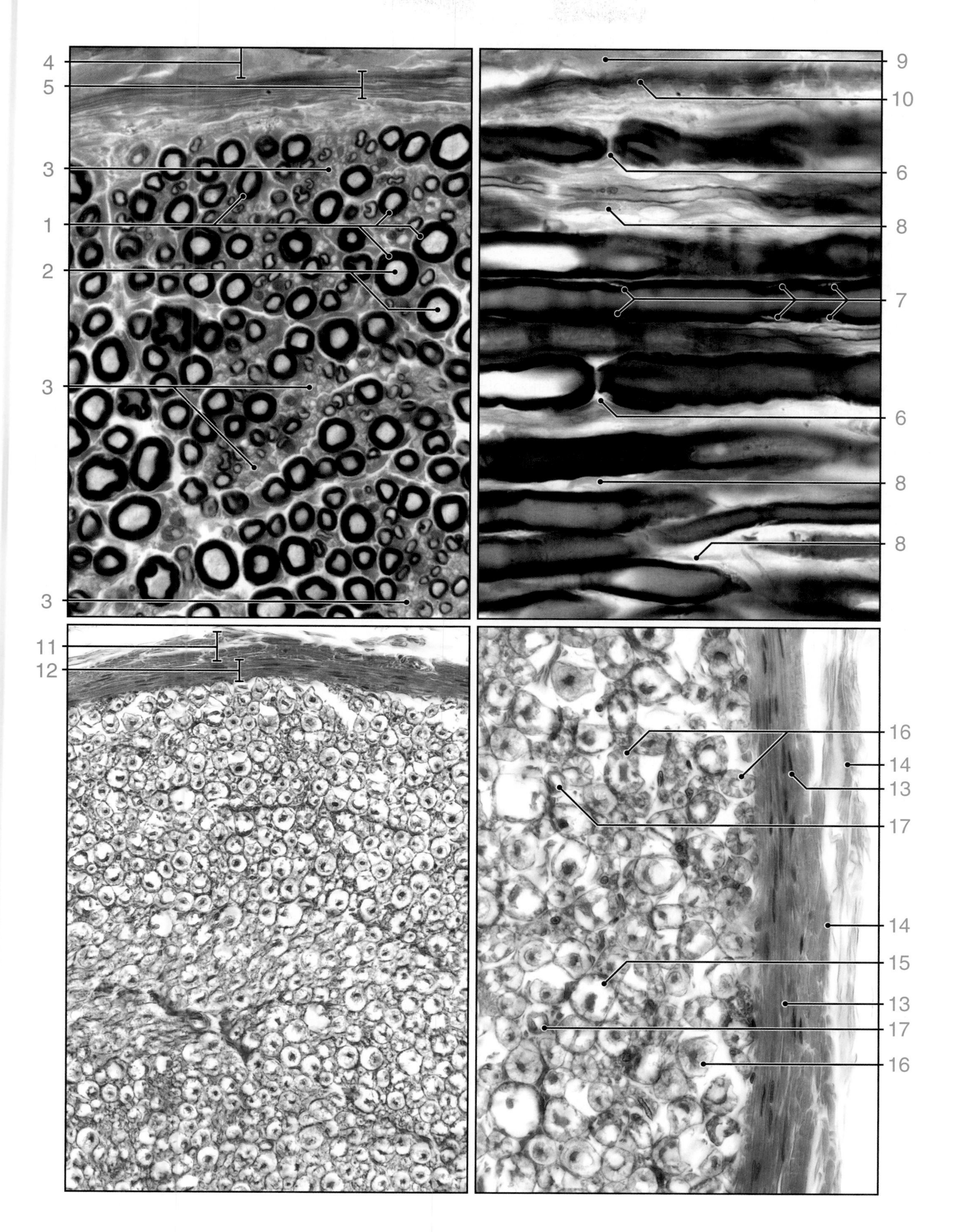

4
5
3
1
2
3
3
9
10
6
8
7
6
8
8
11
12
16
14
13
17
14
15
13
17
16

PLATE 44. PERINEURIUM, ELECTRON MICROSCOPY

Peripheral nerve, human, electron micrograph, x7000; insets x30,000.

This micrograph reveals the outer portion of a small, longitudinally sectioned nerve. An obliquely cut **axon** (1) and its **myelin sheath** (2) are present in the upper right corner. The **perineurium** (3) is to the left of the axon and appears as an orderly series of cell layers. To the left of the perineurium is the epineurium, which contains a moderate amount of **collagen** (4), **epineurial fibroblast processes** (5), and **elastic material** (6). The epineurium, unlike the perineurium, is not well defined. It tends to blend with the connective tissue more removed from the nerve; thus, there is no discrete boundary between the epineurium and the surrounding connective tissue.

Examining the perineurium at this low magnification, the most notable feature is the relative uniformity of the individual cellular layers, or lamellae. The cells that constitute each layer, in effect, are contiguous with one another and joined by tight junctions. Typically, the edge of one cell is in apposition to its neighbors, thus forming a series of uninterrupted, concentric cellular sheaths, each of which completely surrounds the nerve bundle.

In contrast to the perineurial cells, a portion of an **endoneurial fibroblast** (7) can be seen along the inner aspect of the perineurium. Although in close apposition to the perineurium, this cell exhibits characteristics of a typical fibroblast and, therefore, represents an endoneurial fibroblast. Note that it does not form a continuous sheathlike structure, rather it exhibits a lateral margin (*arrowhead*) that does not meet or possess junctions with another cell to provide continuity. This endoneurial cell is morphologically identical with the epineurial cell; both are typical fibroblasts.

The cytologic distinction between the fibroblasts of the epineurium and endoneurium and the specialized perineurial cell can be appreciated better at high magnification (**insets**). The **upper inset** reveals a small portion of the perineurial cell; the **lower inset** reveals an epineurial cell. The specific site from which each is taken is indicated by the *small ovals*. Comparing the two cells, note that the perineurial cell exhibits **basal lamina** (8) on both surfaces; the fibroblast typically lacks this covering material. Both cell types reveal pinocytotic vesicles as well as **mitochondria** (9). The remainder of the cytoplasm in these two micrographs, however, shows an important difference. The perineurial cell reveals numerous fine filaments, that represent myofilaments, and **cytoplasmic densities** (10). Both are features characteristic of the smooth muscle cell. Again, the fibroblasts lack these elements. Although not evident here, the perineurial cells also exhibit profiles of **rough endoplasmic reticulum** (11), but fewer than are seen in the fibroblast. The presence of rough endoplasmic reticulum in the perineurial cell suggests that it, like the fibroblast, produces collagen and is thus responsible for the deposition of collagen between the cellular laminae. The contractile nature of the perineural cells, as evidenced by the myofilaments and cytoplasmic densities, explains the shortening of a nerve when it is accidentally or surgically cut. Finally, the epithelial-like arrangement of the perineurial cells is regarded as a selective permeability barrier that contributes to the blood-nerve barrier.

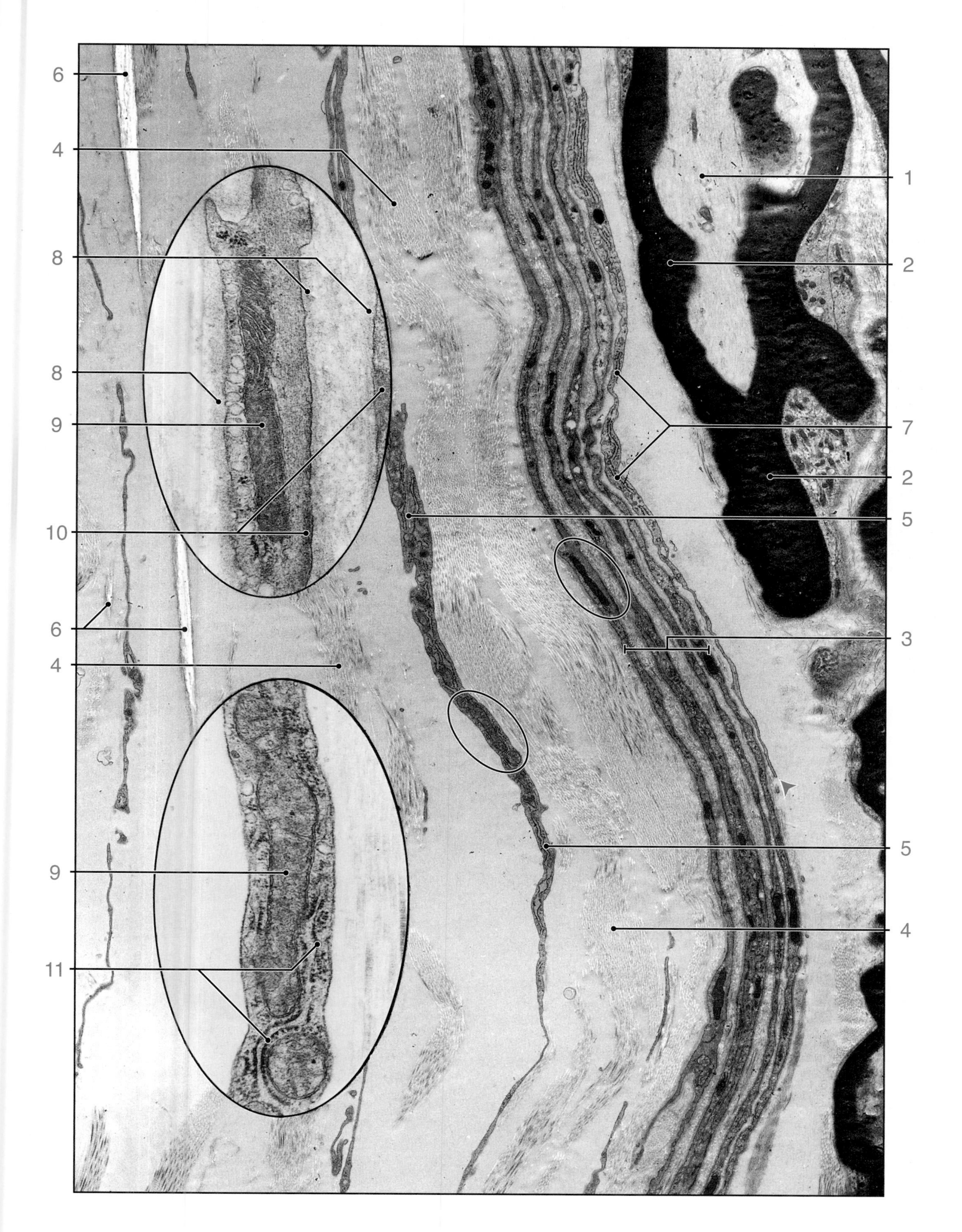

PLATE 44. PERINEURIUM, ELECTRON MICROSCOPY

PLATE 45. SYMPATHETIC AND DORSAL ROOT GANGLIA

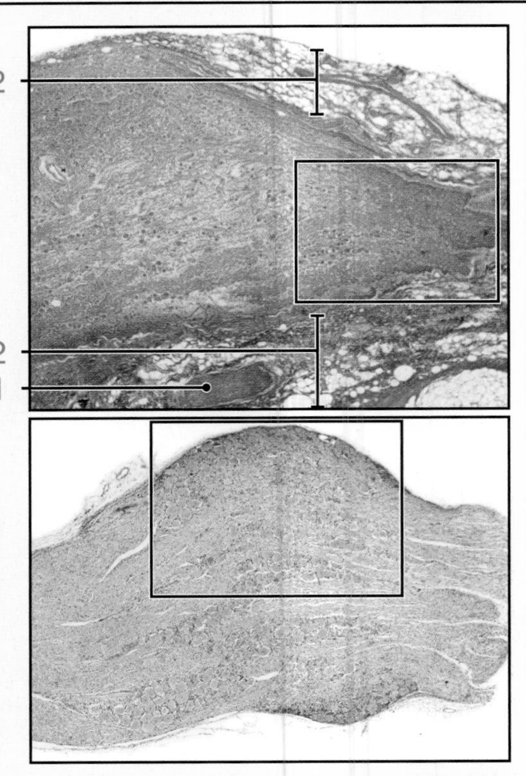

Ganglia are groups of neuron cell bodies that occur outside of the central nervous system (CNS). (Within the CNS, a group of nerve cell bodies is referred to as a nucleus.) Based on structure as well as function, ganglia can be divided into sensory ganglia, with pseudounipolar or bipolar neurons, and autonomic ganglia, with sympathetic and parasympathetic multipolar neurons. The sensory ganglia lie just outside of the CNS and contain the cell bodies of sensory nerves, which carry impulses into the CNS. Autonomic ganglia are peripheral motor ganglia of the autonomic nervous system and contain cell bodies of post-synaptic neurons. They conduct nerve impulses to smooth muscle, cardiac muscle, and glands. Sympathetic ganglia constitute a major subclass of autonomic ganglia; parasympathetic ganglia and enteric ganglia constitute other subclasses. Sympathetic ganglia are located in the sympathetic chain (paravertebral ganglia) and on the anterior surface of the aorta (prevertebral ganglia). They possess long post-synaptic axons that go to the viscera. Parasympathetic ganglia (terminal ganglia) are located in the organs innervated by their postsynaptic neurons. The enteric ganglia are located in the submucosal plexus and the myenteric plexus of the alimentary canal (see Plate 91). They receive parasympathetic, presynaptic input as well as intrinsic input from other enteric ganglia and innervate smooth muscle of the alimentary canal.

ORIENTATION MICROGRAPH: The *upper micrograph* reveals a sympathetic ganglion, an adjacent **nerve** (1), and surrounding **adipose and connective tissue** (2). The right side of the micrograph shows several nerves entering or leaving the ganglion. On the left side of the micrograph, a part of the ganglion is not visible. The *lower micrograph* reveals a dorsal root ganglion. Most of the surrounding connective and adipose tissues have been removed. The ganglion cells are responsible for the expanded form of the ganglion.

Sympathetic ganglion, human, Bodian method with counterstains, x64.

This micrograph shows the boxed area from the upper orientation micrograph. It reveals a portion of a sympathetic ganglion and several **nerves** (1) leaving the ganglion. The ganglion is covered by **dense connective tissue** (2), the epineurium, and an underlying perineurium; both are continuous with the epineurium and perineurium of the nerves entering and leaving the ganglion. The **nerve cell bodies** (3), clustered in groups between **bundles of nerve fibers** (4), are easily seen at this low magnification because of their large size.

Sympathetic ganglion, human, Bodian method with counterstains x256.

In this higher power micrograph of the boxed area on the top left micrograph, the nerve cell bodies are readily apparent. Most exhibit a spherical **nucleus** (5) with a discrete nucleolus, which appears as a dot-like structure in the nucleus. Some of the **cell bodies** (6) do not exhibit a nucleus, a factor attributable to the plane of section through these large cells. Each cell body is surrounded by **small satellite (capsule) cells** (7). Typically, however, they are not readily apparent in this type of ganglion. A few of the **nerve cell bodies** (8) display several processes joined to them. These cell bodies are multipolar and possess many dendrites that are stained in this preparation by silver, a component of the Bodian staining method. The wavy stringlike structures between the ganglion cells are **nerve bundles** (9).

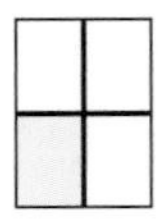

Dorsal root ganglion, human, H&E, x64.

This micrograph shows the boxed area on the lower orientation micrograph. Dorsal root ganglia differ from autonomic ganglia in that the latter contain multipolar neurons and have synaptic connections, whereas dorsal root ganglia contain pseudounipolar sensory neurons and have no synaptic connections in the ganglion. The ganglion, even at this low magnification, shows **nerve bundles** (10) and clusters of **nerve cell bodies** (11).

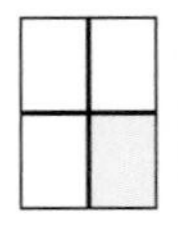

Dorsal root ganglion, human, H&E, 256.

The boxed area on the bottom left micrograph is shown here at higher magnification. Some of the cell bodies exhibit a discrete **nucleus** (12), whereas others (13) appear to lack a nucleus—again, a factor relating to the plane of section of the cell. A layer of **satellite (capsule) cells** (14) surrounds each cell body. They represent a continuation of the Schwann cells that cover the axons. The majority of the nerve fibers leading to and extending from the nerve cell bodies in dorsal root ganglia are myelinated. A single, well-defined **nerve fiber** (15) is indicated.

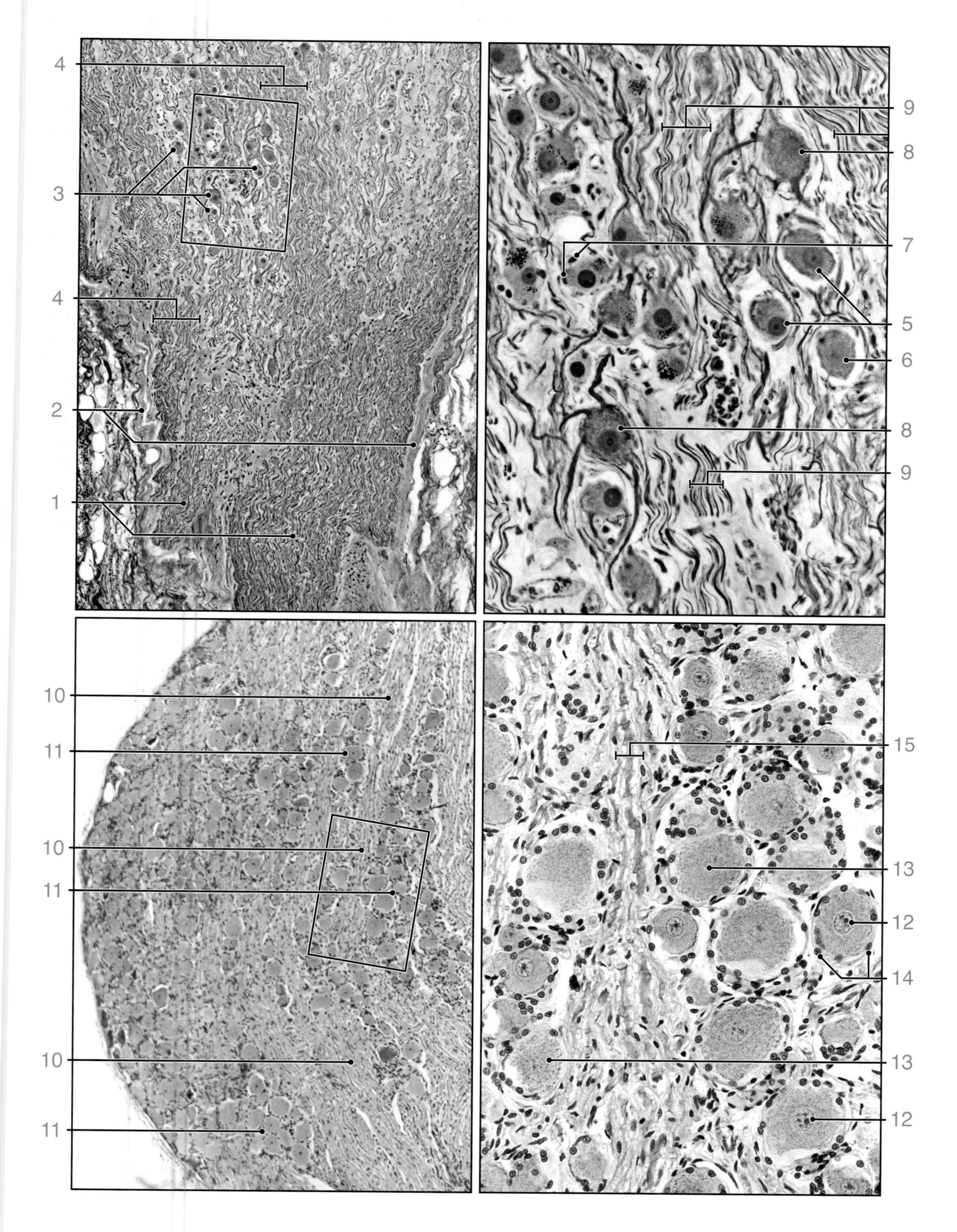

PLATE 45. SYMPATHETIC AND DORSAL ROOT GANGLIA

CHAPTER 9
Cardiovascular System

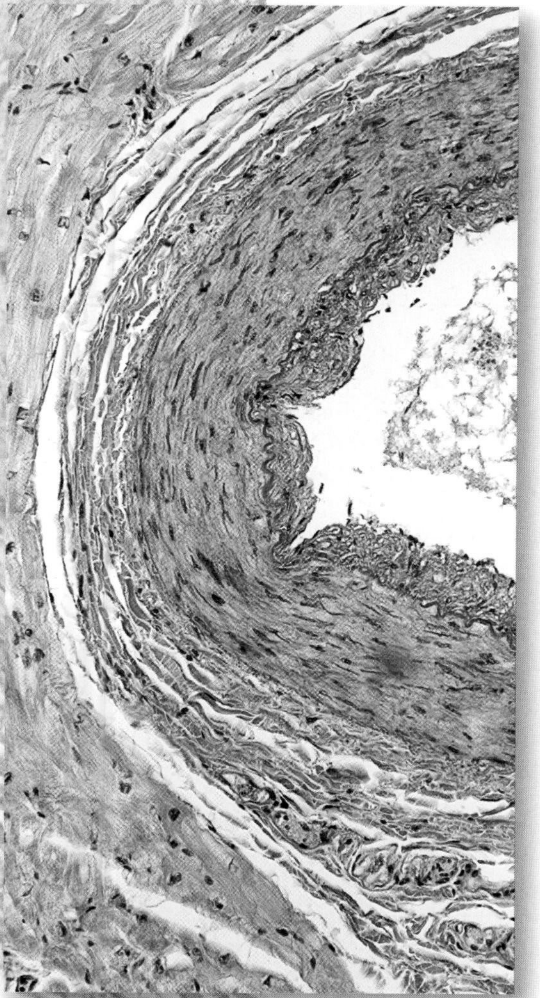

PLATE 50. HEART, ATRIOVENTRICULAR WALL

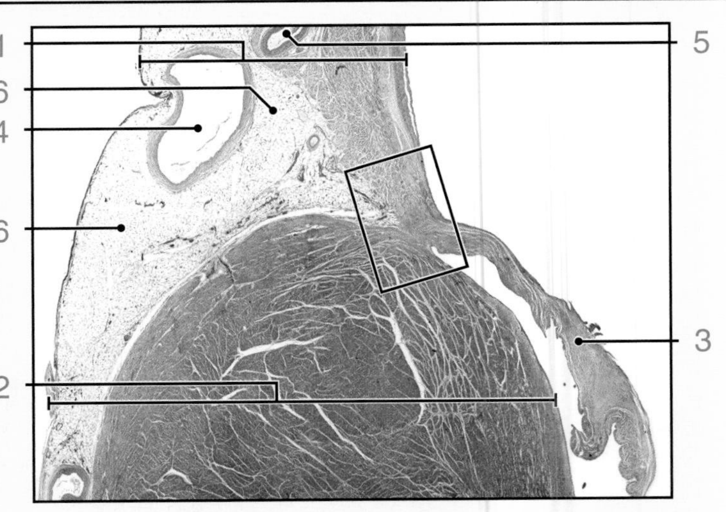

The cardiovascular system is comprised of a pump, namely, the heart, blood vessels that provide the means to circulate blood to and from all parts of the body, and lymphatic vessels, which carry tissue-derived fluid called lymph back to the blood vascular system and transport lymphocytes to and from lymphatic organs.

The heart is a muscular pump that maintains unidirectional flow of blood. It consists of four chambers: a right and left atrium and a right and left ventricle. The right atrium receives blood from the body. From the atrium blood enters the right ventricle and is pumped to the lungs for oxygenation. Blood returns from the lungs to the left atrium, after which it enters the left ventricle and is pumped to the rest of the body. During embryonic development, the heart differentiates from a straight vascular tube into the chambered structure. Its walls have the same basic three-layered structure as the blood vessels with the exception of the smaller capillaries and post-capillary venules. The layers of the heart consist of endocardium, made up of endothelium, connective tissue, and smooth muscle cells; myocardium, made up of cardiac muscle; and epicardium, consisting of a layer of mesothelial cells on the outer surface of the heart and connective tissue underneath.

ORIENTATION MICROGRAPH: This micrograph shows a sagittal section of the posterior wall of the **left atrium** (1) and **left ventricle** (2) at the level of the atrioventricular septum. Evident in this section is part of the **mitral valve** (3). The section includes the coronary (A–V) groove, which contains the **coronary sinus** (4), a vein that receives most of the cardiac veins and empties into the right atrium. It also shows the **circumflex branch of the left coronary artery** (5), which lies in the **epicardium** (6), surrounded by adipose tissue.

Heart, atrioventricular septum, human, H&E, x45. The higher magnification shown in this micrograph is from the boxed area on the orientation micrograph. Both chambers and the valve are lined with simple squamous endothelium of the **endocardium** (1). The section reveals **Purkinje fibers** (2) of the cardiac conduction system in the atrial wall. They are positioned beneath the thin **subendocardial connective tissue** (3). **Dense fibrous connective tissue** (4), continuous with that of the septum and the subendocardial layers of the atrium and ventricle, extends from the root of the **mitral valve** (5) into the valve leaflet. The wall of the ventricle consists mostly of **cardiac muscle** (6). Between the cardiac muscle of the ventricle and the subendothelial connective tissue is a relatively thick layer of **dense connective tissue** (7) that extends from the connective tissue septum and thins out as it continues into the ventricular wall. **Adipose tissue** (8) is present within the epicardium of the coronary groove.

Heart, atrioventricular septum, human, H&E, x125. This micrograph is a higher magnification of the boxed area on the top micrograph. It reveals the nuclei of the **endothelial lining** (9) of the endocardium. A thin layer of **smooth muscle** (10) is seen between the **dense connective tissue** (11) underlying the endocardium and the dense connective tissue of the **subendocardium** (12). Beneath the subendocardial connective tissue are the **Purkinje fibers** (13). Careful examination reveals some of the **intercalated discs** (14). These modified cardiac muscle cells are much larger than the ordinary cardiac muscle cell. In H&E sections, clear areas are typically seen, usually around the nucleus. They represent glycogen-rich regions of the cell.

Heart, atrioventricular septum, human, H&E, x125. This micrograph shows a section through part of one of the two cusps of the mitral valve, near its attachment to the fibrous ring. The surface of the valve, as seen here, faces the ventricle and is covered by **endothelium** (15). A section through the valve typically reveals three layers. The **fibrosa** (16) is one of the layers. It forms the core of the valve and contains fibrous extensions from the dense irregular connective tissue of the skeletal rings of the heart. The **ventricularis** (17) is immediately adjacent to the ventricular surface of the valve. It contains dense connective tissue with many layers of elastic fibers. The fibers of the ventricularis continue into the **chordae tendineae** (18). They are the threadlike chords covered with endothelium that extend from the free edge of the A–V valves to the muscular projections from the wall of the ventricles, the papillary muscles. The third layer, the spongiosa, is not seen in this micrograph. It is composed of loosely arranged collagen and elastic fibers infiltrated with a large amount of proteoglycans. The spongiosa acts as a shock absorber that dampens vibrations associated with the closing of the valve.

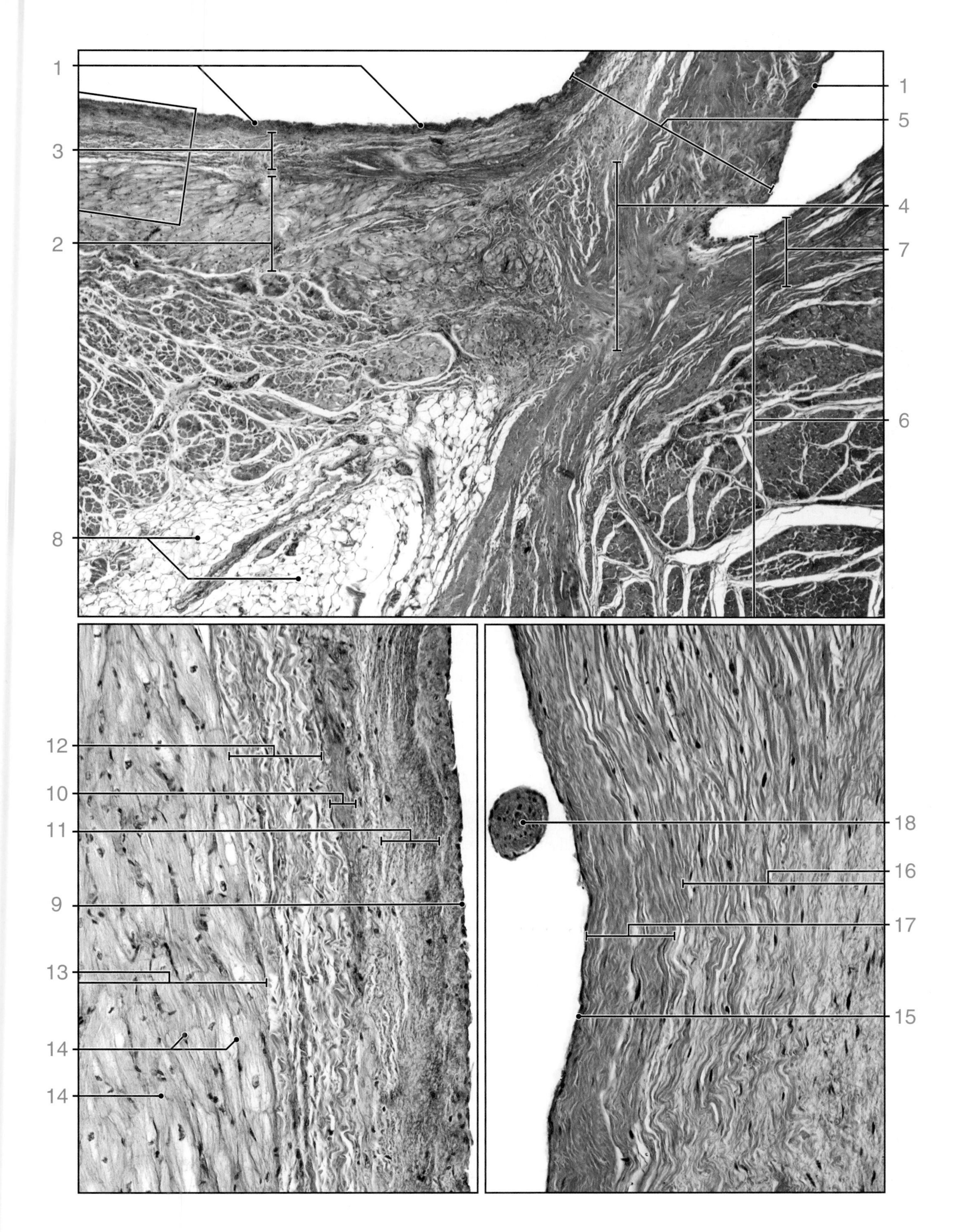

PLATE 50. HEART, ATRIOVENTRICULAR WALL

PLATE 51. **CORONARY ARTERIES AND CORONARY SINUS**

Arteries and veins, like the heart, are composed of three major layers. They are classified both by size and by characteristics of component parts of the vessel wall. Thus, arteries are classified as large or elastic arteries, medium or muscular arteries, and small arteries and arterioles. The large and medium arteries are also named vessels, (e.g., aorta, femoral artery). Similarly, veins are classified by size—large, medium, and small; large and medium veins are also named. Some of the smaller venous vessels are also called muscular venules; the smallest venules are referred to as postcapillary venules. The three major layers of the vascular wall, in both arteries and veins, from the lumen outward are the tunica intima, the tunica media, and the tunica adventitia. The tunica intima is further divided into an endothelium, a basal lamina, belonging to the endothelial cells, and a subendothelial layer, consisting of loose connective tissue. The tunica media, the middle layer, consists primarily of circumferentially arranged smooth muscle cells. The tunica adventitia, the outermost layer, consists of connective tissue, which merges with the loose connective tissue surrounding the vessel. The tunica adventitia is relatively thin in most of the arteriole system, but tends to be a major component, in terms of thickness, in the venules and veins.

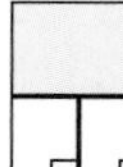

Heart, coronary artery and coronary sinus, human, H&E, x30.

The coronary artery and the coronary sinus seen in the atrioventricular septum orientation micrograph on Plate 50 are shown here at higher magnification. The coronary artery is in the lower left of the micrograph. The **tunica intima** (1) of this vessel appears as the darker-stained band lining the vessel. The **tunica media** (2) is the thickest portion of the wall. The **tunica adventitia** (3), comprised of dense connective tissue, is the outer layer. Adjacent to the coronary artery is a band of **conduction fibers** (4), consisting of Purkinje cells. The larger vessel, the coronary sinus, has a thin wall relative to its size, a feature typical of veins compared to arteries. The **tunica intima** (5) again appears as a darker layer. At this magnification it is not possible to distinguish the tunica media from the adventitia.

Heart, coronary artery and coronary sinus, human, H&E, x125; inset x250.

This micrograph is a higher magnification of the coronary artery in the boxed area on the top micrograph. This magnification allows for distinction between the three layers of the vessel wall. The **tunica intima** (6) is moderately thick and is separated from the **tunica media** (7) by the **internal elastic membrane** (8). When the vessel wall is seen in cross-section, the internal elastic membrane appears as a thin, undulating band of elastic material. Its undulating nature is due to contraction of smooth muscle in the tunica media. The smooth muscle cells are oriented in a circumferential pattern as indicated by the elongate profiles of their nuclei. The **tunica adventitia** (9) consists of dense connective tissue. Note the relative paucity of nuclei in this tissue. Also included in this micrograph is a band of conduction fibers consisting of **Purkinje cells** (10). The **inset** shows to advantage the **tunica intima** (11). The **endothelial cells** (12) are seen along the luminal surface of the vessel. The **subendothelial layer** (13) of the tunica intima contains collagenous fibers, fibroblasts, and some smooth muscle cells. The **tunica media** (14) consists of smooth muscle cells. Collagenous fibers produced by the smooth muscle cells are present, but fibroblasts are absent.

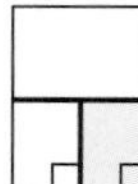

Heart, coronary artery and coronary sinus, human, H&E, x125; inset x250.

The portion of the wall of the coronary sinus in the boxed area on the top micrograph is shown here at higher magnification. The **tunica intima** (15) is relatively thin. The **tunica media** (16) contains smooth muscle, but unlike the artery, there is also a significant amount of dense connective tissue. The **tunica adventitia** (17), like that of the artery, consists of dense connective tissue, but it is notably thicker relative to the thickness of the wall. The tunica media and the tunica adventitia appear somewhat similar, however, the tunica media exhibits more nuclei per unit area, a reflection of the number of smooth muscle cells that are present in this layer. The **inset** shows the thin **tunica intima** (18) and the adjacent **tunica media** (19). The endothelium is represented by the nuclei that are located on the luminal surface. The fibrous nature of the tunica media is evident here. Also note the just perceptible **elastic fibers** (20) present in this layer.

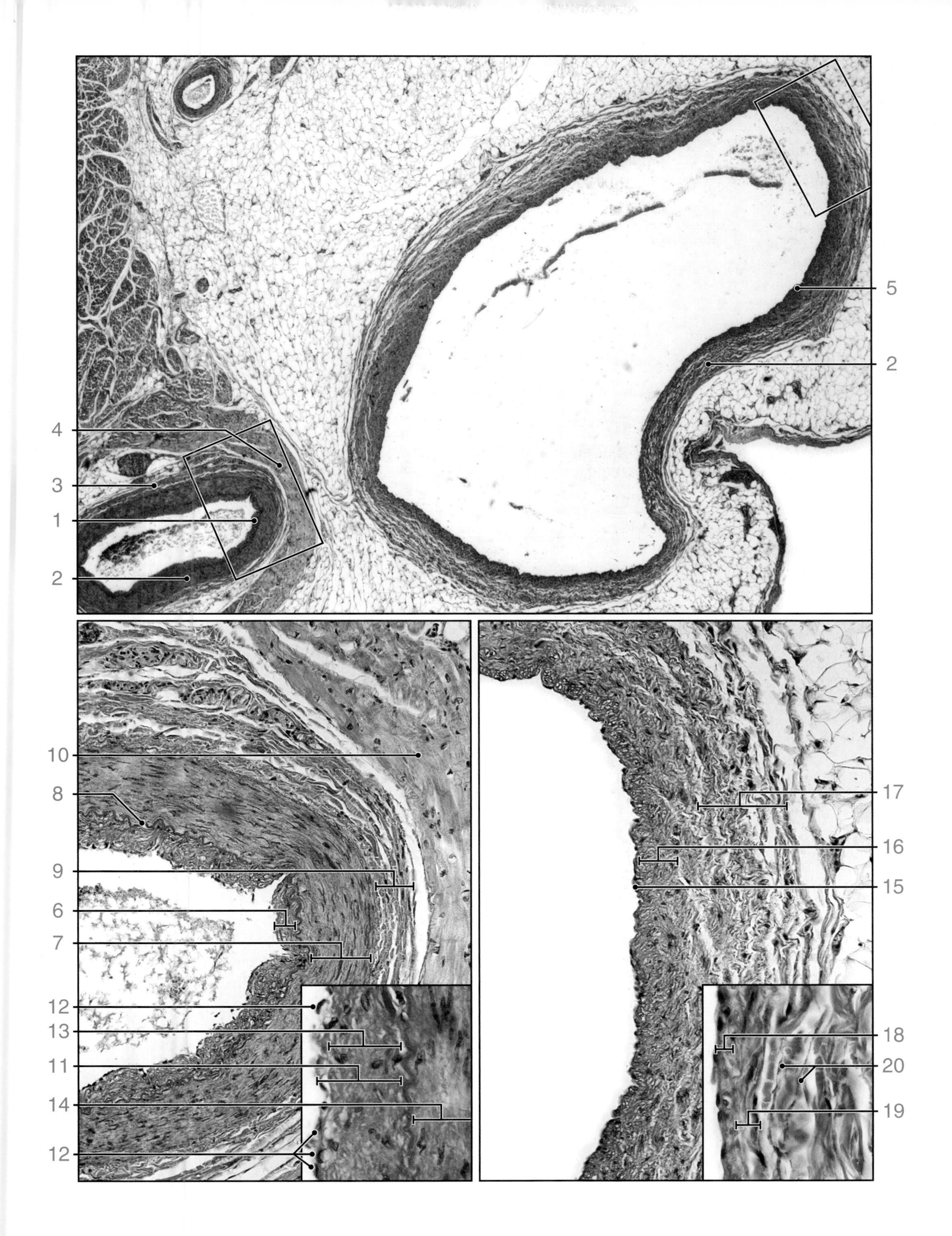

PLATE 51. CORONARY ARTERIES AND CORONARY SINUS

PLATE 52. AORTA

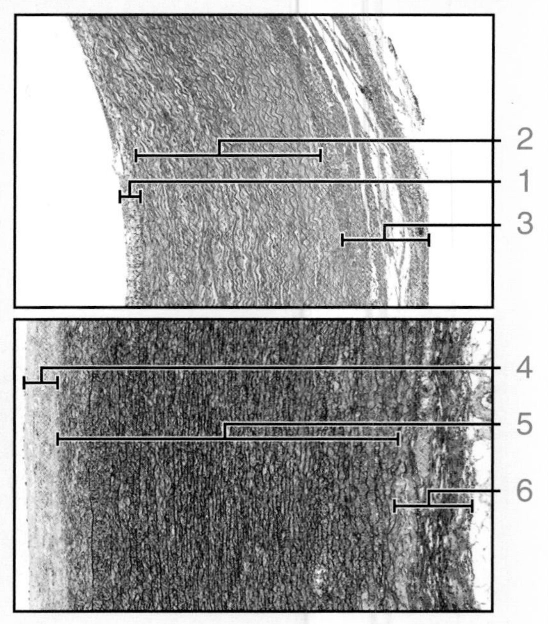

The aorta, the main systemic artery of the body, is the largest arterial vessel. It is an elastic artery. The presence of numerous fenestrated elastic lamellae allow it to resist the pressure variations caused by beating of the ventricle, as well as to create a relatively steady flow of blood into the arterial system. The intima is, comparatively, much thicker than that seen in muscular arteries. The subendothelial layer of the intima consists of connective tissue with both collagen and elastic fibers. The cellular component consists of smooth muscle cells and fibroblasts. The external border of the intima is bounded by an internal elastic lamina or membrane that is not always readily distinguishable as an individual entity, but represents the first layer of the many concentric fenestrated laminae in the media of the vessel.

The media constitutes the bulk of the wall. Between the elastic laminae are collagen fibers and smooth muscle cells. The latter are responsible for the production and maintenance of elastic material as well as the collagen and proteoglycans. With age, the number of elastic laminae in the wall increases. By 35 years, as many as 60 laminae are found in the thoracic aorta. The individual laminae also increase in thickness with age, but at approximately 50 years they begin to show signs of degeneration and gradually become replaced by collagen leading to a gradual loss of elasticity, but an overall increase in strength of the aortic wall.

The adventitia consists of irregular dense connective tissue with intermixed elastic fibers that tend to be organized in a circumferential pattern. It also contains small blood vessels that supply the outer portion of the media. They are the vasa vasorum of the aorta. Also present in the adventitia are lymphatic capillaries. The inner portion of the media and the intima rely on the diffusion of oxygen and nutrients from the lumen of the vessel.

ORIENTATION MICROGRAPHS: The *upper micrograph* shows a part of a cross section of an H&E stained human aorta from a child. The **intima** (1) stains considerably lighter than the adjacent **media** (2). The **adventitia** (3) contains an abundance of collagenous fibers and stains more densely than that of either the media or intima. The *lower micrograph* is from an adult and has been stained to reveal the elastic component of the vessel wall. The **intima** (4) is very lightly stained due to the paucity of elastic material. The **media** (5) is heavily stained due to the presence of large amounts of elastic laminae. The **adventitia** (6) contains a moderate amount of elastic fibers in addition to the dense collagenous tissue.

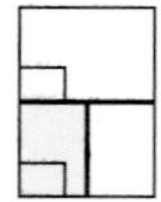

Aorta, human, H&E, x365; inset x700.

This micrograph shows the layers of the aortic wall. The intima consists of an **endothelium** (1) overlying **loose connective tissue** (2). The thickest portion of the vessel wall is the **media** (3). The wavy eosinophilic material is the collagenous fibers. The H&E stain does not reveal the elastic laminae. In the media, the nuclei are of smooth muscle cells. Fibroblasts are absent. The outer layer of the vessel wall is the **adventitia** (4). The eosinophilic material here consists of dense connective tissue. In the adventitia, the nuclei that are evident belong to fibroblasts. Also, note the small **blood vessel** (5) in the adventitia. The **inset** shows the intima at higher magnification and includes part of the media. Note the **endothelium** (6). The eosinophilic material in the intima consists of **collagenous fibers** (7). The elastic fibers are not stained. The prominent cell type here is the **smooth muscle cell** (8).

Aorta, human, iron hematoxylin and aniline blue, x255; inset x350.

The specimen shown here has been stained to distinguish collagen from elastic material. The **intima** (9) consists mostly of collagenous fibers. The **endothelium** (10), represented by several nuclei, is just barely evident. The **media** (11) contains numerous elastic lamellae that appear as black wavy lines. The intervening, blue-stained material consists of collagen fibers. Careful examination of the media reveals nuclei dispersed between the elastic lamellae. These are nuclei of smooth muscle cells. The **inset** shows the intima from the boxed area on this micrograph at higher magnification. Note the nuclei of the **endothelial cells** (12) at the luminal surface. The remainder of the intima consists mostly of collagenous fibers (stained blue) with occasional **elastic fibers** (13) identified by their darker coloration. The nuclei of the fibroblasts and occasional **smooth muscle cells** (14) appear randomly arranged; the two types of cells are difficult to distinguish from one another in this preparation.

Aorta, human, iron hematoxylin and aniline blue, x255.

This micrograph shows the outer portion of the **media** (15) with its elastic lamellae. The major portion of the micrograph is the **adventitia** (16). The thick **collagenous fibers** (17) are readily recognized. The outer portion of the adventitia contains numerous elastic fibers. These elastic fibers are arranged in a circumferential pattern, which appear as black, dot-like structures when cross-sectioned.

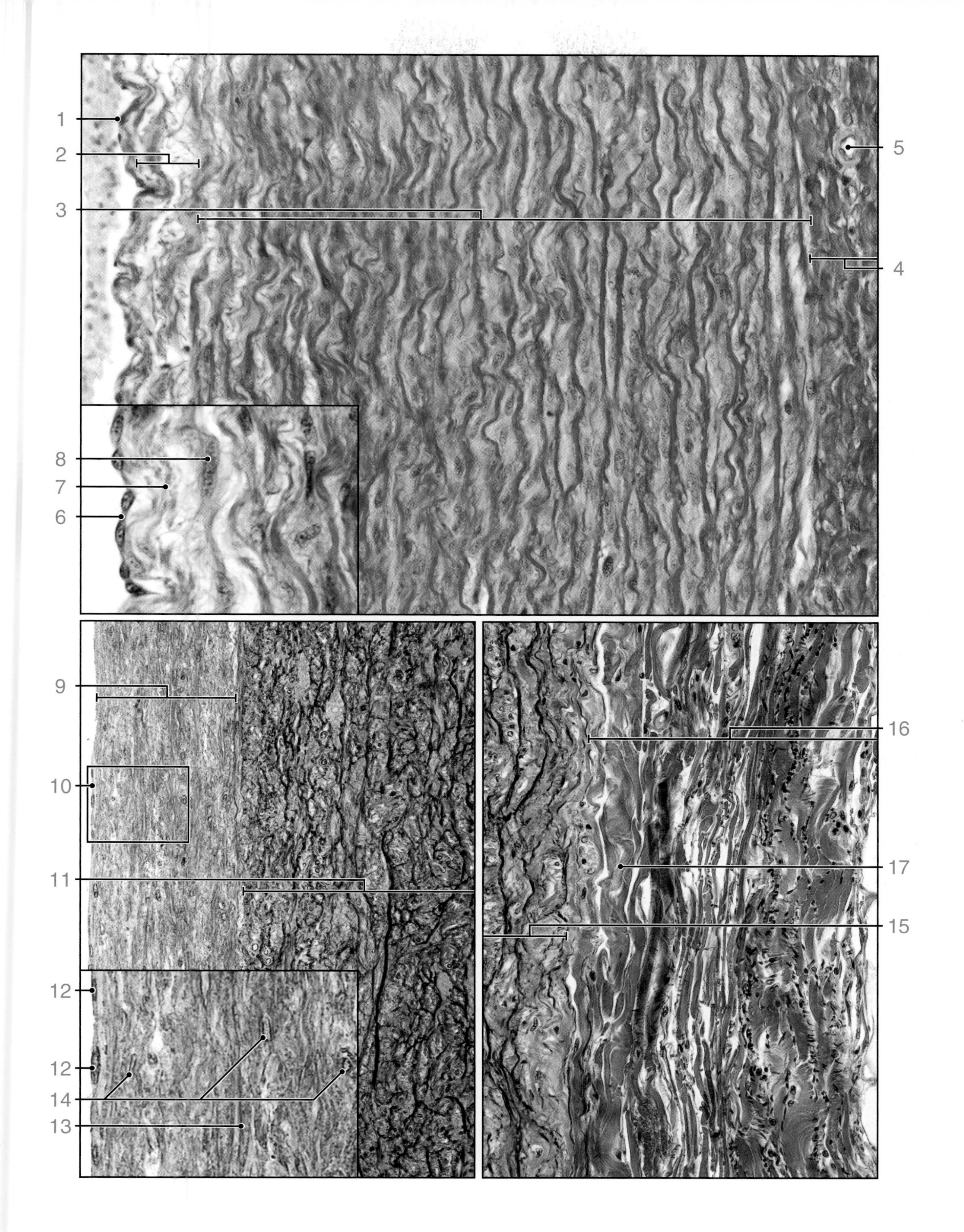

PLATE 52. AORTA

PLATE 53. MUSCULAR ARTERIES AND VEINS

Muscular arteries, also referred to as distributing arteries, constitute the majority of the named arteries in the body. The large to medium sized blood vessels encountered in routine sections are either muscular arteries or veins. As the arteriole tree is traced further from the heart, the elastic tissue in the vessel wall is considerably reduced, and smooth muscle becomes the predominant component of the tunica media. In almost all cases, muscular arteries exhibit a distinctive internal elastic membrane, separating the tunica intima from the tunica media. Sometimes, an external elastic membrane separates the tunica media from the tunica adventitia. Veins usually accompany arteries as they travel in the connective tissue. The accompanying veins have the same three layers in their walls, but the tunica media is much thinner than in the artery, and the tunica adventitia is the predominant layer in the vein wall. The veins usually have the same name as the artery they accompany.

ORIENTATION MICROGRAPH: This micrograph shows a muscular **artery** (1) and a portion of its accompanying **vein** (2). Where the two vessels lie in immediate proximity to one another, the **adventitia** (3) of each vessel is conjoined. Also showing in the micrograph is a part of a **lymph node** (4).

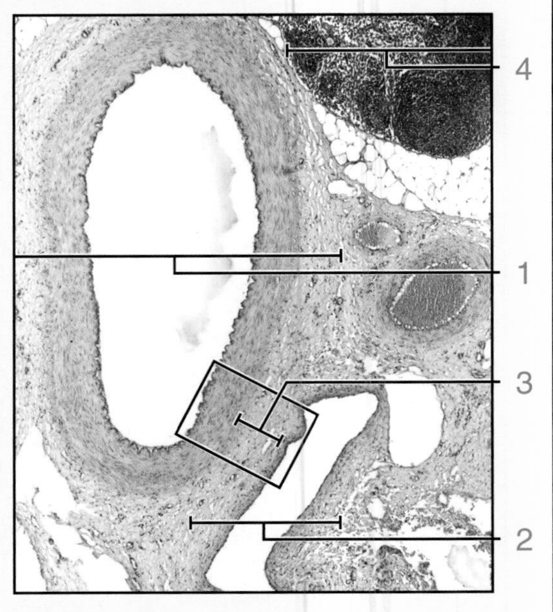

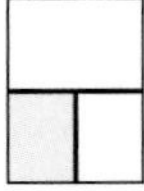

Muscular artery and vein, monkey, H&E, x365.
This micrograph is a higher magnification of the boxed area on the orientation micrograph. As noted in the orientation micrograph, the adventitia of the artery and vein are in apposition. The thickness of the wall of the artery is best determined where the two vessels are not in apposition. In the region shown here, the thickness of the **arteriole wall** (1), by extrapolation, is approximately the same as the thickness of the **wall of the vein** (2). The major distinction in terms of the components of each vessel is the relatively thick **tunica media of the artery** (3) compared to the **tunica media of the vein** (4). Conversely, the **tunica adventitia of the vein** (5) is considerably thicker than the **tunica adventitia of the artery** (6).

Muscular artery, monkey, H&E, x545.
This is a higher magnification micrograph of the boxed area on the top micrograph. At this magnification, the **endothelial cell nuclei** (7) are readily evident. The **internal elastic membrane** (8), which separates the tunica intima from the tunica media, appears as a poorly stained, undulating, homogeneous layer. Beneath this are the **smooth muscle cells** (9) of the tunica media. Note the corkscrew appearance of the contracted smooth muscle cells. Their contraction accounts for the undulating appearance of the internal elastic membrane. The lower portion of the micrograph includes the **tunica adventitia** (10).

Muscular vein, monkey, H&E, x600.
In this higher magnification of the vein, the **endothelial cells** (11) are readily seen lining the internal surface of the vessel. Beneath the endothelium is a thin layer of **connective tissue** (12), belonging to the intima. The **tunica media** (13) is relatively thin. Again note the general corkscrew shape of some of the smooth muscle cells in this region. The remainder of the micrograph consists of **tunica adventitia** (14). The few nuclei that are present in the tunica adventitia belong to fibroblasts.

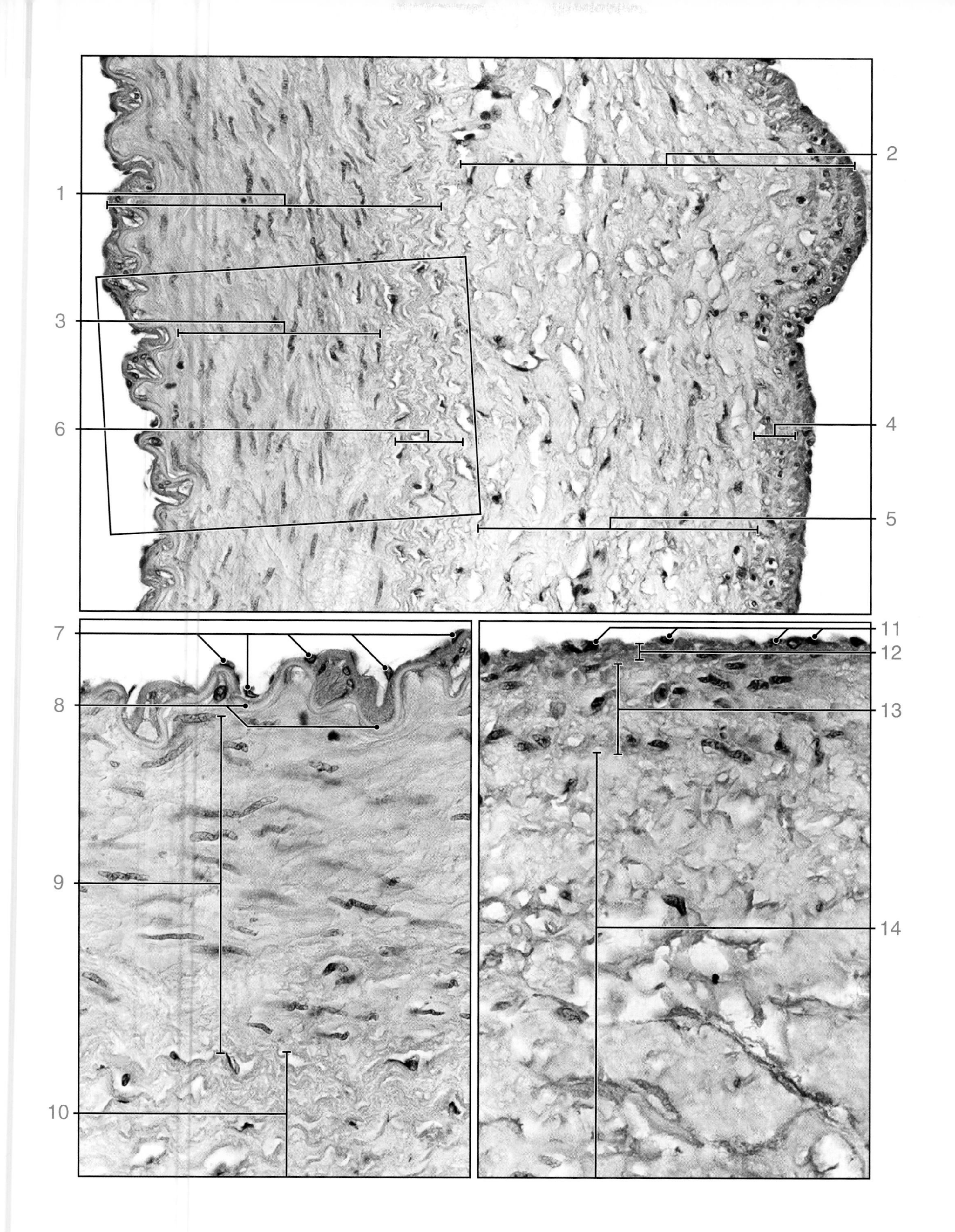

PLATE 53. MUSCULAR ARTERIES AND VEINS

PLATE 56. ARTERIOLE, ELECTRON MICROSCOPY

Arteriole, electron micrograph, x12,000; inset, H&E, x875.

The portion of the vessel wall seen in this electron micrograph corresponds to the area indicated by boxed area on the **inset** light micrograph. The electron micrograph reveals a portion of the wall of a longitudinally sectioned arteriole. The vessel is comparable in size and general structural organization to the arteriole shown in the inset.

Comparing the light and electron micrographs, note that in the light micrograph the smooth muscle cell nuclei are not evident, a result of the site at which the smooth muscle cells were sectioned. Also, the cytoplasm of the individual smooth muscle cells blends together; only their nuclei, when present in the section, are readily discernible. In contrast, the electron micrograph reveals the individual **smooth muscle cells** (1) of the vessel wall. Each cell is separated from its neighbor by a definable intercellular space. Evident in the micrograph are portions of two **endothelial cells** (2), one of which includes part of its nucleus. At the site of apposition, the two endothelial cells are separated by a narrow (~20 nm) intercellular space (*arrows*).

Between the endothelium and the smooth muscle cells is an **internal elastic membrane** (3). This is typically absent in the smallest arterioles. The elastic membrane in this vessel is a continuous sheath, permeated by irregularly shaped openings (*arrowheads*). Together with the endothelium, it forms the **tunica intima** (4). In sectioned material, the openings give the appearance of a discontinuous membrane rather than a true sheath that simply contains fenestrations. The basal lamina of the endothelial and smooth muscle cells is closely applied to the elastic membrane; consequently, other than the elastic material, little extracellular matrix is present in the tunica intima of this vessel.

The tunica media is represented by the single layer of smooth muscle cells. The muscle cells are circumferentially disposed and appear here in-cross section; none includes a nucleus. Each muscle cell is surrounded by **basal lamina** (5); however, there may be focal sites where neighboring smooth muscle cells come into more intimate apposition to form a nexus or gap junction. At these sites of close membrane apposition, the basal lamina is absent. The cytoplasm of the muscle cell displays aggregates of mitochondria along with occasional profiles of rough endoplasmic reticulum. These organelles tend to be localized along the central axis of the cell at both ends of the nucleus. In addition, numerous **cytoplasmic densities** (6) are present. They are adherent to the plasma membrane and extend into the interior of the cell, forming a branching network. The cytoplasmic densities are regarded as attachment devices for the contractile filaments, analogous to the Z lines in striated muscle.

The **tunica adventitia** (7) is composed of **elastic material** (8), numerous **collagen fibrils** (9), and **fibroblasts** (10). The elastic material is somewhat more abundant nearest the smooth muscle. If the vessel shown here could be traced back toward the heart, the elastic material would quantitatively increase, ultimately forming a more continuous sheathlike structure, namely, an external elastic membrane. The bulk of the adventitia consists of collagen fibrils, arranged in small bundles, separated by fibroblast processes.

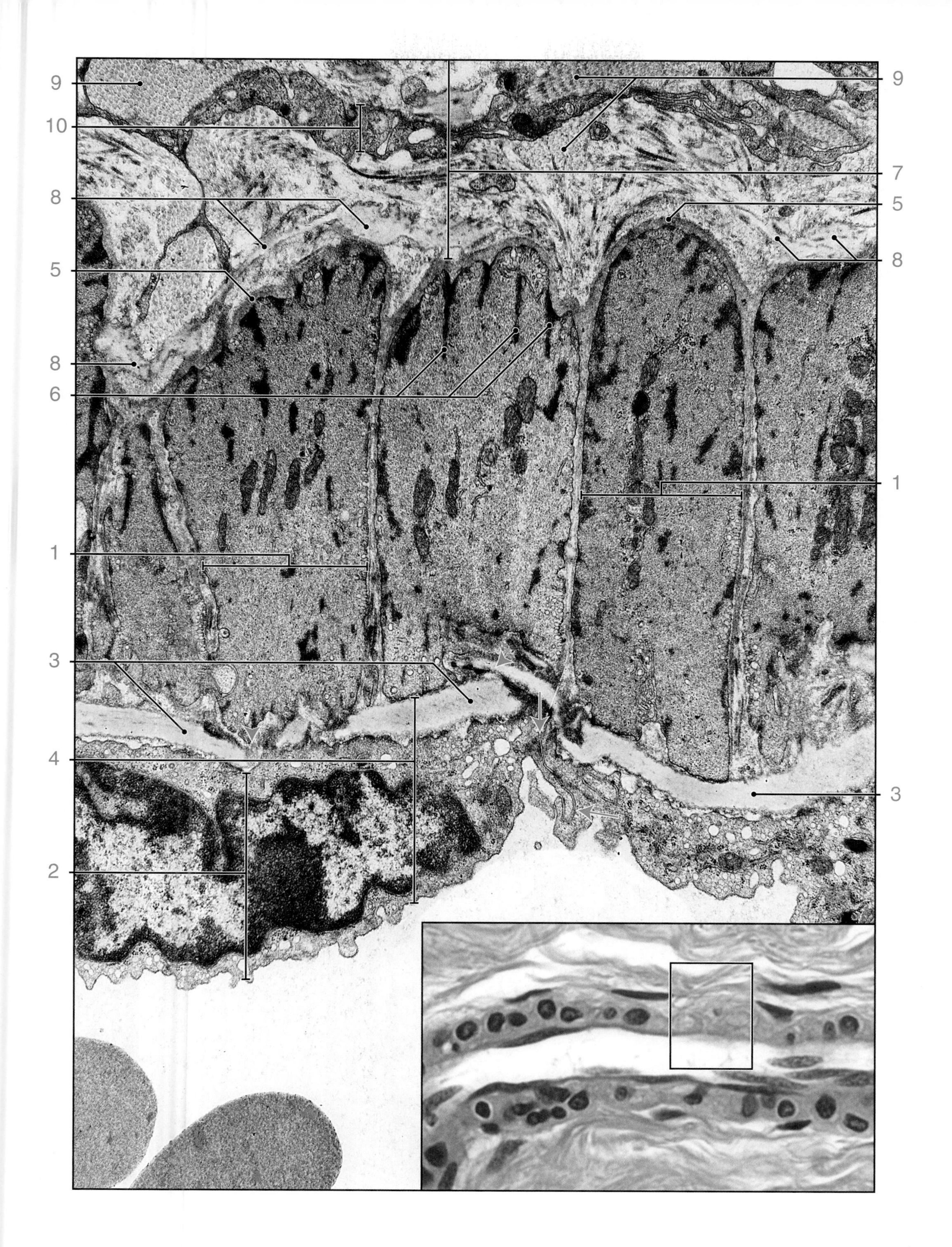

PLATE 56. ARTERIOLE, ELECTRON MICROSCOPY

CHAPTER 10

Lymphatic Tissue and Organs

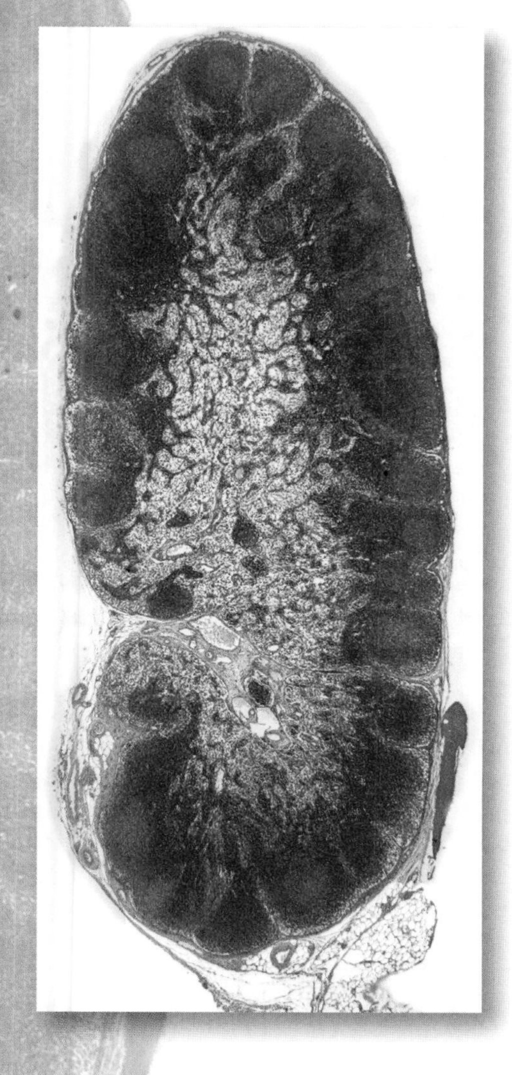

PLATE 57. **PALATINE TONSIL**

The palatine tonsils are paired structures, consisting of masses of lymphatic tissue, located on either side of the pharynx. They, along with the pharyngeal tonsils (adenoids) and lingual tonsils, form a ring at the entrance to the oropharynx (Waldeyer's ring). Structurally, the tonsils contain numerous lymphatic nodules located in the mucosa. The stratified squamous epithelium that covers the surface of the palatine tonsil (and pharyngeal tonsil) dips into the underlying connective tissue, forming many crypts—the tonsillar crypts. The walls of these crypts contain lymphatic nodules. The epithelial lining of the crypts is typically infiltrated with lymphocytes and often to such a degree that the epithelium may be difficult to discern. Although the nodules principally occupy the connective tissue, the infiltration of lymphocytes into the epithelium tends to mask the epithelial–connective tissue boundary. The tonsils guard the opening of the pharynx, the common entry to the respiratory and digestive tracts. The palatine and pharyngeal tonsils can become enflamed due to repeated infection in the oropharynx and nasopharynx, and they can harbor bacteria that cause repeated infections if they are overwhelmed. When this occurs, the inflamed tonsils are removed surgically (tonsillectomy and adenoidectomy). Tonsils, like other aggregations of lymphatic nodules, do not possess afferent lymphatic vessels. Lymph, however, does drain from the tonsillar lymphatic tissue through efferent lymphatic vessels.

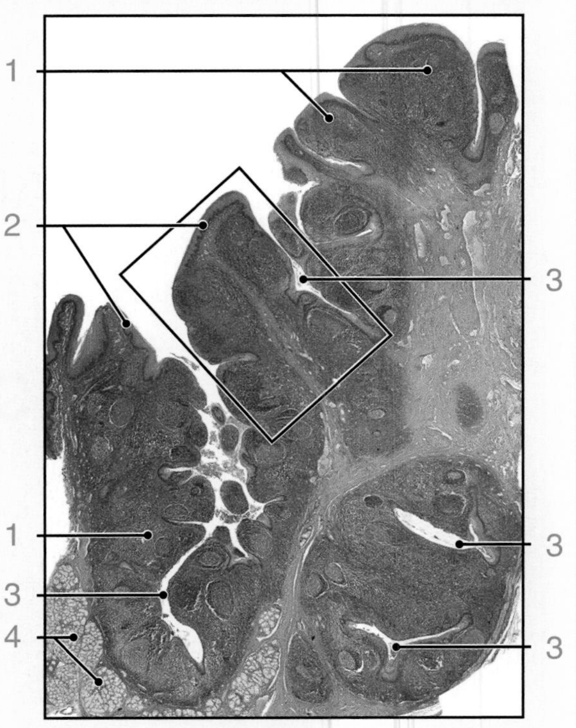

ORIENTATION MICROGRAPH: This low magnification micrograph is a section through a palatine tonsil. The hematoxylin staining areas represent the **lymphatic tissue** (1). The tonsil is surfaced by **stratified squamous epithelium** (2), which dips into the underlying connective tissue forming the **tonsillar crypts** (3). At the base of one of the crypts are a number of **mucus-secreting glands** (4).

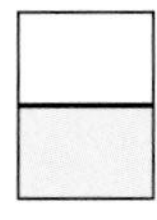

Tonsil, human, H&E, x47.

This micrograph is from the boxed area on the orientation micrograph. At this higher magnification, part of the **surface epithelium** (1) of the tonsil can be readily identified. In other sites, the **lymphocytes** (2) have infiltrated the epithelium to such an extent that the epithelium is difficult to identify. The body of each **nodule** (3) lies within the mucosa; the bodies merge because of their close proximity. Several of the nodules have been cut in a plane that includes their **germinal center** (4). Note the eosinophilic staining in these areas. Beneath the nodules is the **submucosa** (5) consisting of dense connective tissue, which is continuous with the dense connective tissue beyond the tonsillar tissue.

Tonsil, human, H&E, x365.

At the higher magnification of this micrograph, the characteristic invasiveness of the lymphocytes into the overlying epithelium is readily evident. Note, on the lower left side of the micrograph, a clear boundary between the epithelium and the underlying lamina propria. The **basal cells** (6) of the stratified squamous epithelium can be recognized. The underlying lamina propria is occupied by numerous lymphocytes; only a few have entered the epithelial compartment. Also note the thin band of **collagen fibers** (7) that can be seen at the boundary between the epithelium and lamina propria. In contrast, the lower right side of the micrograph displays numerous lymphocytes that have invaded the epithelium. More striking is the presence of what appear as isolated **islands of epithelial cells** (8) within the periphery. The thin band of **collagen** (9) lying at the interface of the epithelium is so disrupted in this area that it appears as small fragments. In effect, the small portion of the nodule seen in the right side of the micrograph has literally grown into the epithelium with the consequent disappearance of the well-defined epithelial–connective tissue boundary.

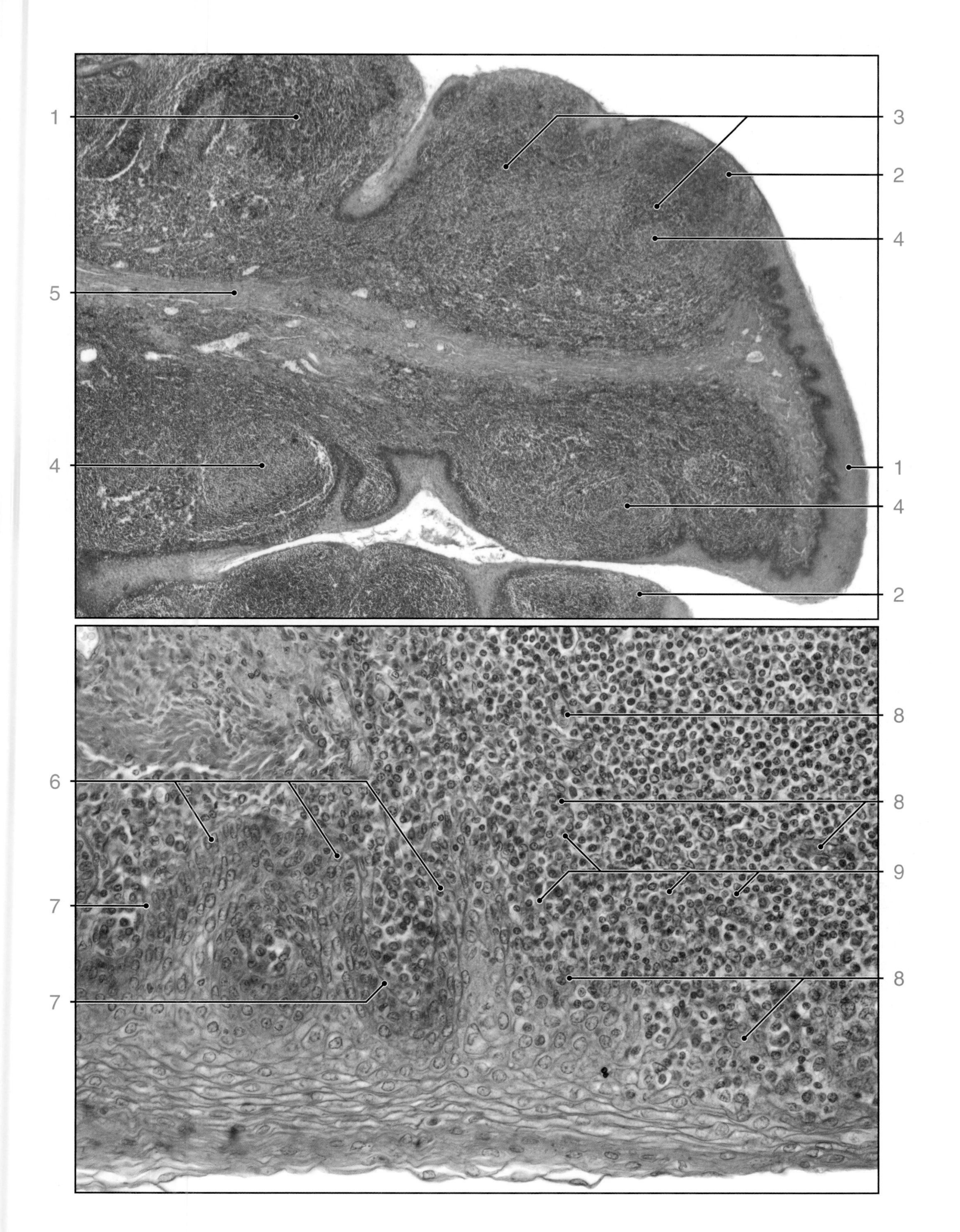

PLATE 57. PALATINE TONSIL

PLATE 58. **LYMPH NODE I**

Lymph nodes are small organs distributed along the course of lymph vessels. They are usually kidney shaped with a small indentation on one side, the hilus, where blood vessels enter and efferent lymphatic vessels leave the node. Lymph nodes receive lymph fluid from afferent lymphatic vessels that enter the node at the convex surface. The lymph fluid percolates through the node where it is filtered by the action of macrophages and reticular cells. The nodes are covered by a connective tissue capsule; beneath this is a cortical or subcapsular sinus, which receives the lymph that has been delivered by the lymph vessels that enter the node. The lymph fluid percolates through the sinus and into the outer portion, or cortex, of the node via the trabecular sinuses to enter the sinuses of the medulla in the inner part of the node. The lymph fluid leaves the medulla through the efferent lymph vessels at the hilus. The cortex largely consists of lymphatic nodules. Typically the nodules present a germinal center in which new lymphocytes are produced in response to an antigen stimulus. The nodules contain mainly B-lymphocytes. Beneath the nodules is the part of the cortex referred to as the deep cortex, or juxtamedullary cortex. The degree of development of this region is dependent upon the number of T-lymphocytes that it contains. The framework of the lymph node, exclusive of the capsule, is made up of reticular cells. These cells produce the collagenous fibers of the node and bear a special relationship to the fibers that form this framework. The cytoplasm of these cells completely covers the collagen framework, isolating it from other components of the node. The sinuses of the node are lined by endothelium that is continuous with the endothelium of the afferent and efferent lymphatic vessels.

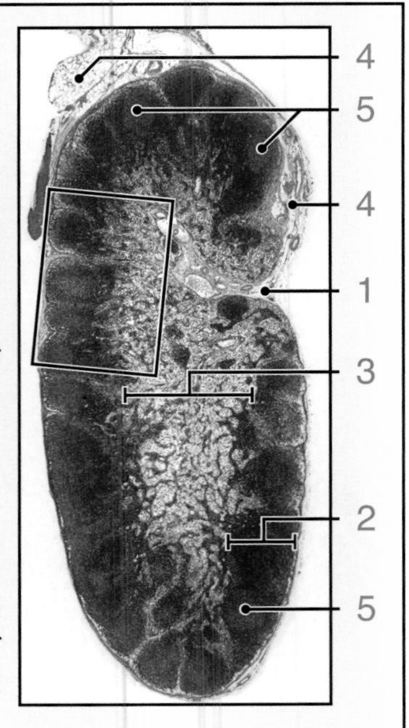

ORIENTATION MICROGRAPH: This micrograph shows a sagittal section through a lymph node. It reveals the **hilus** (1) of the node, the **cortex** (2) of the node, and the much lighter appearing inner portion of the node, the **medulla** (3). Typically, **adipose tissue** (4) surrounds part or all of the node. The lighter-staining regions of the outer cortex are the **germinal centers** (5).

Lymph node, human, H&E, x90.

The boxed area on the orientation micrograph is shown here at higher magnification. The **connective tissue capsule** (1) is composed of dense connective tissue from which **trabeculae** (2) extend into the substance of the organ. Several of the lymphatic nodules exhibit a lighter-staining, more eosinophilic region, which represents the **germinal center** (3). In some instances, the nodule may appear to lack a germinal center (4). This is usually due to the section passing through the outer portion of the nodule, thereby not including the germinal center. The **deep cortex** (5) is variable in thickness. Beneath the deep cortex is the **medulla** (6), which is largely made up of **cords of lymphatic tissue** (7) separated by **medullary sinuses** (8), the almost empty appearing spaces. In the lower right corner of the micrograph is the beginning of an **efferent lymphatic vessel** (9), which is shown in more detail in Plate 59.

Lymph node, human, H&E, x365; left inset x700; right inset x615.

This micrograph is a higher magnification of the boxed area on the top micrograph. It reveals the **capsule** (10) and, beneath it, the **subcapsular sinus** (11). The remainder of this micrograph is occupied by a lymphatic nodule. Its **germinal center** (12) contains mostly medium and large-sized lymphocytes as well as some plasma cells; compared this to the outer portion of the nodule, which contains small, densely packed lymphocytes. The **right inset** shows the subcapsular sinus at much higher magnification. The sinus is lined by endothelial cells; the nucleus of one is seen here (13). Spanning the sinus are the **reticular cells** (14), whose processes form a meshwork within the sinus. Their nuclei typically exhibit a nucleolus. Also present within the sinus are **circulating lymphocytes** (15). They are recognized by their dense, round nuclei. The **left inset** is a higher magnification of part of the germinal center. It reveals several **dividing cells** (16) identifiable by their mitotic figures. The small, dense, round nuclei belong to **lymphocytes** (17) that are just outside of the germinal center. Also identifiable are a few **reticular cells** (18).

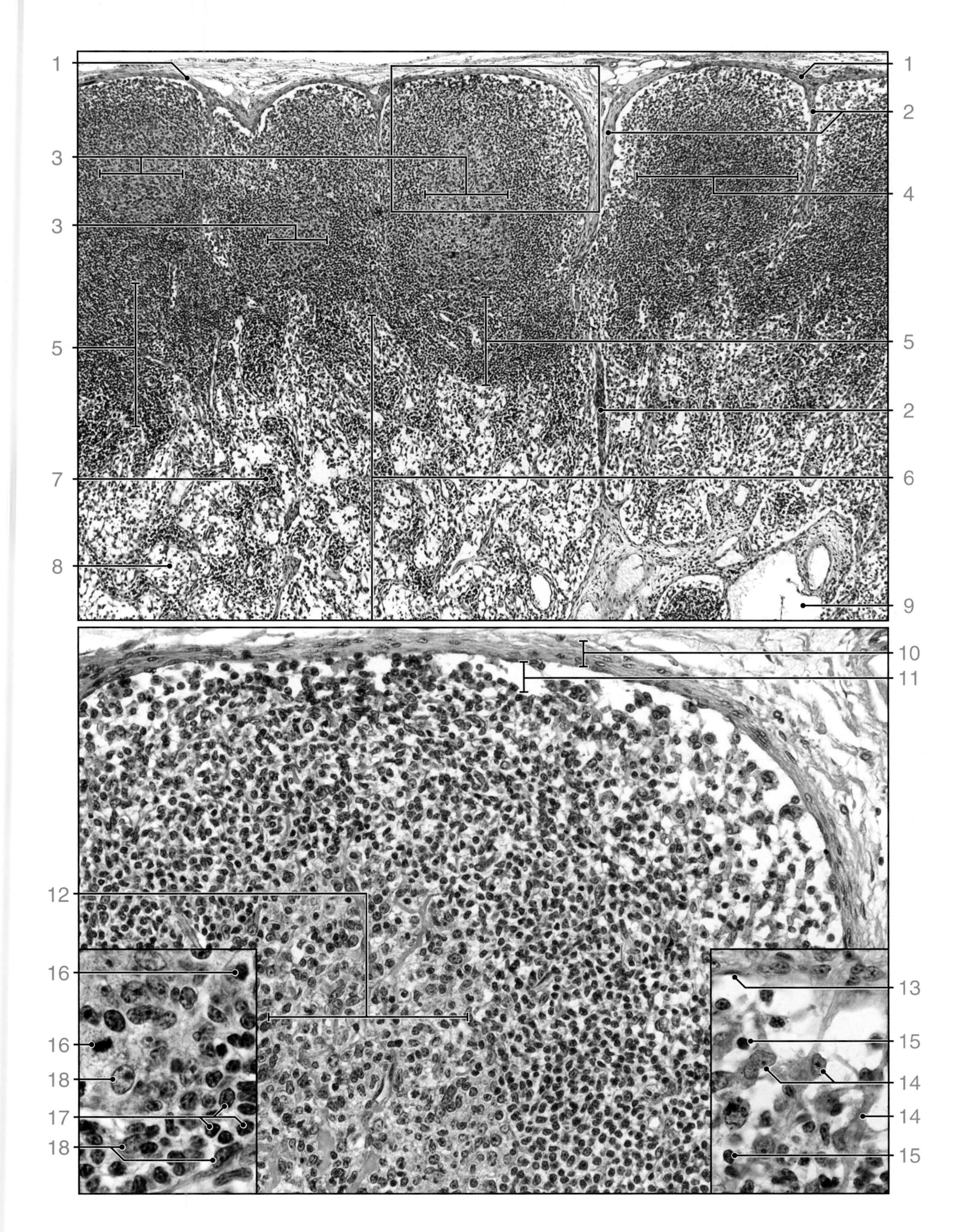
1
1
2
3
4
3
5
5
2
7
6
8
9
10
11
12
16
13
16
15
18
14
17
14
18
15

PLATE 59. **LYMPH NODE II**

As noted in the introductory text of the Plate 60, the cortex of a lymph node is divided into a superficial, or nodular cortex, which contains the lymphatic nodules, and a deep cortex, or paracortex, which contains most of the T-cells in the lymph node. The deep cortex is also regarded as the thymus-dependent cortex. This name is based on the observation that experimental thymodectomy in animals results in a poorly developed deep cortex. An important characteristic of the deep cortex, other than its mass of T-cells, is the presence of specialized postcapillary venules. These venules are lined by cuboidal or columnar endothelial cells and are referred to as high endothelial venules (HEV). The endothelial cells of these vessels play two important roles. They have the ability to transport fluid and electrolytes, which have entered the lymph node via afferent lymph vessels, directly into the blood stream. The HEV cells have a high concentration of water channels that facilitate the rapid resorption of interstitial fluid via the water channels into the blood stream. This causes lymph entering through the afferent lymph vessels to be drawn into the deep cortex. These endothelial cells also possess receptors for antigen-primed lymphocytes. They signal those lymphocytes to leave the circulation and migrate into the deep cortex. Both B- and T-cells leave the venules by diapedesis. The T-cells remain in the thymus-dependent cortex, and the B-cells further migrate to the nodular cortex. Those lymphocytes that are destined to leave the lymph node first enter lymphatic sinuses and flow to an efferent lymphatic vessel near the hilus.

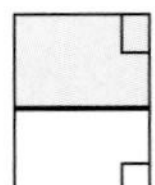

Lymph node, deep cortex, human, H&E, x365; inset x700.

The deep cortex of a lymph node is shown here. The micrograph includes a small area of the **medulla** (1). The deep cortex contains large numbers of lymphocytes, most of which are T-cells. In contrast, the medulla consists mostly of lymphatic channels known as **medullary sinuses** (2). The sinuses are lined by endothelial cells, and within these sinuses are lymphocytes, which will leave the node by emptying into an efferent lymphatic vessel near the hilus. Of special interest in the deep cortex are the **high endothelial venules (HEV)** (3). These vessels are lined by an endothelium consisting of cuboidal cells. A **typical venule** (4) entering the deep cortex is included in this section. This vessel will become an HEV. Note that the endothelium displays typical squamous cells in contrast to the HEVs. The **inset** reveals an HEV cut in cross-section at higher magnification. Note that the **endothelial cell nuclei** (5) tend to have a round profile. Also evident in the micrograph are several **lymphocytes** (6) that are migrating from the lumen and passing between the endothelial cells.

Lymph node, hilar region, human, H&E, x175; inset x350.

This micrograph reveals medullary tissue near the hilus. The **medullary tissue** (7) shows the medullary sinuses. The **inset** shows the sinuses that occupy the bulk of the tissue at higher magnification. The micrograph also includes **loose connective tissue** (8) of the hilar region as well as the beginning of several **lymphatic vessels** (9). Also present within the hilar connective tissue is a **vein** (10) and an **artery** (11). The uppermost lymphatic vessel has a discontinuous wall at one site (*arrows*). At this site, the medullary sinuses are emptying into the lymphatic vessel. Also note that this vessel and the lymphatic vessel below exhibit a set of **valves** (12). These vessels will leave the node as efferent lymphatics.

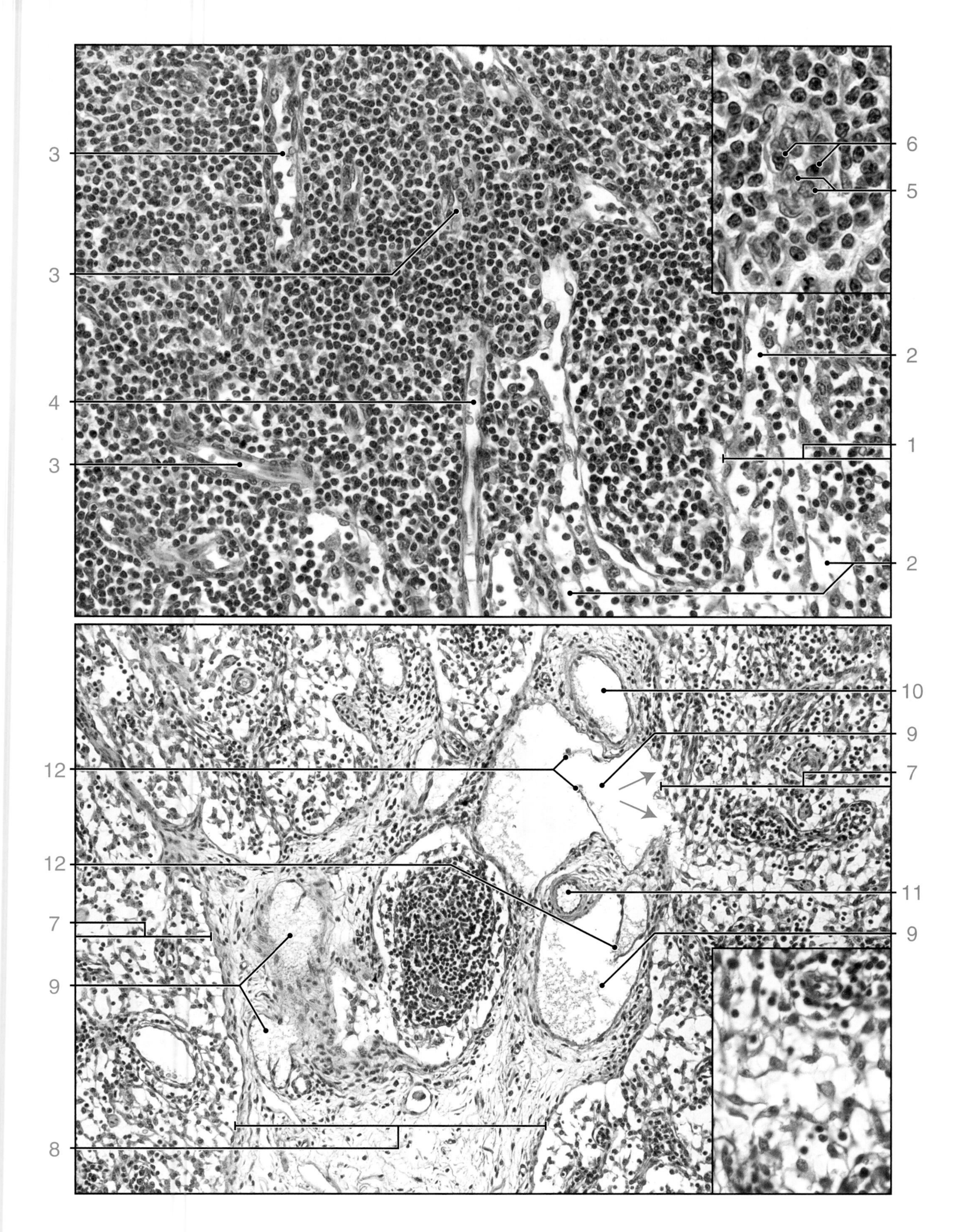

PLATE 59. LYMPH NODE II

PLATE 62. **SPLEEN II**

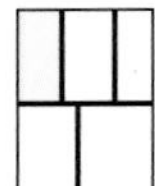

Spleen, human, H&E, x360.
As previously noted, the red pulp consists of the **venous sinuses** (1) and the area between the venous sinuses, the **splenic cords (of Bilroth)** (2). In this specimen, the red blood cells have been lysed, leaving only a clear outline of the individual cells. Thus, the relatively clear spaces with scattered nuclei represent the lumen of the venous sinus; the nuclei are those of white blood cells. When the wall of a **venous sinus** (3) is tangentially sectioned, as in this figure, the endothelial cells, which are rodlike in shape, appear as a series of thin, linear bodies.

Spleen, human, H&E, x1200.
This micrograph is a high magnification of the boxed area on the top left micrograph. The venous sinus in the center of this micrograph has been cut in cross-section. Other than the lysed red blood cells, which appear as empty circular profiles, a number of **lymphocytes** (4) are present in the lumen. The wall of the sinus, as seen here, consists of **rodlike endothelial cells** (5) that have been cut in cross-section. A narrow, but clearly visible, intercellular space is present between adjacent cells. These spaces allow blood cells to pass readily into and out of the sinuses. Processes of macrophages, located outside of the sinuses in the splenic cords, also pass between the endothelial cells, and they extend into the lumen of the sinuses to monitor the passing blood for foreign antigens. The **endothelial cell nuclei** (6) project into the lumen of the vessel and appear to be sitting on top of the cell. A **macrophage** (7), identified by residual bodies in its cytoplasm, is seen just outside of the sinus.

Spleen, human, H&E, x160.
This figure shows a **trabecular vein** (8) and surrounding red pulp. At the top of the micrograph, two venous sinuses (*arrows*) can be seen emptying into the trabecular vein. These small trabecular veins converge into larger veins, which eventually unite, giving rise to the splenic vein.

Spleen, human, silver preparation, x128.
This micrograph shows a **splenic nodule** (9) occupying the upper portion of the micrograph and **red pulp** (10) below. The components that can be identified are a **germinal center** (11), a **central artery** (12), and **venus sinuses** (13) in the red pulp. The structural elements that are stained by the silver in the nodule consist of reticular fibers. Note their paucity within the germinal center. The fine, stained, threadlike material that encircles the venous sinuses is a usual modification of basement membrane.

Spleen, human, silver preparation, x515.
This micrograph reveals several **venous sinuses** (14). Where the vessel wall has been tangentially sectioned, the **basement membrane** (15) appears as a ladderlike structure. Where the vessel has been cut deeper along its long axis (16), the basement membrane appears as dot-like structures. A three-dimensional reconstruction of the basement membrane would reveal it as a series of ringlike structures.

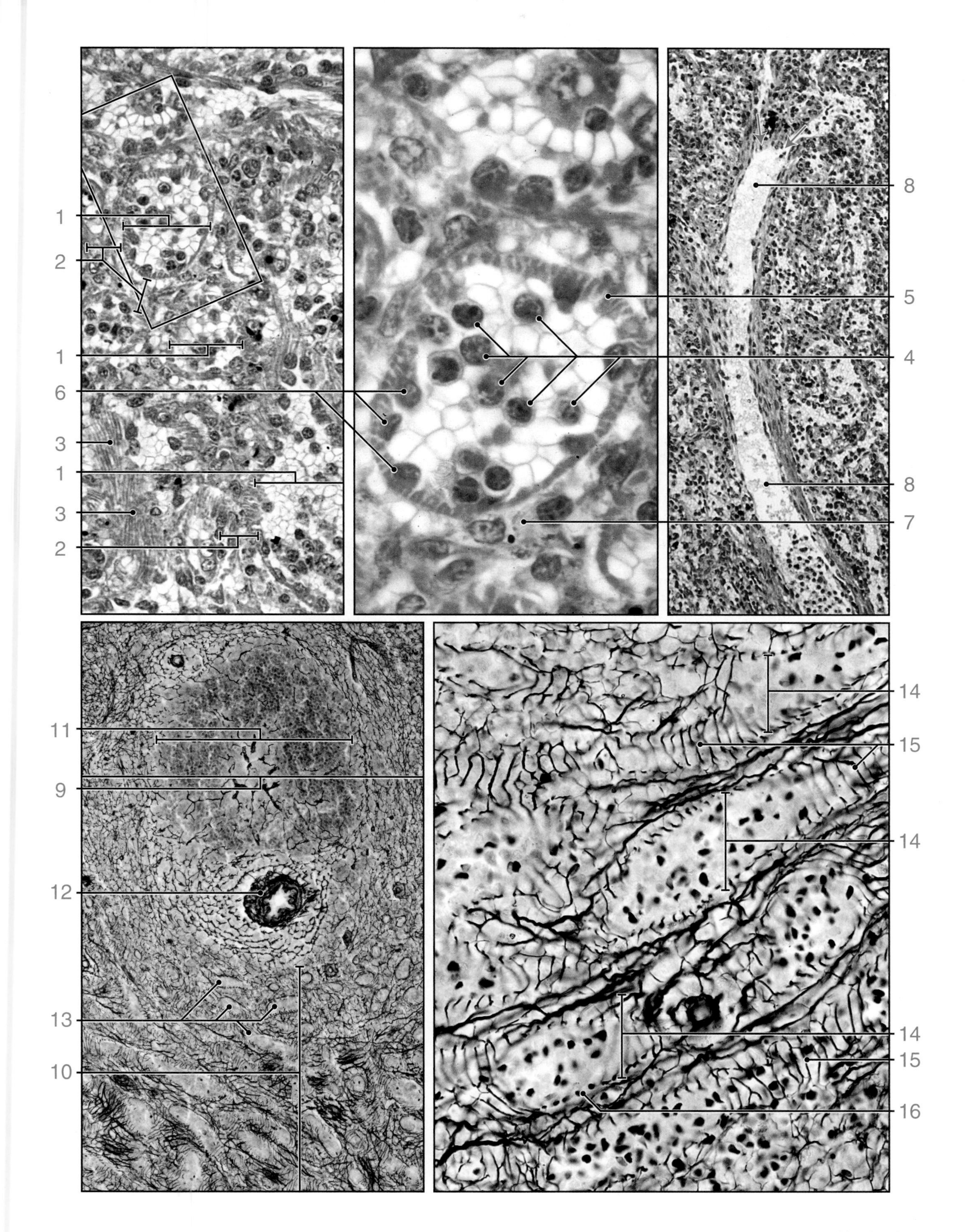
1
2
1
6
3
1
3
2
5
4
7
8
8
11
9
12
13
10
14
15
14
14
15
16

PLATE 64. **THYMUS, ADOLESCENT AND MATURE**

As previously noted, the thymus is fully formed and functional at birth. At this time, it weighs about 10 to 15 grams. From the age of about 5 to 6 years, there is a slow reduction in the number of cortical lymphocytes. However, the organ continues to grow until the age of puberty, when its weight is between 30 and 40 grams. After puberty, there is a progressive and rapid reduction in both cortical lymphocytes and epithelial cells. These elements are replaced, in part, by adipose tissue. By mid-adult life, the thymus may weigh only about 10 grams. It continues to decline with age; and by the later years of life, the thymus is difficult to identify, consisting of groups, or nests, of corpuscles with relatively few lymphocytes and Hassel's corpuscles.

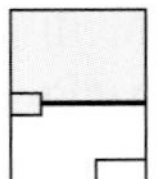

Thymus, adolescent, human, H&E, x45; inset x350.

This section of the adolescent thymus reveals identifiable **cortical** (1) and **medullary tissue** (2). Surrounding the thymus is its **capsule** (3). A substantial portion of the thymus is occupied by **adipose tissue** (4). **Connective tissue septa** (5) are prominent, some of which contain identifiable **blood vessels** (6). Also evident within the medullary tissue are **Hassall's corpuscles** (7) of various sizes. The **inset** is a higher magnification of a portion of a Hassall's corpuscle and surrounding lymphocytes. At the periphery of the corpuscle, several **nuclei** (8) of the epithelioreticular cells are readily recognized. Deeper within the corpuscle, the **nuclei** (9) of these cells have become pycnotic.

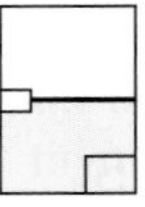

Thymus, mature, human, H&E, x60; inset x365.

The thymus shown in this micrograph is from an individual at a late stage in life. The **thymic tissue** (10) is reduced to small islands that appear disconnected. Several **veins** (11) and **arteries** (12) are also readily recognized. The bulk of the thymus, however, consists of adipose tissue. The **inset** provides a higher magnification of one of the thymic tissue islands. It reveals a **Hassall's corpuscle** (13) and surrounding lymphocytes. At the stage of involution of the thymus in this micrograph, its identity as thymus at the gross level would be very difficult. Thus, in preserved cadavers of the aged, it is rarely identifiable.

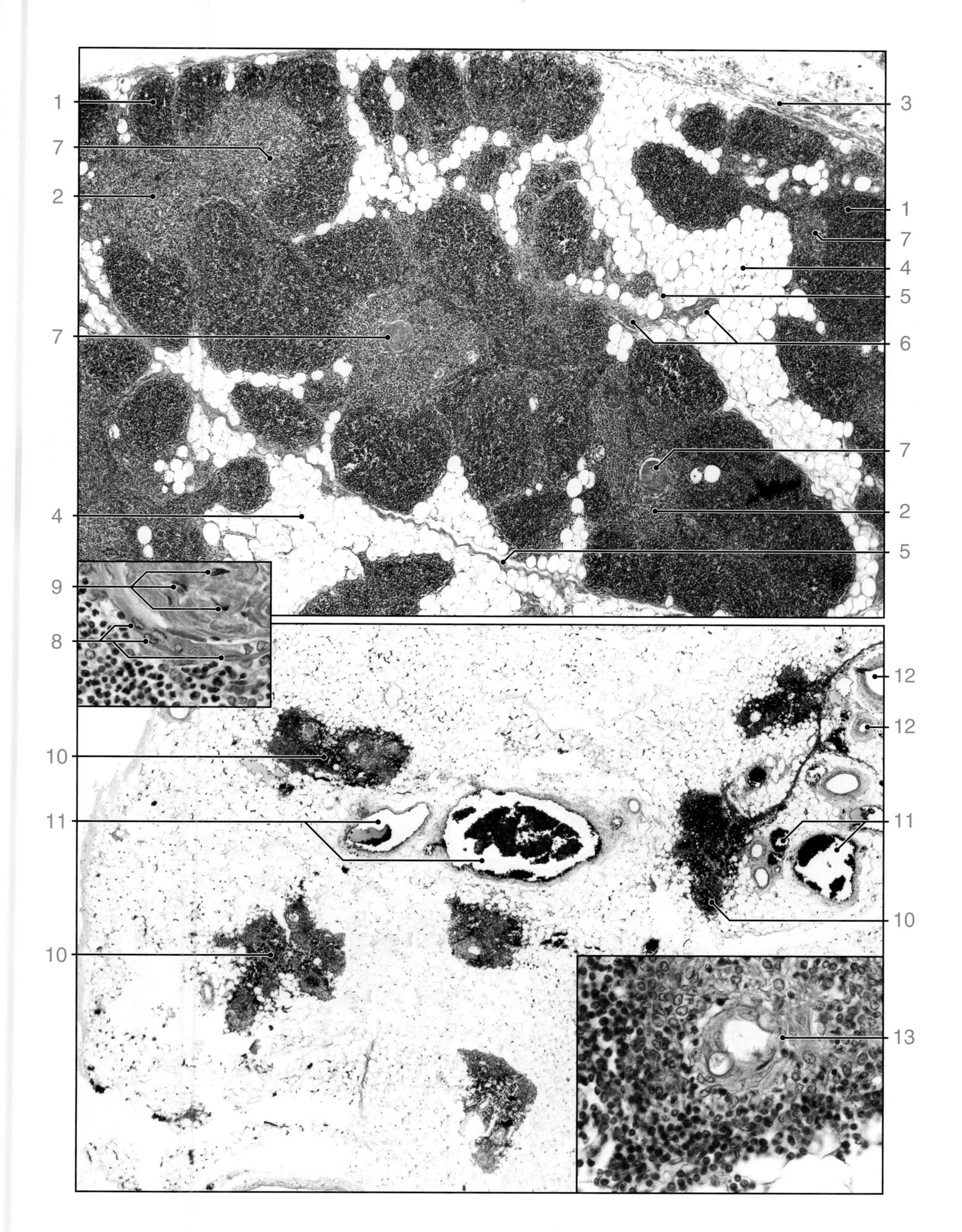
1
7
2
7
4
9
8
10
11
10
3
1
7
4
5
6
7
2
5
12
12
11
10
13

CHAPTER 11

Integumentary System

PLATE 65. **THICK SKIN**

Skin or integument covers the external surface of the body. It consists of two principle layers:

- Epidermis, a keratinized stratified squamous epithelium
- Dermis, the underlying connective tissue component

Beneath the dermis is the hypodermis, referred to as subcutaneous tissue or, by gross anatomists, superficial fascia. It is a looser connective tissue than the dermis. At many sites, it contains an abundance of adipose tissue.

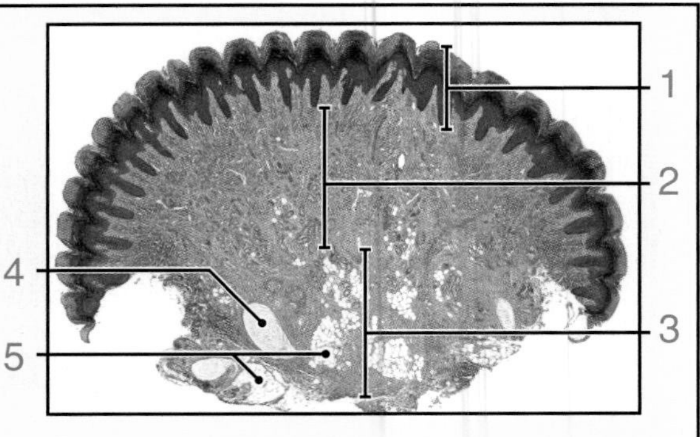

The epidermis, through the downgrowth of the epithelial basal layer into the connective tissue, gives rise to hair follicles, sweat glands, and sebaceous glands. By modifying its cellular metabolism to create a hard keratin, the epithelium also gives rise to the nails. The thickness of the epidermis varies from as little as 0.06 mm on the eyelid to as much as 0.12 mm on the back. Exceptions are the palmar and plantar epidermis, where the thickness may reach 0.8 mm and 1.4 mm, respectively. The skin of the palms and soles is usually referred to as thick skin, in contrast to the skin at other sites on the body, referred to as thin skin.

ORIENTATION MICROGRAPH: An example of thick skin, a section through the pad of a fingertip, is shown in the orientation micrograph. The thick, multicolored band at the surface with peg-like profiles is the **epidermis** (1). Below is the **dermis** (2) and, in the deepest region, the **hypodermis** (3), containing an example of a **Pacinian corpuscle** (4), a pressure receptor (described below), and **adipose tissue** (5).

Thick skin, fingertip, human, H&E, x40.
This micrograph is a low power micrograph from the specimen in the orientation micrograph above. It shows the **epidermis** (1), the underlying dense connective tissue of the **dermis** (2), and the **hypodermis** (3), which occupies most of the lower half of the micrograph. At its surface, the epidermis is characterized by a series of **epidermal ridges** (4); these collectively form what are known as dermatoglyphic patterns, or fingerprints. Each epidermal ridge exhibits, at its base, a prominent peg-like projection, the **rete ridge** (5), which interfaces with the dermal connective tissue. These peg-like structures, the rete, are true ridges and appear as pegs only when the section is cut at right angles to the epidermal ridges. In the deeper region of the dermis and hypodermis are **sweat glands** (6). Conspicuous aggregates of **adipose tissue** (7) and a **Pacinian corpuscle** (8) are also seen in the hypodermis; the Pacinian corpuscle, a pressure receptor, is shown in Plate 73. The sweat glands are shown in Plates 69 and 70.

Thick skin, fingertip, human, H&E, x160.
The boxed area in the top left micrograph is shown here at higher magnification and reveals the various epidermal layers. The deepest layer, the **stratum basale** (9), is only a single layer thick and thus is difficult to resolve even at this magnification. It is also called the germinative layer because it contains the stem cells of the epidermis. Above this layer is the **stratum spinosum** (10). It includes the cells that make up the rete ridges. Above this level is a dark basophilic-staining region, the **stratum granulosum** (11). The cells in this layer contain keratohyalin granules that produce the basophilia. The most superficial layer is the **stratum corneum** (12). The keratinocytes at this level have lost their nuclei and most organelles. Keratin filaments, produced during the cytomorphosis of the keratinocyte, pack the cell and give this layer an almost homogenous appearance. The micrograph also reveals several profiles of eccrine sweat gland ducts. One **duct** (13) is seen just entering a rete ridge; another **duct** (14) is seen at a higher level within the stratum spinosum. Also evident is a **Meissner's corpuscle** (15).

Thick skin, foot, human, H&E, x40.
This specimen, from the sole of the foot, is another example of thick skin. The thickness of the **epidermis** (16) is similar to that of the fingertip. An obvious difference is the lack of pronounced epidermal and rete ridges. Also, the **stratum corneum** (17) of the foot is significantly thicker than that of the fingertip. The **dermis** (18) is composed of dense connective tissue. In contrast, the underlying **hypodermis** (19) contains abundant adipose tissue. The primary role of adipose tissue at this site is cushioning. The **sweat glands** (20) within the hypodermis are conspicuous because of the surrounding adipose tissue.

Thick skin, foot, human, H&E, x125.
This higher magnification of the boxed area on the bottom left micrograph emphasizes the **epidermis** (21) and two of the **sweat gland ducts** (22) passing from the dermis through the epidermis. The coiling nature of the ducts within the epidermis is responsible for their seeming discontinuity. The **stratum basale** (23), again, is barely discernable. The **stratum spinosum** (24) is bounded above by the **stratum granulosum** (25), which exhibits distinct basophilia. An additional layer not usually seen in the fingertip or thin skin is the **stratum lucidum** (26), a narrow, homogenous, pale-staining band. The stratum lucidum consists of several layers of compacted, anuclear cells that appear to have different staining and refractive properties than those of the **stratum corneum** (27).

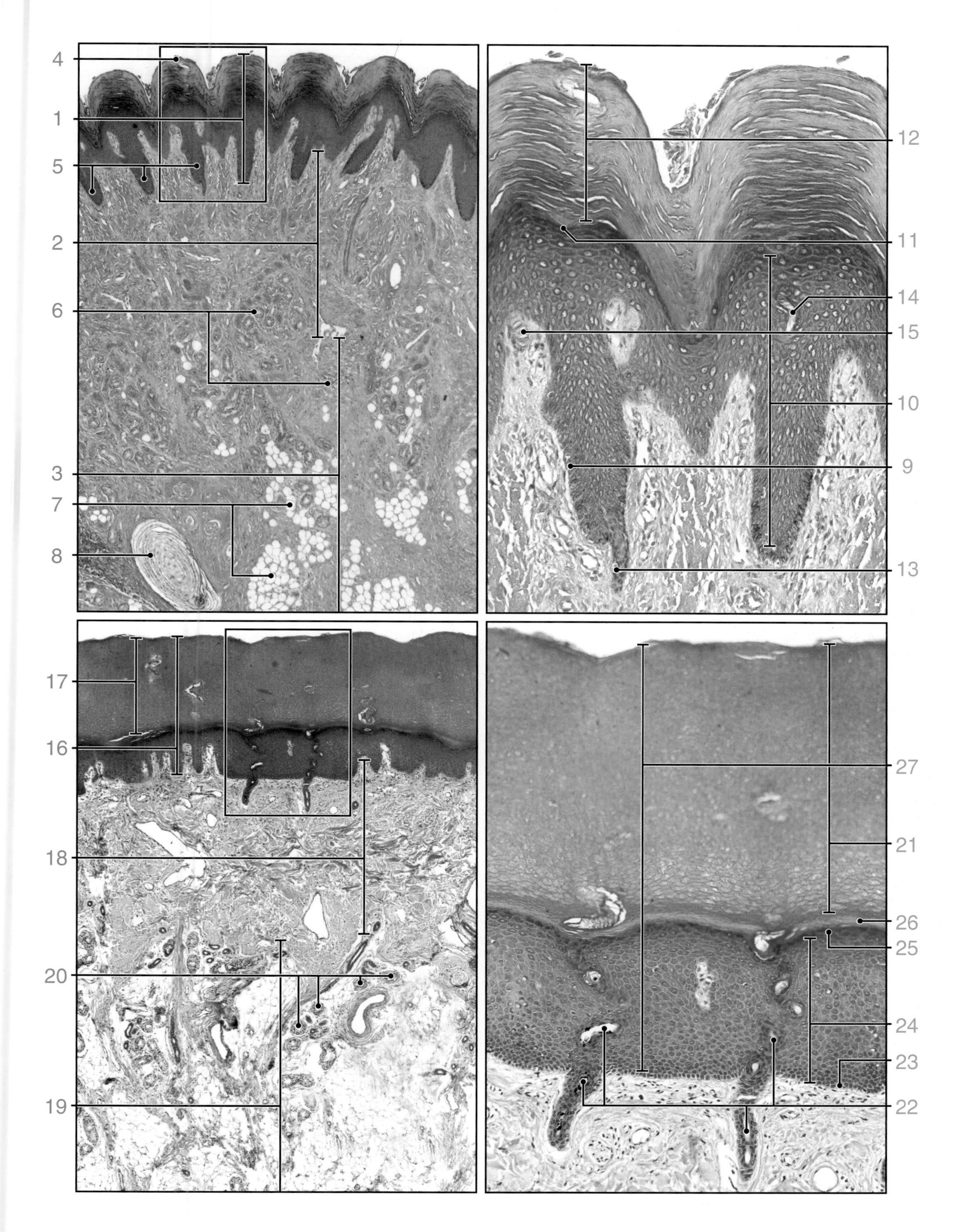

PLATE 65. THICK SKIN

PLATE 66. **THIN SKIN**

Thin skin covers most of the body surface. Variations occur from one site to another, but its common features throughout are a thinner epidermis and dermis compared to thick skin. In thin skin, the stratum corneum is particularly thin, usually representing a very small portion of the epidermal thickness; whereas in thick skin, the stratum corneum is very prominent and contributes to more than half the thickness of the epidermis. In areas subject to abrasion, such as the cheeks, shoulders and surface of the arms, the cornified layer is somewhat thicker than in more protected areas, such as the eyelid. Similarly, the dermis is also thinner than that of thick skin, but there is variation here, too. For example, the skin of the back has a very thick dermis, thicker than that found in thick skin, though its epidermis is similar to that of other sites of thin skin. Lastly, in most locations, thin skin possesses sebaceous glands and hair, ranging from fine, vellus hairs to the coarse hairs of the scalp. Both are absent in thick skin.

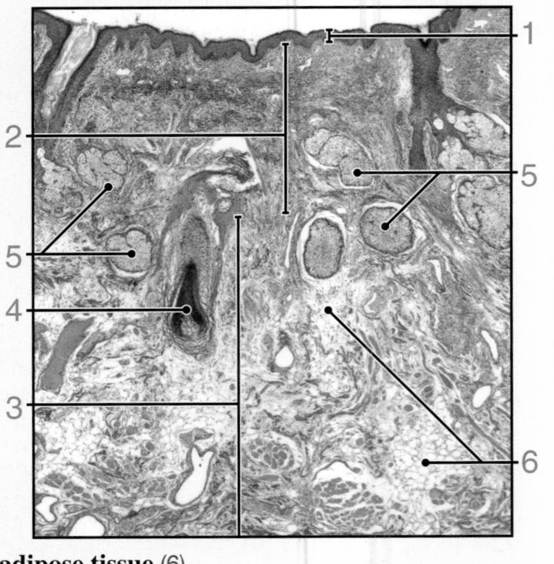

ORIENTATION MICROGRAPH: This micrograph of facial skin reveals the thin **epidermis** (1), the underlying **dermis** (2), and the **hypodermis** (3). The root of a **hair follicle** (4) and associated **sebaceous glands** (5) are also evident. The light-appearing areas in the hypodermis represent **adipose tissue** (6).

Thin skin, with sebaceous glands, H&E, x175.

The appearance of the **epidermis** (1) in thin skin is very different from that in thick skin (compare with Plate 65). The stratum corneum is almost undetectable at this magnification. Also, note that the epidermis is continuous with the epithelium of the **hair follicles and sebaceous gland follicles** (2). The surface of the epidermis reveals minor contours; however, the contour between the epidermal–dermal interface is somewhat more pronounced, producing shallow **connective tissue papillae** (3). Within the **dermis** (4) are a number of **sebaceous glands** (5). They open into the hair follicle. Note the presence of secreted **sebum** (6) in the follicle on the right. Another notable feature is the presence of the many **venous vessels** (7) that are part of the superficial vascular plexus.

Thin skin, epidermis and dermis, H&E, x725.

In this higher magnification of the epidermis, the epidermal layers are readily discernable. The low, columnar-shaped cells of the **stratum basale** (8) form a single cell layer and lie on the basement membrane. They contain melanin granules, the pigment that gives the brown color to the cytoplasm of these basal keratinocytes. Cells in this layer serve as **stem cells** from which all of the keratinocytes are derived. Also scattered in this layer are **melanocytes** (9), the cells that produce the melanin. They possess nuclei that are slightly smaller and more dense than the basal keratinocytes' nuclei, and their cytoplasm is typically clear. The keratinocytes that make up the **stratum spinosum** (10) form the thickest layer in thin skin and vary in shape as they are moved upward. The most superficial cells in this layer acquire a squamous shape. The cells of the **stratum granulosum** (11) retain the squamous shape, but are identified by the keratohyalin granules within the cytoplasm. The granules readily define this layer and stain deeply with the hematoxylin stain. The cells of the **stratum corneum** (12) are anucleate and form a thin layer that looks almost homogenous. Occasional wisp-like profiles seen at the surface represent **keratinized cells** (13) that are desquamating (i.e., sloughing off). The superficial part of the underlying connective tissue is relatively cellular and is called the **papillary layer** (14) of the dermis. In addition to numerous fibroblasts, it contains **macrophages** (15), some of which can be identified here by the melanin granules they contain. Variable numbers of other cells of the immune system, such as lymphocytes, may also be present.

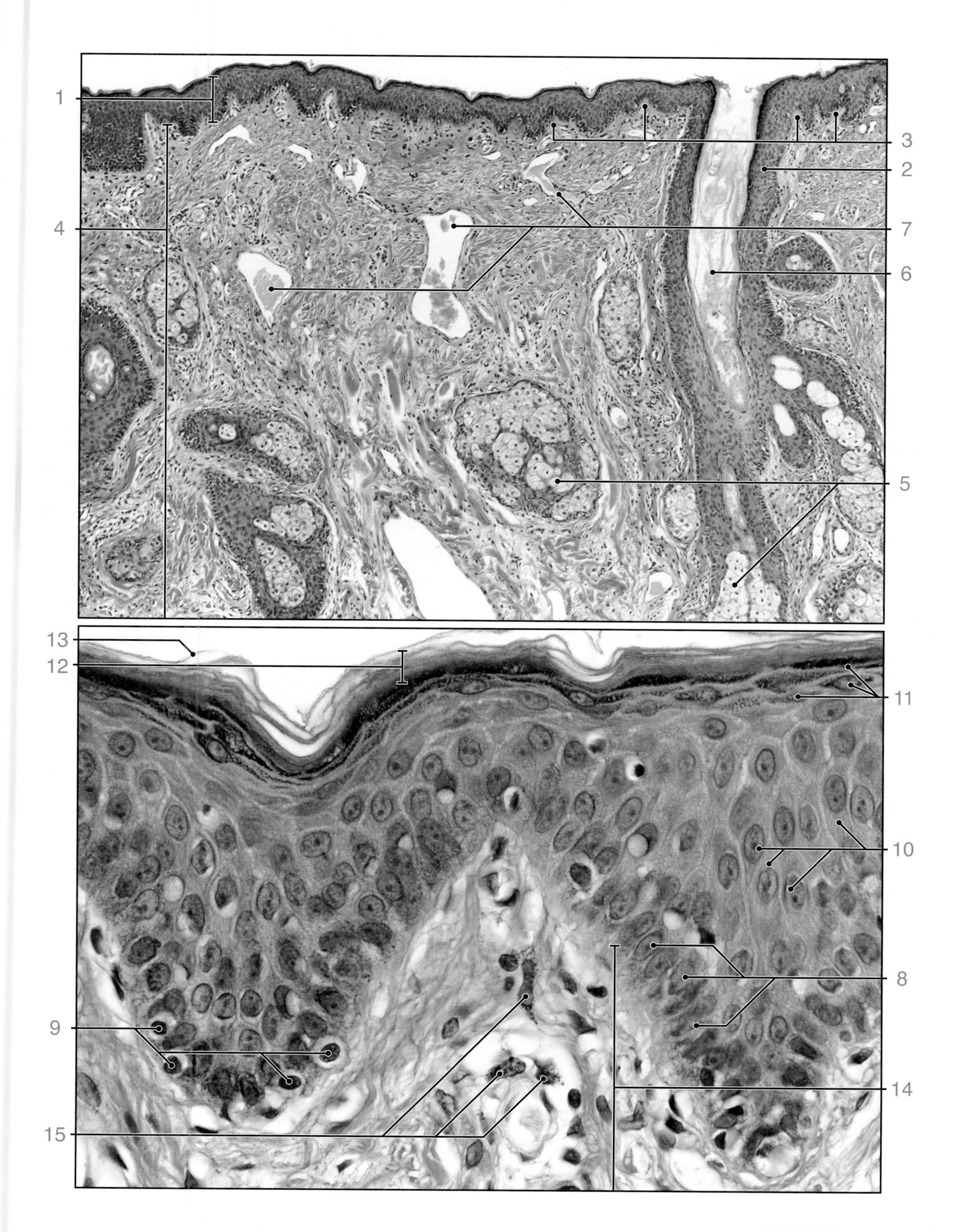

PLATE 66. THIN SKIN

PLATE 67. EPIDERMAL CELLS

Epidermis consists of four cell types: keratinocytes, melanocytes, Langerhans' cells, and Merkel's cells.

Keratinocytes constitute about 93% of epidermal cells. They arise from stem cells in the basal layer and maintain epithelial integrity by a high density of desmosomal attachments. They also produce tonofilaments (bundles of keratin intermediate filaments) in the stratum basale and stratum spinosum—initially expressed as keratin types 5 and 14 in the basal cells and keratin types 1 and 10 in cells above this layer.

Membrane-coating granules, containing packed lipid lamellae, are synthesized and discharged into the extracellular space by keratinocytes in the upper stratum spinosum and in the stratum granulosum, creating a barrier to keep foreign bodies from invading the skin. Tight junctions are also present in the stratum granulosum and contribute to the barrier system.

Keratohyalin granules, first produced in the deep cells of the stratum granulosum, contain the intermediate filament-associated protein filaggrin that aggregates the keratin (tono) filaments into an organized compact mass that fills the organelle-depleted, cornified cells.

Finally, cleavage of desmosomes by cholesterol sulfate is implicated in the desquamation of the surface cornified cells.

Melanocytes are melanin-producing cells located in the epidermal basal layer; about 2 to 10% of basal cells are melanocytes. Melanin, produced by oxidation of tyrosine, is contained in ellipsoid membrane-bounded granules, melanosomes. They rapidly pass to the ends of long dendritic processes and are then phagocytized by the adjacent keratinocytes. The granules accumulate in a supranuclear site in the keratinocytes, protecting the nucleus from UV radiation. As the keratinocytes move toward the surface, the melanosomes degrade, having accomplished their protective role.

Langerhans' cells constitute about 4% of the epidermal cell population. They are scattered among nonkeratinized keratinocytes. They have clear cytoplasm but are best identified in the TEM by their racket-shaped Birbeck's granules. Langerhans' cells function as antigen-presenting cells, notably in hypersensitivity reactions.

Merkel cells are located in the basal layer, attached to adjacent keratinocytes by desmosomes. Transmission electron microscopy reveals small granules in the basal cytoplasm. The base of the cell lies in close apposition to an expanded, afferent (unmyelinated) axonal nerve ending. The Merkel cell is thought to provide cutaneous sensation.

Thin skin, human, H&E, x1400.
This is a high magnification micrograph of the epidermis, showing the **stratum corneum** (1), **stratum granulosum** (2), and the upper part of the **stratum spinosum** (3). The stratum corneum often appears homogeneous. Desquamation (*arrows*) is usually evidenced by individual cells or sheets of cornified cells separating from the surface. Cells of the granular layer exhibit increased staining as they mature partly because of an increase in the number of the keratohyalin granules and partly because of condensation of the cell from the loss of its organelles and nucleus. Note that the granules are distinct at first, but as the cell matures, the granules condense into a dark-staining mass. In this same layer, secretion of membrane-coating granules results in sealing and obscuring the intercellular space. Compare this with the appearance of the intercellular space immediately below the stratum granulosum.

Thin skin, human, H&E, x1400; inset x2100.
This micrograph of the stratum spinosum reveals the spinous processes of the keratinocytes. Conspicuous "bridges" crossing the intercellular space are clearly evident. The **inset** shows thickenings (*arrows*) in many of the intercellular bridges. This is the site of the desmosome.

Thin skin, human, H&E, x1400.
This micrograph shows the basal cell layer, the stratum germanitivum, and, above, the stratum spinosum. The keratinocytes in the basal layer contain numerous **melanin granules** (4), which produce the brown/black coloring of the cytoplasm. **Melanocytes** (5) are also seen in the basal cell layer. Their cytoplasm stains poorly, often appearing as a clear area around the nucleus. In well-preserved specimens, melanin granules can sometimes be seen in the melanocyte. The processes that extend from the cell body and deliver the melanosomes to the keratinocytes are not visible in typical H&E preparations.

Thin skin, human, electron micrograph, x6000.
This low power micrograph is comparable to the upper portion of the top left micrograph. The cells of the **stratum corneum** (6) are exceedingly dense, reflecting the packed intermediate filaments that they contain. The cell below, in the stratum granulosum, shows a **nucleus** (7) that is about to degenerate. Note its irregular surface contour and **heterochromatin** (8) aggregated against the nuclear membrane. The **nucleolus** (9) is still prominent, but condensed. The **keratohyalin granules** (10) appear as irregular, dense bodies.

Thin skin, human, electron micrograph, x6000.
This micrograph, from the same specimen as the above micrograph, shows several cells in the stratum spinosum. The **nucleus** (11) has a smooth surface contour, and its **nucleoli** (12) are not condensed. Note that there is little heterochromatin along the nuclear membrane. The cytoplasm contains numerous **tonofibrils** (13) that extend into the **cytoplasmic bridges** (14). Desmosomes are not seen at this magnification.

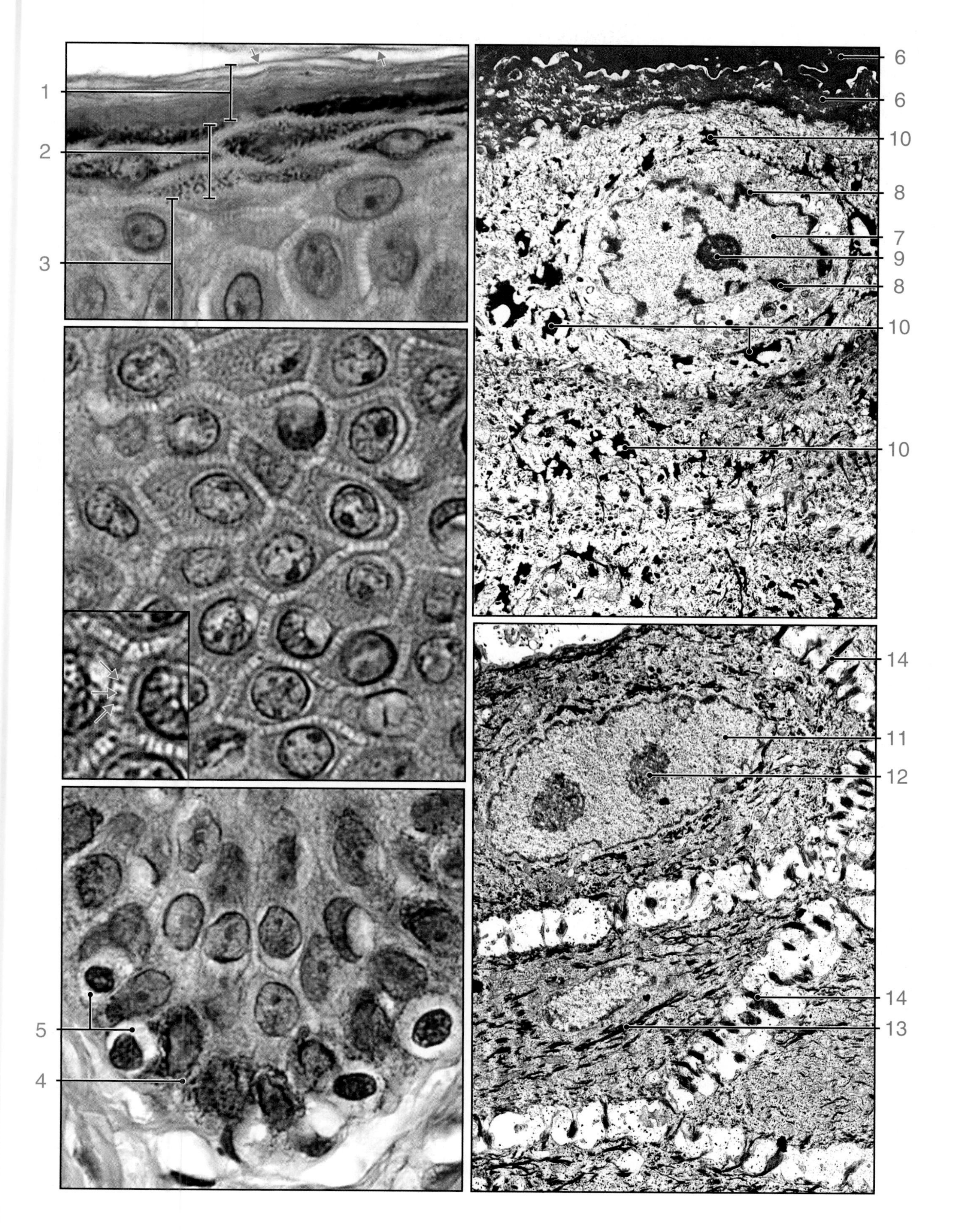

PLATE 67. EPIDERMAL CELLS

PLATE 68. FETAL SKIN AND DERMIS

The primordial epidermis of the fetus first appears as a single layer of squamous cells. Around the fourth week, this simple epithelium develops into a stratified double cell layer. The surface cells constitute the periderm, a permeable but protective barrier. Around 8 weeks, melanocytes derived from the neural crest migrate into the basal layer. They begin to deliver melanosomes to the basal cells later, around the mid to late second trimester. By the end of the first trimester, an intermediate zone is formed between the periderm and the basal cell layer. The cells in this zone are rich in glycogen and contain bundles of intermediate filaments. Small desmosomes are also present. Functional Langerhans' cells appear in the intermediate zone between 12 and 14 weeks, having migrated from hematopoietic stem cells in the yolk sac and liver. Later, they arise from bone marrow. Merkel cells are believed to arise from undifferentiated cells within the early epidermis. They are first identified in the beginning of the second trimester. After the fifth month, keratohyalin granules are found in the upper cell layer of the intermediate zone. This is followed by the loss of nuclei in these cells and evidence of keratinization. By the seventh month, cornification of the epidermis is complete, and the periderm is sloughed off. At term, the epidermis is fully functional as an impermeable barrier.

The dermis is divided into two regions: a superficial part, the papillary layer, which includes the dermal papillae; and below, the reticular layer, which is much thicker and extends to the hypodermis. The papillary layer is a relatively loose connective tissue that consists of thin collagen and elastic fibers, with a cell population that includes fibroblasts, macrophages, mast cells, and occasional lymphocytes. In contrast, the reticular layer is composed of thick bundles of collagen fibers and relatively thick elastin fibers. The principal cell type in this layer is the fibroblast, but there are fewer per unit area than in the papillary layer. The reticular layer provides the integument's strength. Its thickness varies; for example, it is thickest in the upper back and thinnest in the eyelids, where there is little stress. Blood is supplied to the dermis by a deep, vascular plexus that parallels the surface of the skin in the lower part of the reticular layer. Communicating vessels ascend from this plexus through the reticular layer to form a superficial plexus in the upper part of the reticular layer. From this plexus, capillary networks loop upward, reaching the dermal papillae to provide a rich vascular supply throughout the papillary layer.

Fetal skin, human, H&E, x130.
This low power micrograph is of fetal skin at approximately 8 weeks. The **epidermis** (1) is exceedingly thin and consists of only two cell layers. The primitive **dermis** (2) at this stage of development consists of mesenchyme. The interface between epidermis and dermis is smooth; no papillae are evident. Small **blood vessels** (3) are present, but there are no epidermal derivatives (e.g., sweat glands, hair follicles) developing at this stage.

Fetal skin, human, H&E, x515.
At this higher magnification, the two cell layers of the epidermis can be seen. The **periderm** (4) is the name given to the surface cell layer. Beneath the epidermis, the **mesenchymal cells** (5) exhibit cytoplasmic processes of varying lengths. Extracellular fibers are sparse, giving the matrix an empty appearance.

Fetal skin, human, Mallory, x130.
This is a low power micrograph of fetal skin early in the third trimester. The **epidermis** (6) is considerably thicker and appears similar to mature epidermis. There is evidence of cornification at the surface as well as a distinctive basal cell layer. The dermis exhibits a number of obliquely sectioned **hair follicles** (7). A cellular **papillary layer** (8) can be distinguished from the **reticular layer** (9), which stains blue because of the abundant collagen fibers.

Fetal skin, human, Mallory, x515.
Examination of the epidermis in the top right micrograph at higher magnification reveals the typical four strata seen in mature epidermis. The **basal layer keratinocytes** (10) appear as a uniform row of columnar cells. Above this is the **stratum spinosum** (11). The spinous processes are not prominent, but desmosomes are well developed. **Keratohyalin granule-containing cells** (12) can be seen, but they do not appear as a continuous cellular layer. The densely stained surface layer represents keratinized cells, the **stratum corneum** (13).

Skin, dermis, human, H&E/Verhoeff, x90; inset x350.
An important feature of the dermis is its connective tissue fiber organization. The tautness of skin is attributed to dermal elastic fibers. The elastin in the papillary layer is in the form of very thin branching fibers. In the reticular layer, the elastic fibers also branch, but they are considerably thicker and more numerous. At low magnification, only the thick **elastin fibers** (14) of the reticular layer are seen. The higher magnification of the **inset** reveals both the **fine elastin fibers** (15) of the papillary layer and the coarse elastin fibers (14) of the reticular layer.

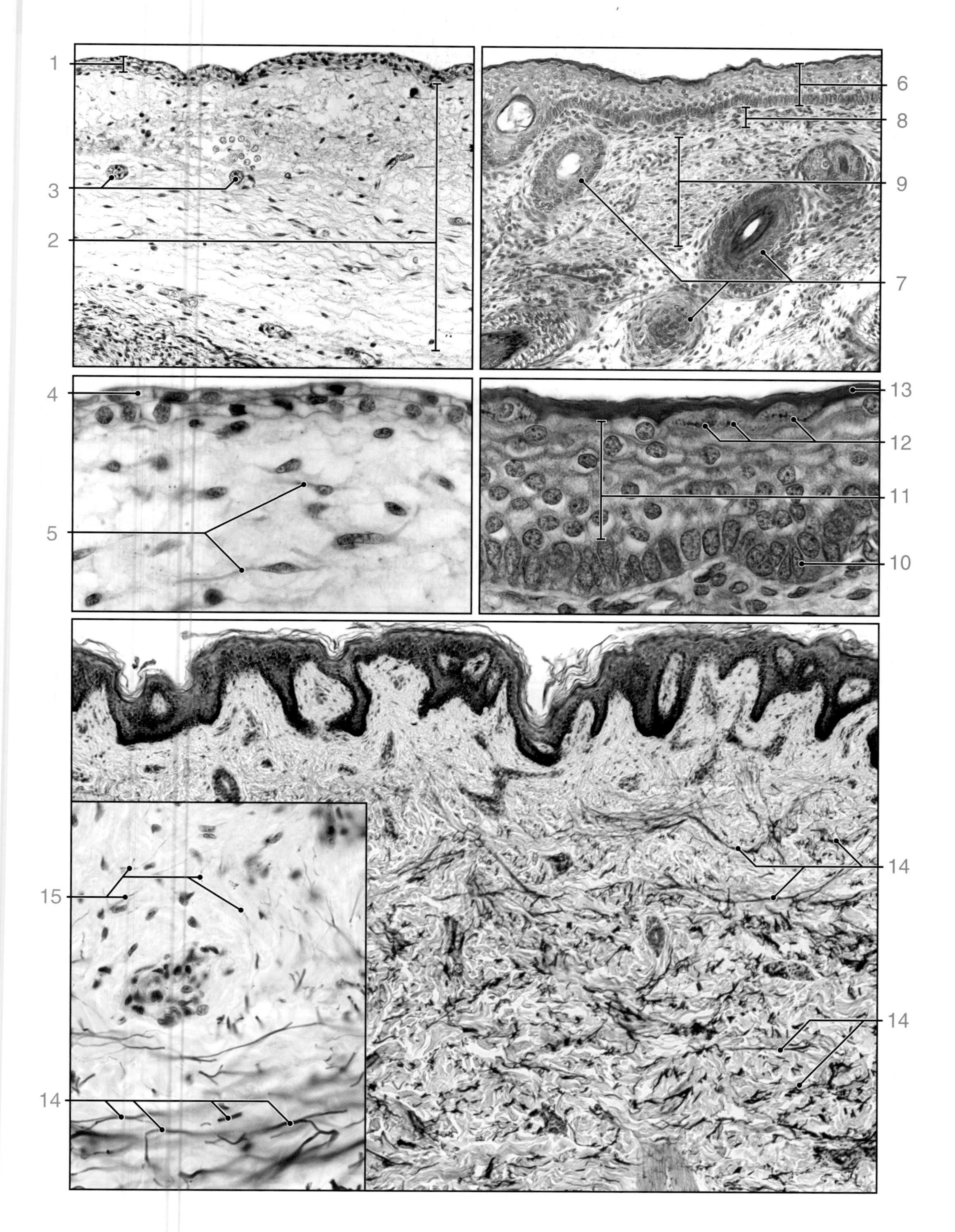

PLATE 68. FETAL SKIN AND DERMIS

PLATE 73. SENSORY ORGANS OF THE SKIN

The skin is endowed with numerous sensory receptors of various types. They are the peripheral terminals of sensory nerves whose cell bodies are in the dorsal root ganglia. The receptors in the skin are described as free nerve endings and encapsulated nerve endings. Free nerve endings are the most numerous. They subserve fine touch, heat, and cold. They are found in the basal layers of the epidermis and as a network around the root sheath of hair follicles. Encapsulated nerve endings include Pacinian corpuscles (deep pressure), Meissner's corpuscles (touch, especially in the lips and thick skin of the fingers and toes), and Ruffini endings (sustained mechanical stress on the dermis).

Motor endings of the autonomic nervous system supply the blood vessels, the arrector pili muscles, and the apocrine and eccrine sweat glands.

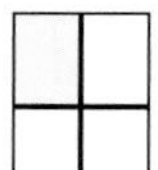

Skin, human, H&E, x20.

This specimen is a section of thick skin from the fingertip, showing the **epidermis** (1), the **dermis** (2), and a portion of the **hypodermis** (3). The thickness of the epidermis is largely due to the thickness of the stratum corneum. This layer is more lightly stained than the deeper portions of the epidermis. Notable even at this low magnification are the thick collagenous fibers in the reticular layer of the dermis. **Sweat glands** (4) are present in the upper part of the hypodermis, and several **sweat ducts** (5) are seen passing through the epidermis.

This specimen depicts those sensory receptors that can be recognized in a routine low power, H&E-stained paraffin section. They are Meissner's corpuscles and **Pacinian corpuscles** (6). Pacinian corpuscles are found in the lower part of the hypodermis. These corpuscles are large, slightly oval structures, and even at low magnification, a layered, or lamellate, pattern can be discerned. Several **nerve bundles** (7) are seen in proximity to the Pacinian corpuscles. Meissner's corpuscles are in the upper part of the dermis, in the dermal papillae immediately under the epidermis. These corpuscles are small and difficult to identify at this low magnification; however, their location is characteristic. Knowing where they are located is a major step in finding Meissner's corpuscles in a tissue section; they are shown at higher magnification in the bottom left micrograph.

Skin, human, H&E, x330.

At this higher magnification, the concentric layers, or lamellae, of the Pacinian corpuscle can be seen to be made of flattened cells. These are fibroblast-like cells, and although not evident in this tissue section, these cells are continuous with the perineurium of the nerve fiber. The space between the cellular lamellae contains mostly fluid. The neural portion of the Pacinian corpuscle travels longitudinally through the center of the corpuscle. In this specimen, the corpuscle has been cross-sectioned; an *arrowhead* points to the centrally located nerve fiber.

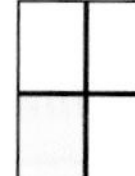

Skin, human, H&E, x175.

This high magnification micrograph shows the upper left portion of the top left micrograph where two **Meissner's corpuscles** (8) are in direct proximity to the undersurface of the epidermis in adjacent dermal papillae. The section shows the long axis of the corpuscles. A Meissner's corpuscle consists of an axon (sometimes two) taking a zigzag or flat spiral course from one pole of the corpuscle to the other. The nerve fiber terminates at the superficial pole of the corpuscle. Consequently, as seen here, the nerve fibers and supporting cells are oriented approximately at right angles to the long axis of the corpuscle. Meissner's corpuscles are particularly numerous near the tips of the fingers and toes.

Skin, human, H&E, x650.

At the higher magnification of this micrograph, the close apposition of the Meissner's corpuscle to the undersurface of the epidermis is well demonstrated throughout the entire area of the dermal papilla. The flat, spiral path of the neuron (not seen) and its supporting cells are evident here, as is the **fibrous capsule** (9) that surrounds the ending.

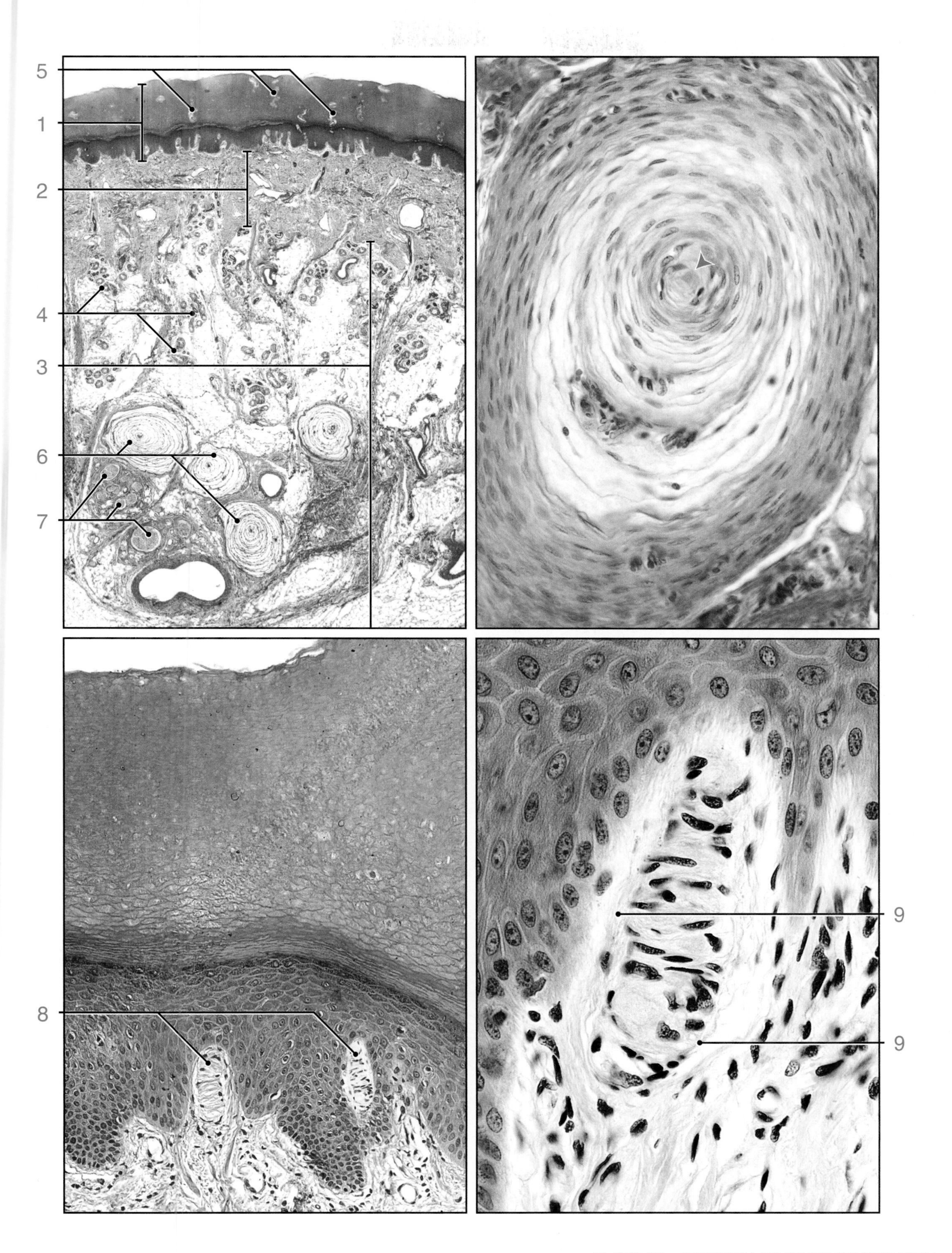
5
1
2
4
3
6
7
8
9
9

CHAPTER 12

Digestive System I: Oral Cavity

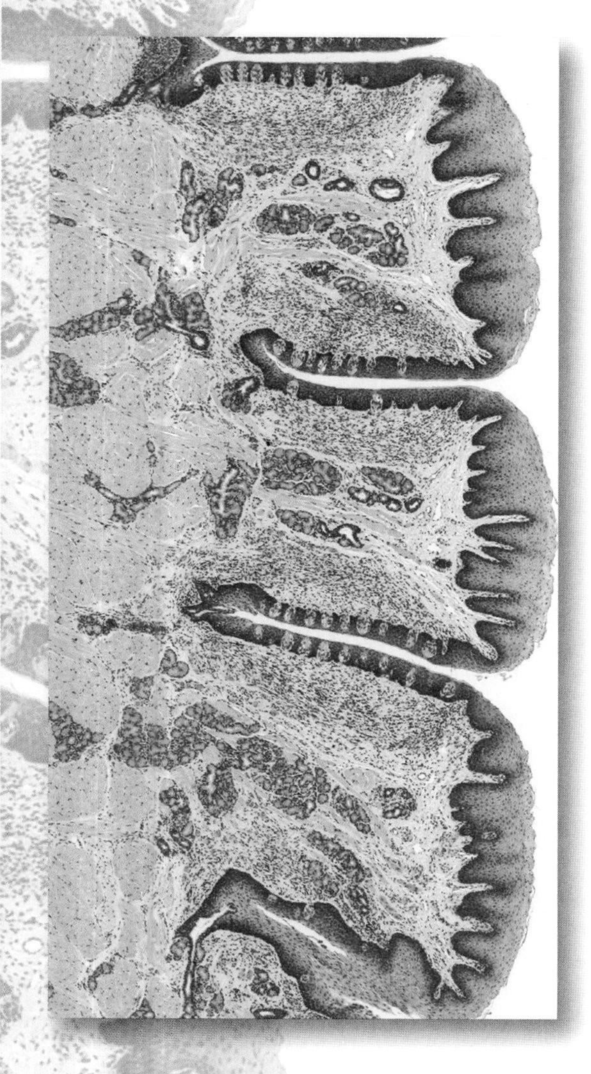

PLATE 74. LIP, A MUCOCUTANEOUS JUNCTION

The lips consist of two folds, the upper and lower labia, surrounding the orifice of the mouth. Externally, they are covered by a keratinized epithelium and internally by a mucous membrane. Each labium contains the orbicularis oris muscle, which is embedded in dense connective tissue. Contraction of the muscle causes a narrowing or shortening of the lips into the smallest possible circle, as in the expression of whistling. The labial blood vessels are numerous, and the venous component is somewhat unusual in that most of these vessels are of large diameter. They are presumed to play a role in heat exchange.

ORIENTATION MICROGRAPH: This micrograph shows a sagittal section through the lower lip. It reveals the **skin of the face** (1), the **vermilion border** (2), also known as the red margin, and the **mucosa of the posterior oral surface** (3). Also evident is the **orbicularis muscle** (4). Note the change in thickness of the epithelium of the skin, the vermilion border, and the mucosal surface. The clear spaces are the large venous **blood vessels** (5) beneath the vermilion border of the lip.

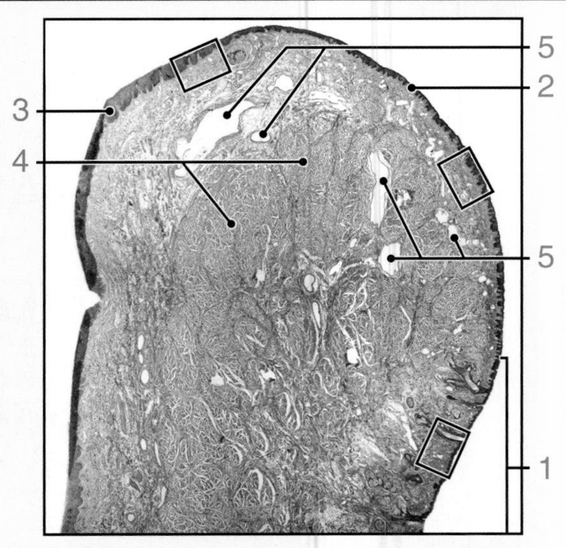

Skin, face, human, H&E, x100.
The area shown here is a higher magnification of the area in the lower box on the orientation micrograph. This region is immediately below the vermilion border and is typical of thin skin. Note the **sebaceous glands** (1). The epithelium is relatively uniform in thickness with the exception of intermittent, shallow **dermal papillae** (2).

Skin, face, human, H&E, x800.
In this higher magnification micrograph of the boxed area on the top left micrograph, the features that characterize the epidermis of thin skin are evident. Note the brown colored **melanin granules** (3) in the basal cell layer. Four or five cell layers are present in the **stratum spinosum** (4), and above that, dark, blue-staining keratohyalin granules are evident in the several layers of the **stratum granulosum** (5). The keratinized cells at the surface constitute the **stratum corneum** (6).

Lip, vermilion border, human, H&E, x100.
The region of the lip shown here is the red margin. Its reddish coloration in life, compared to the adjacent skin, is due to the presence of numerous large **veins** (7) combined with the presence of the deep, dermal **papillae** (8) that occur at this location. Compare the depth of the dermal papillae here with the more shallow papillae of the skin on the top left micrograph.

Lip, vermilion border, human, H&E, x800.
This higher power micrograph of the boxed area on the middle left micrograph shows one of the dermal **papillae** (9), along with the **epidermis** (10). A significant feature observed in this part of the lip, particularly in light skinned individuals, is the virtual absence of melanin granules in the **basal cells** (11). Compare with (3) on the micrograph immediately above. In contrast, the **stratum granulosum** (12) and the **stratum corneum** (13) are identical to that of the skin.

Lip, mucosal portion, human, H&E, x100.
This micrograph shows part of the vermilion border of the **lip** (14) and the continuing **mucosal portion** (15). The rectangle encloses the transition between the two sites. Note the much thicker epithelium (far left) of the mucosal portion compared to the epithelium of the vermilion border (far right). The mucosal epithelium of the lip is continuous with, and has the same structure as, the lining of the walls of the oral cavity.

Lip, mucosal portion, human, H&E, x800.
The transition between the vermilion border and the mucosal part of the lip is characterized by the disappearance of cells of the **stratum granulosum** (16) and the appearance of nuclei in the **surface cells** (17) of the mucosal epithelium. At this transitional site there is some keratinization of the nucleated surface cells, but after a very short distance, the surface cells become typical stratified squamous nonkeratinized cells of the oral mucosa.

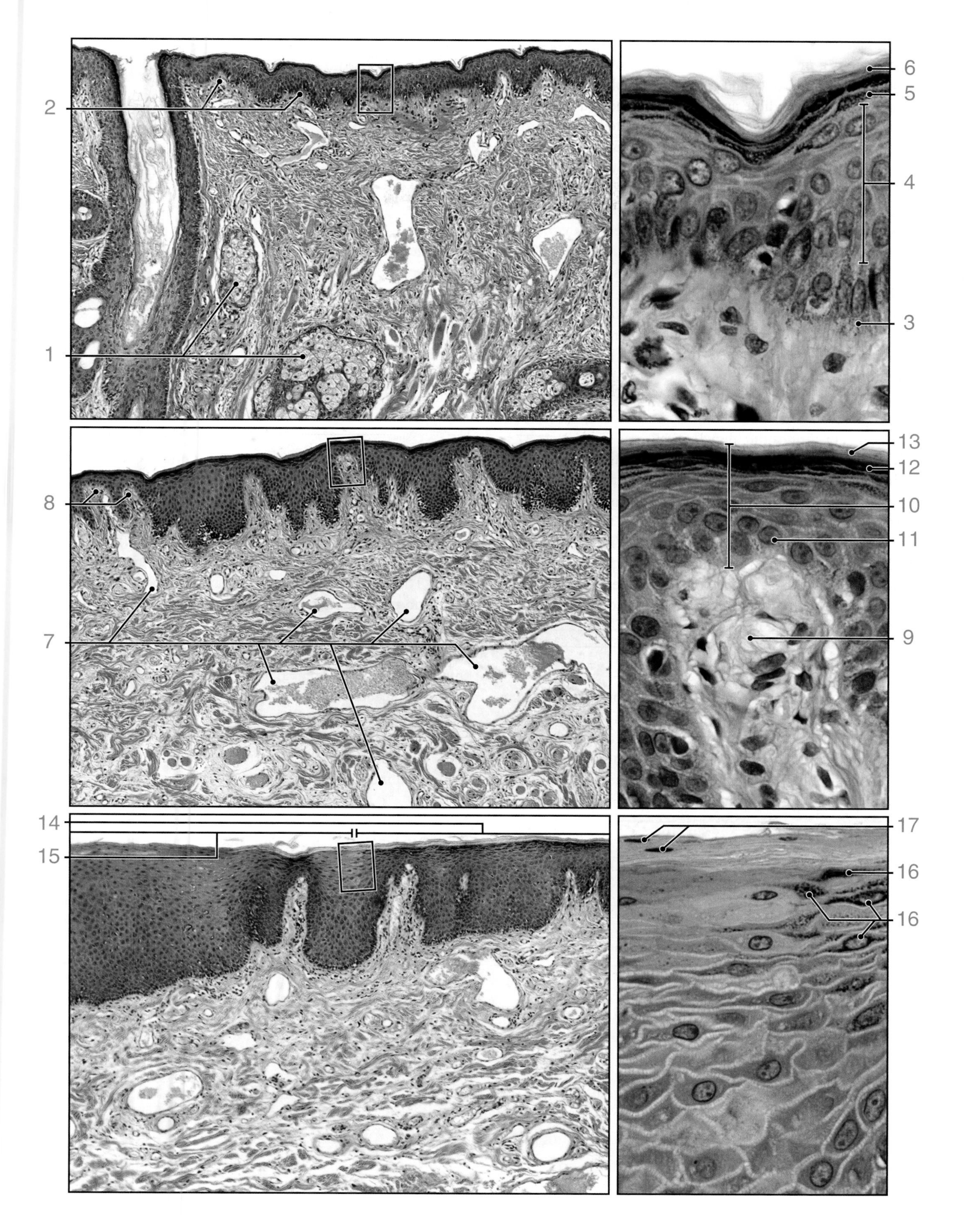

PLATE 74. LIP, A MUCOCUTANEOUS JUNCTION

PLATE 76. **TONGUE II, MUSCULATURE AND FILIFORM PAPILLAE**

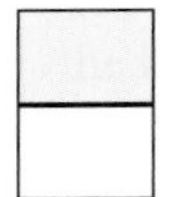

Tongue, lingual muscle and serous salivary glands, human, H&E, x175.

This medium power micrograph reveals the organization of the musculature of the tongue. The tongue is a unique organ in several respects. First, it is divided into two halves by a median connective tissue septum, therefore all muscles of the tongue are paired. Second, muscles of the tongue are divided into extrinsic muscles, in that they have one attachment outside of the tongue, and intrinsic muscles, in that they lack an external attachment. Third, as noted previously, the musculature is arranged in bundles that course in three planes. In this micrograph, there are **cross-sectioned muscle bundles** (1), **longitudinally sectioned bundles** (2) that run horizontally, and **vertically arranged bundles** (3). Each directional plane is at right angles to the other two planes. This arrangement of the muscle bundles allows flexible movement of the tongue in almost all directions (downward movement of the tongue is restricted by its position lying in the floor of the mouth). The nature of the innervation of the tongue musculature allows for extremely precise movement, an essential feature that facilitates human speech and the swallowing process. Another notable feature is the presence of minor **serous salivary glands** (4) interposed between the muscle bundles. These glands contribute to the formation of saliva and play a role relative to taste buds of some of the papillae (see Plate 77).

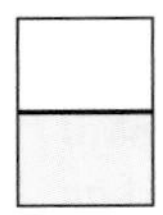

Tongue, filiform papillae, human, H&E, x90.

This micrograph shows a portion of the dorsal surface of the tongue with filiform papillae. The papillae are conical, elongated projections distributed over the anterior dorsal surface of the tongue. The tips of the papillae point posteriorly. Each papilla is comprised of a core of **connective tissue** (5) and a covering of **epithelium** (6). In humans, the epithelium constituting the anterior surface of the papilla is generally **stratified squamous** (7). However, that part of the papilla that faces posteriorly is keratinized (8). The appearance of **connective tissue surrounded by epithelium** (9) in one of the papilla is due to the plane of section at its base rather than the presence of a true island of connective tissue. These papillae rest on a layer of **dense connective tissue** (10), beneath which is **striated muscle** (11).

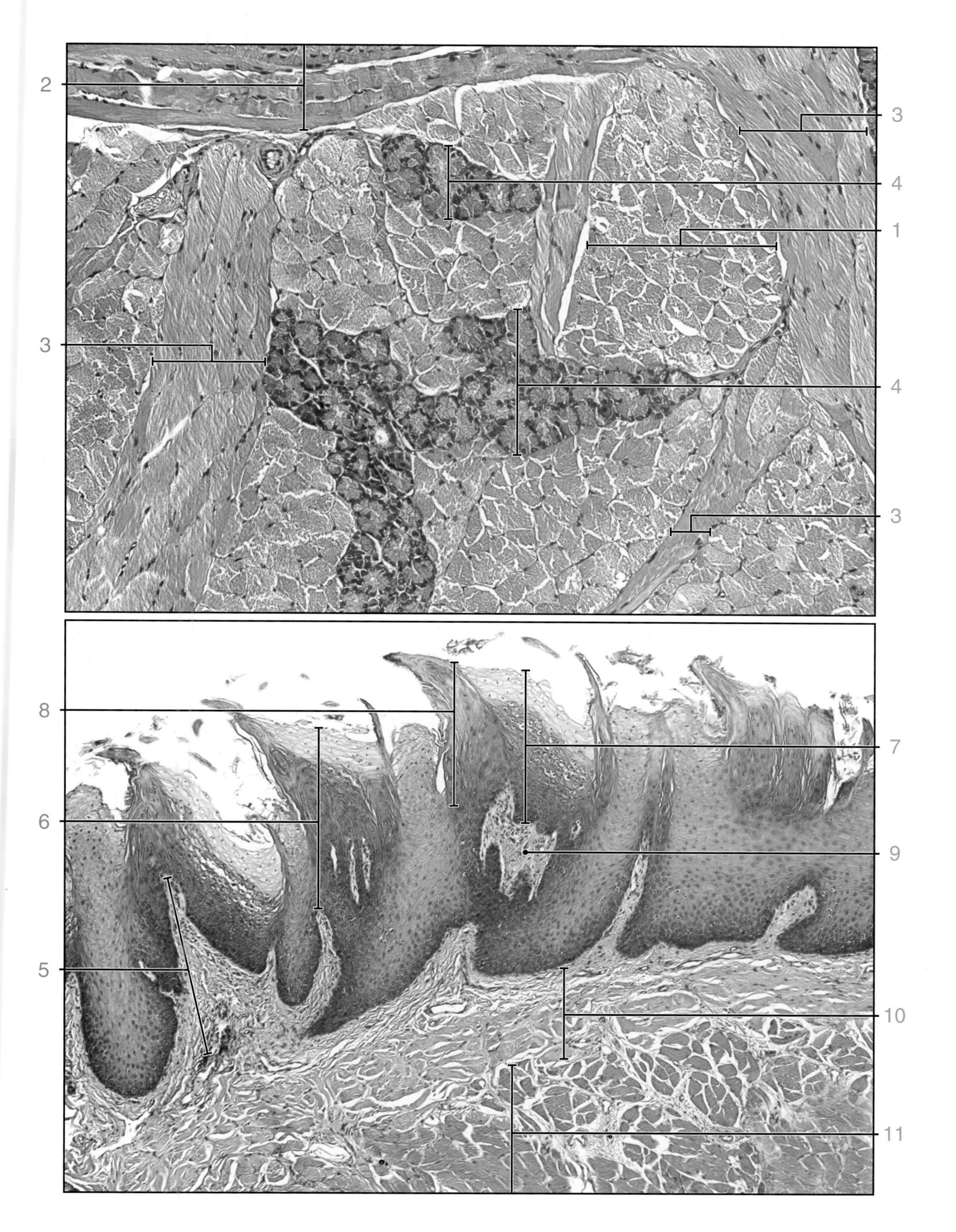

PLATE 76. TONGUE II, MUSCULATURE AND FILIFORM PAPILLAE

PLATE 77. TONGUE III, FUNGIFORM AND CIRCUMVALATE PAPILLAE

Tongue, fungiform papillae, human, H&E, x50; inset x270.

Fungiform papillae, as the name implies, are mushroom-shaped projections located on the dorsal surface of the tongue. They are more concentrated on its anterior half. Two **fungiform papillae** (1) are present in this micrograph. The fungiform papilla on the right has been sectioned vertically, showing its inner **connective tissue core** (2) and its **connective tissue papillae** (3) that penetrate upwards into the epithelium. Taste buds (*arrows*) are just barely perceptible near the top of the papillae. The fungiform papilla on the left has been obliquely sectioned near its periphery and consequently shows part of the connective tissue core as an island of tissue. Despite the plane of section, its domed shape and similarity of the epithelium to the fungiform papillae on the right aid in its identification. Note that adjacent to these fungiform papillae there are examples of **filiform papillae** (4). The **inset** reveals a fungiform papilla at higher magnification. It shows a **taste bud** (5) in one of the connective tissue papillae. Also evident is the thin layer of **keratinized epithelium** (6) at the surface of the papilla.

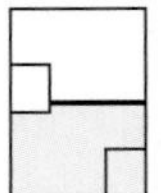

Tongue, circumvallate papillae, human, H&E, x50; inset x240.

Circumvallate papillae are large, dome-shaped structures that reside in the mucosa anterior to the sulcus terminalis. In humans, there are eight to twelve papillae arranged in an inverted V-shape across the tongue. Each papilla is surrounded by a **moatlike invagination** (7) lined by **nonkeratinized stratified squamous epithelium** (8). The surface of the papilla, like the fungiform papillae, is covered by **keratinized stratified squamous epithelium** (9). The epithelium lining the papilla itself contains numerous **taste buds** (10). In the immediate region of these papillae are numerous **lingual serous salivary glands (von Ebner's glands)** (11) that lie in the dense connective tissue between the lamina propria and the underlying muscle of the tongue. Some of the salivary glands also extend into the muscle between fiber bundles. Their secretions enter the base of the moats. The area in the rectangle on the micrograph is shown at higher magnification in the **inset**. A **duct of von Ebner's gland** (12) is seen entering the base of the moat. Also note in the inset the nonkeratinized stratified squamous epithelium. In addition to other functions, it is believed that the salivary gland secretions flush material from the moat so that the taste buds can respond rapidly to changing stimuli.

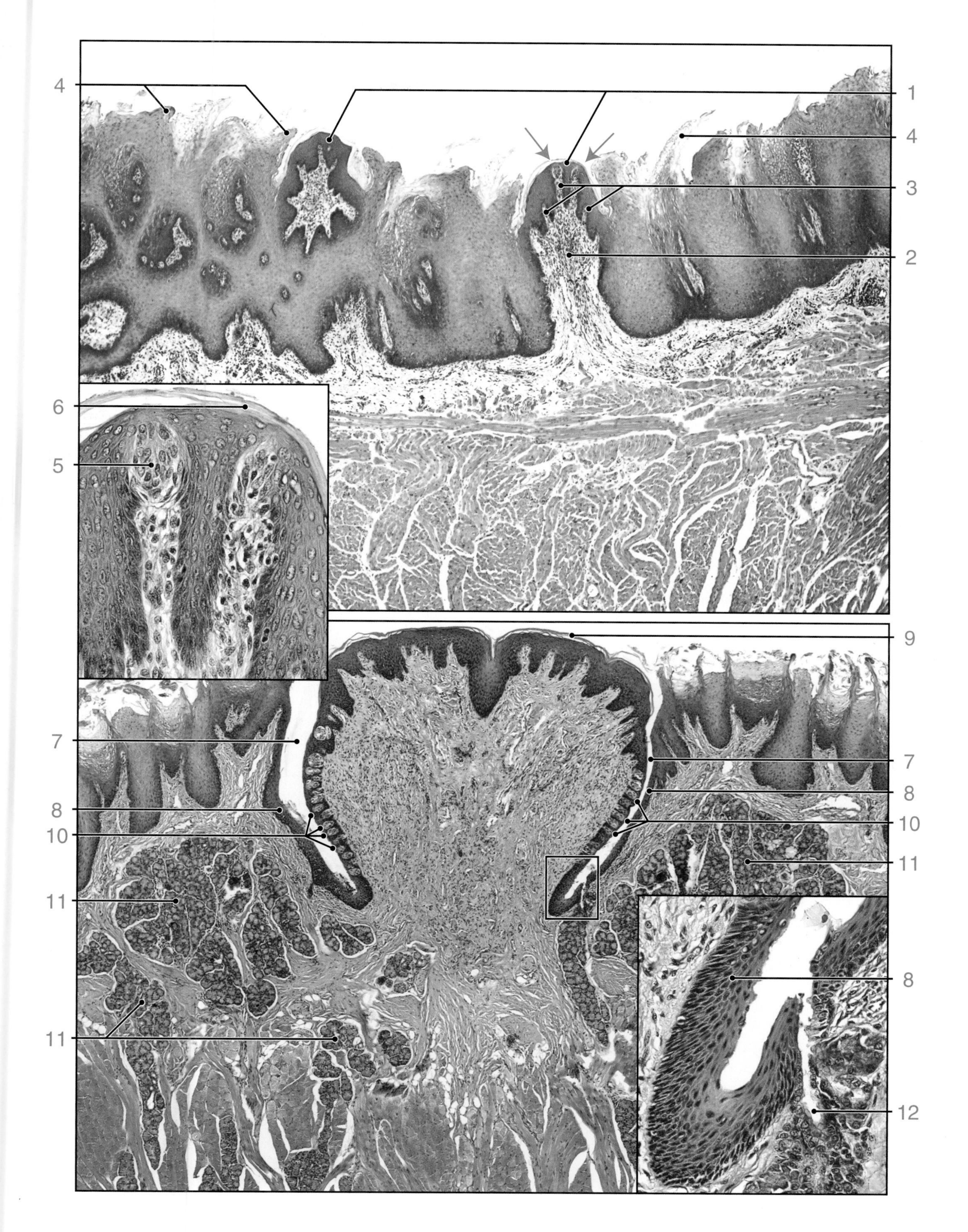

PLATE 77. TONGUE III, FUNGIFORM AND CIRCUMVALATE PAPILLAE

PLATE 78. TONGUE IV, FOLIATE PAPILLAE AND TASTE BUDS

Tongue, foliate papillae, human, H&E, x50.
Foliate papillae consist of a series of parallel ridges that are separated by narrow, deep mucosal clefts (see orientation photograph, Plate 75). They are aligned at right angles to the long axis of the tongue on its posterior lateral edge. In younger individuals, they are readily observed by gross inspection. However, with age, foliate papillae may not be recognized. This micrograph shows three papillae, each is separated from its neighbor by a narrow **cleft** (1). The surfaces of these papillae are covered by a thick **nonkeratinized stratified epithelium** (2). The basal surface of the epithelium is extremely uneven due to the presence of deep, penetrating **connective tissue papillae** (3). In contrast, the **epithelium lining the clefts** (4) is relatively thin and uniform. It contains numerous taste buds (the light-staining structures seen in the cleft epithelium). Underneath the epithelium are a layer of **loose connective tissue** (5) and a central core of dense connective tissue. Within this core and between bundles of muscle fibers beneath the papillae are **lingual serous glands** (6). These glands, like the serous glands associated with the circumvallate papillae, have **ducts** (7) that empty into the base of the clefts between papillae.

Tongue, taste buds, human, H&E, x500.
This higher magnification micrograph shows the taste buds located within the cleft epithelium. The taste buds typically appear as oval, pale-staining structures that extend through much of the thickness of the epithelium. Beneath the taste bud are **nerve fibers** (8), which are also lightly stained. At the apex of the taste bud is a small opening in the epithelium called the **taste pore** (9).

Tongue, taste bud, human, H&E, x1100.
This micrograph shows to advantage the **taste pore** (10), the cells of the taste bud, and the taste bud's associated **nerve fibers** (11). The cells with large, round nuclei are **neuroepithelial sensory cells** (12). They are the most numerous cells of the taste bud. At their apical surface, they possess microvilli that extend into the taste pore. At their basal surface, they form a synapse with the afferent sensory fibers that make up the underlying nerve. Among the sensory cells are **supporting cells** (13). These cells also contain microvilli on their apical surface. At the base of the taste bud are small cells referred to as **basal cells** (14), one of which is identified here. They are the stem cells for the supporting and neuroepithelial cells: cells which have a turnover of about 10 days.

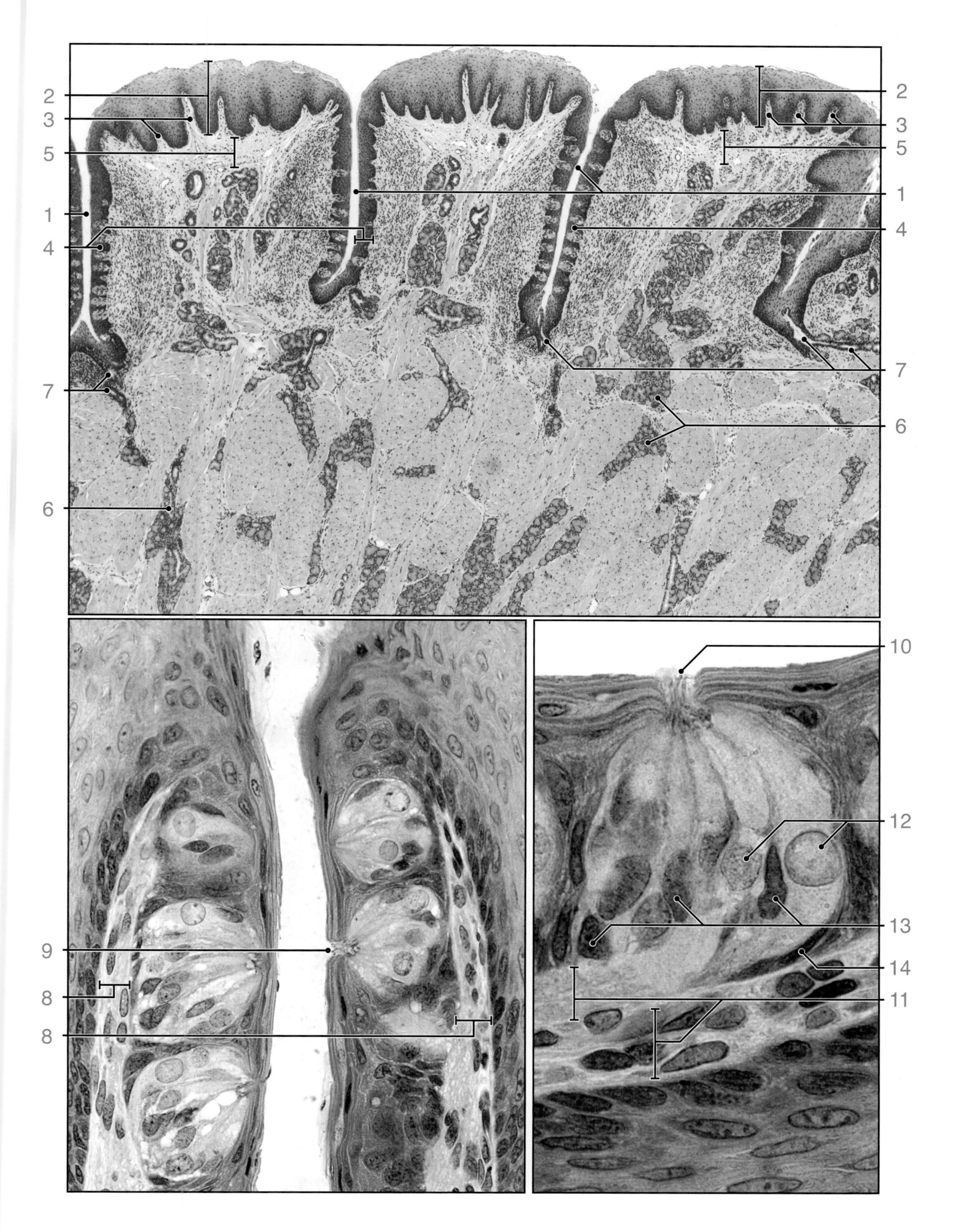

PLATE 78. TONGUE IV, FOLIATE PAPILLAE AND TASTE BUDS

PLATE 79. PAROTID GLAND

The major salivary glands are the parotid, submandibular, and sublingual glands. They, along with a number of minor accessory glands scattered in the oral mucosa, secrete saliva. The parotid glands, described here, are the largest. They are located below and in front of the ear. It is a compound tubular gland, consisting entirely of serous secretory cells. Their duct (Stensen's duct) enters the oral cavity opposite the second upper molar tooth. The secretion is hypotonic, very watery, and rich in enzymes and antibodies. A characteristic feature of the gland, in contrast to the sublingual and submandibular glands, is the presence of varying numbers of adipocytes. The facial nerve (cranial nerve VII) passes through the parotid gland (but does not innervate it), thus sometimes a section of this large nerve may be encountered in a slide section.

The major salivary glands are encapsulated by dense connective tissue from which septa divide the gland into lobes and lobules. The minor salivary glands lack a capsule. Numerous lymphocytes as well as plasma cells can be found in the connective tissue between lobules. The plasma cells are responsible for the production of salivary immunoglobulin A. The secreted IgA binds to receptors of the salivary gland cells and is internalized by endocytosis to be carried through the acinar cell and released into the acinar lumen as secretory sIgA, which, among other functions, serves to prevent dental caries.

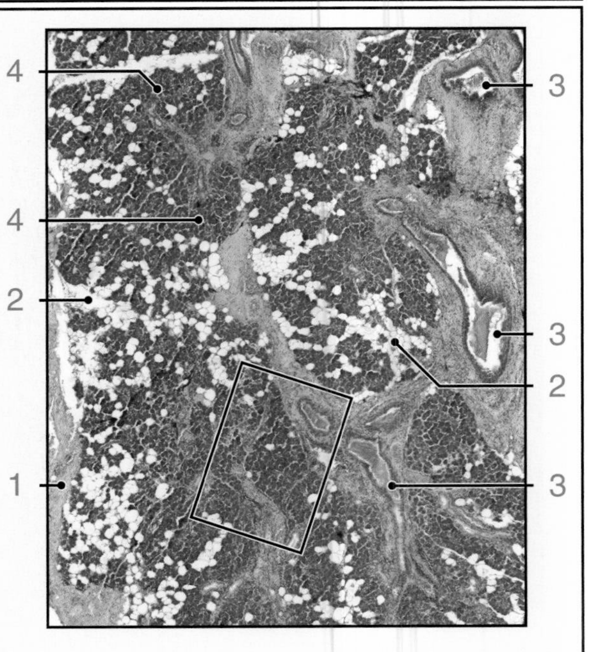

ORIENTATION MICROGRAPH: This micrograph shows a portion of a parotid gland. A part of the **capsule** (1) is present and aggregates of **adipocytes** (2) appearing as clear areas are present. Several large **excretory ducts** (3) lying within connective tissue septa of the gland are also recognizable. The **serous secretory units** (4) are deeply stained.

Parotid gland, human, H&E, x180.
This micrograph is a higher magnification of the boxed area on the orientation micrograph. The parenchyma of the parotid gland consists entirely of **serous acini** (1), made up of pyramidal cells surrounding a narrow lumen. The secretions from these acini enter small ducts, known as **intercalated ducts** (2), which are recognizable even at this fairly low magnification. In this region of the gland, a relatively small number of **adipocytes** (3) are seen interspersed among the acini. An **excretory duct** (4) can be readily recognized by the stratified epithelium of the duct, as evidenced by the multilayering of the nuclei, and its presence within a connective tissue septa. Also evident in the micrograph is a **vein** (5). Adjacent to the vein is a **striated duct** (6), which receives secretions from the intercalated ducts and connects with the larger excretory ducts.

Parotid gland, human, H&E, x365; insets x1000.
The boxed area on the above micrograph is shown here at higher magnification. Careful examination of the **acini** (7) reveals their tubular rather than true sphere-like nature. This micrograph shows to advantage the intercalated ducts, ranging from the **smaller intercalated ducts** (8), whose nuclei are elongate, to the **larger intercalated ducts** (9), whose nuclei tend to be spherical. The **upper inset** shows portions of several acini, with one including its **lumen** (10). Also note that in this specimen the secretory granules, which appear as small, round, densely staining objects, are retained—a feature which is not always seen. The **lower inset** shows an intercalated duct, including its **lumen** (11).

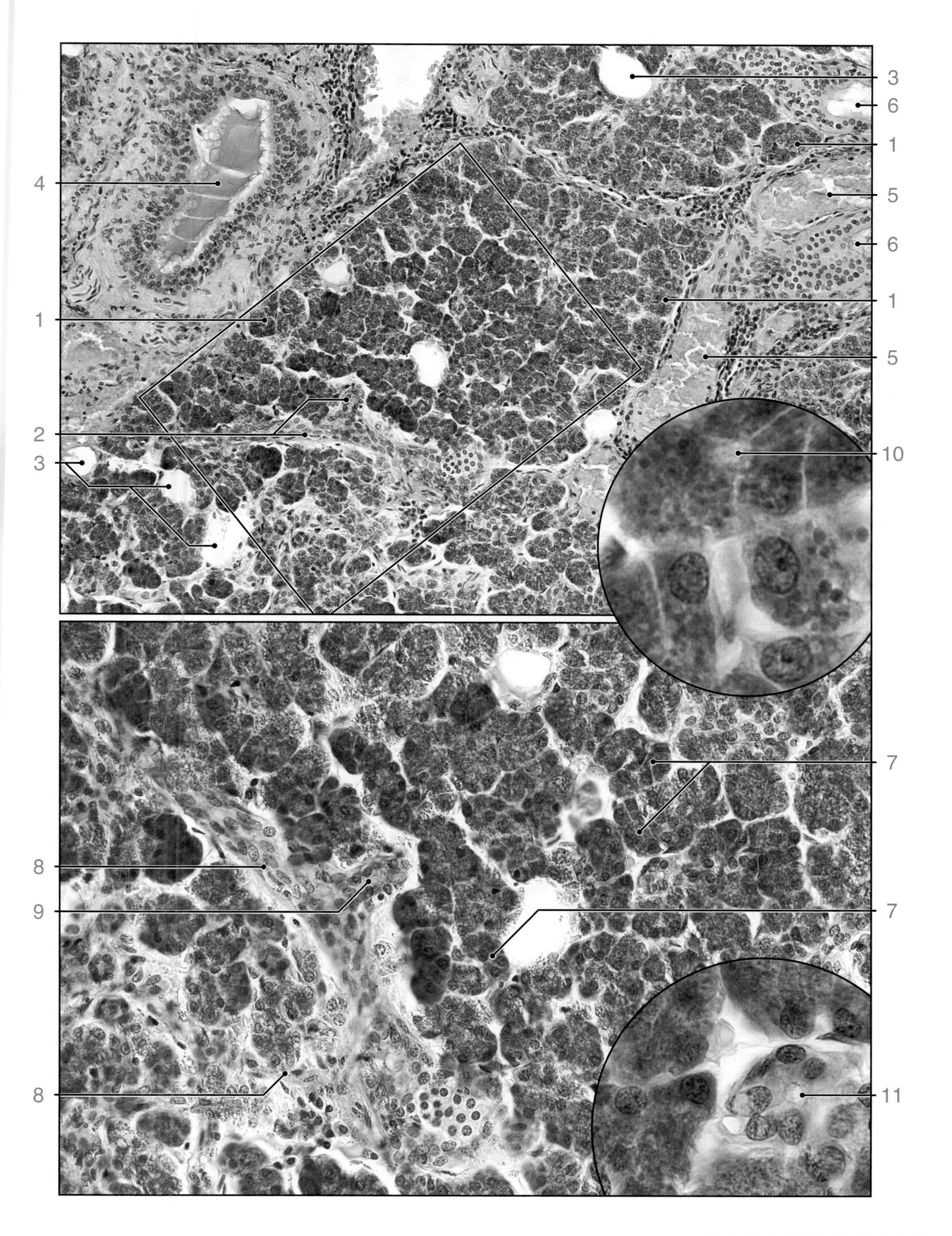
3
6
1
4
5
6
1
1
5
2
3
10
7
8
9
7
8
11

PLATE 80. SUBMANDIBULAR GLAND

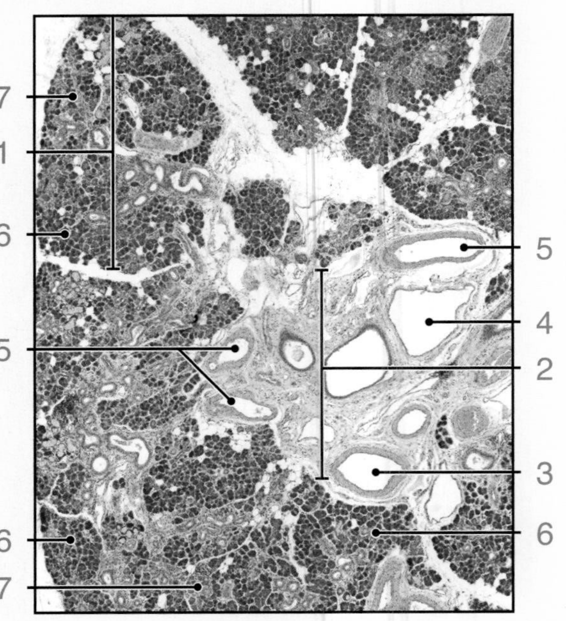

In contrast to the parotid glands, which reside entirely outside the oral cavity, the submandibular glands are located partially outside and partially within the floor of the oral cavity. Their larger portions are located below the mylohyoid muscle near the medial side of the mandible. A duct runs forward and medially from each of the two glands to a papilla located on the floor of the mouth, just lateral to the frenulum of the tongue. The secretory components of the submandibular glands are the acini, which are of three types: serous acini, which are protein-secreting like those of the parotid gland; mucous acini, which secrete mucin; and acini containing both serous and mucus-secreting cells. In the mixed acini, the mucous cells are capped by serous cells, which are typically described as demilumes. Recent studies suggest that the demilume is an artifact of tissue preparation and that all of the cells are aligned to secrete into the acinus lumen. Traditional fixation in formaldehyde appears to expand the mucous cells, consequently squeezing the serous cells to form their cap-like position.

ORIENTATION MICROGRAPH: This micrograph reveals a portion of the submandibular gland. A single, well-defined **lobe** (1) is seen in the upper part of the micrograph. Within the central portion of the gland, there is a **dense connective tissue core** (2) containing the larger **arteries** (3), **veins** (4), and **excretory ducts** (5) of the gland. The submandibular gland is a mixed gland; those regions containing **serous acini** (6) are darkly stained, whereas regions containing **mucous acini** (7) are lighter in appearance.

Submandibular gland, human, H&E, x175.

This micrograph reveals the various components of the submandibular gland. The **serous acini** (1) are darkly stained compared to the lighter-staining **mucous acini** (2). Furthermore, the serous acini are generally spherical in shape. The mucous acini are more tubular or elongate and sometimes can be seen to branch. The secretion from the acini enters an intercalated duct. They are the smallest ducts and are relatively short. They reside within the lobule, but are often difficult to find because of their shortness. These ducts empty into the larger **striated duct** (3). This type of duct is better seen in the micrograph below. The striated duct contents empty into an **excretory duct** (4), which is recognized by a stratified or pseudostratified epithelium. Other features of note in this micrograph are **arteries** (5) and **veins** (6), which also course through the connective tissue with the ducts. Also evident in this micrograph is an area containing an accumulation of **lymphocytes** and **plasma cells** (7).

Submandibular gland, human, H&E, x725.

The boxed area on the above micrograph is shown here at higher magnification. It includes several **mucous acini** (8) on the left side of the micrograph; a number of **serous acini** (9) on the right side of the micrograph; and two **mixed acini** (10) in the center, consisting of mucous secreting cells and serous secreting cells. Characteristically, the mucous secreting cells have a pale staining cytoplasm with their nuclei flattened at the base of the cell. In contrast, the serous secreting cells are deeply stained and exhibit round nuclei. In addition, the **lumen** (11) of the acini associated with the mucous secreting cells is relatively wide; the lumen of the serous acini is relatively narrow and difficult to find. Additionally, the serous cells of the mixed acini generally appear as a cap in relation to the mucous cells. Such an arrangement is referred to as a serous demilume and most likely represents an artifact of tissue fixation. It is possible that some of the acini that appear to be serous in nature are simply a tangential section of a demilume. A **striated duct** (12) is also present in the micrograph. It is so named because of the faint striations seen in the basal cytoplasm. These ducts, as noted above, receive secretions from the intercalated ducts and empty into the larger excretory ducts.

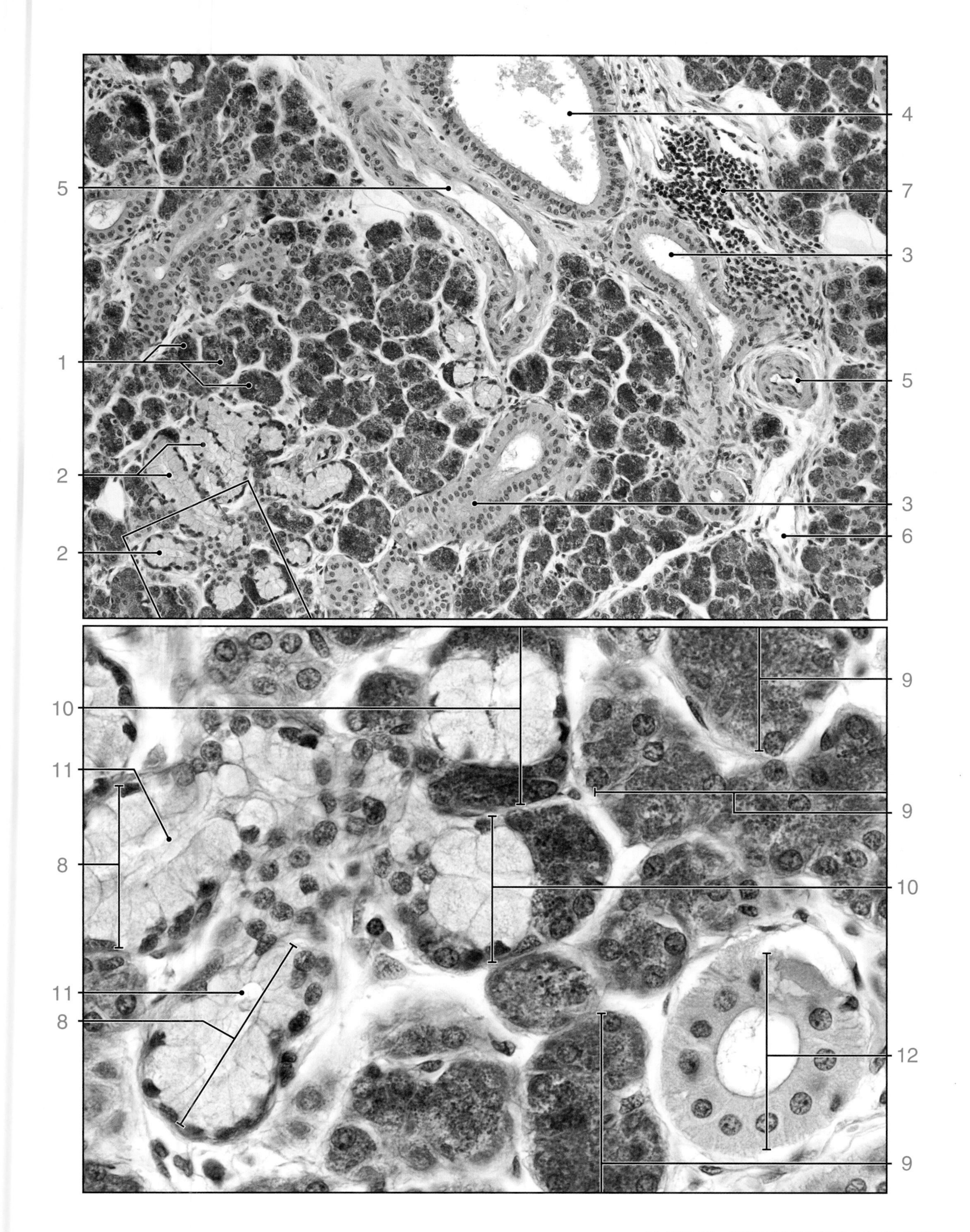

PLATE 80. SUBMANDIBULAR GLAND

PLATE 81. SUBLINGUAL GLAND

The sublingual glands are the smallest of the paired major salivary glands. Located anterior to the submandibular glands on the floor of the oral cavity, they have multiple ducts, some of which empty into the duct of the submandibular gland and others that empty directly onto the floor of the mouth. The secretory portion of the sublingual glands are mixed; mucous cells are considerably more numerous in the sublingual gland than in the submandibular gland. Overall, the mucous acini predominate, although there is considerable variation in the proportion of mucous to seromucous acini in different portions of the gland. Many of the mucous acini exhibit serous demilumes. However, pure serous acini are generally absent from sublingual glands. With respect to the duct system, the intercalated ducts and striated ducts are relatively short, making them difficult to locate.

ORIENTATION MICROGRAPH: This micrograph reveals the kind of variation of mucous and serous components seen in the sublingual gland. The upper, light-staining portion of the micrograph consists of pure **mucous acini** (1). The lower, dark-staining portion consists of **seromucous acini** (2). The serous component is represented by serous demilumes. The gland is encapsulated by **dense connective tissue** (3). The larger ducts, seen at this magnification as residing in the connective tissue septa, are **excretory ducts** (4).

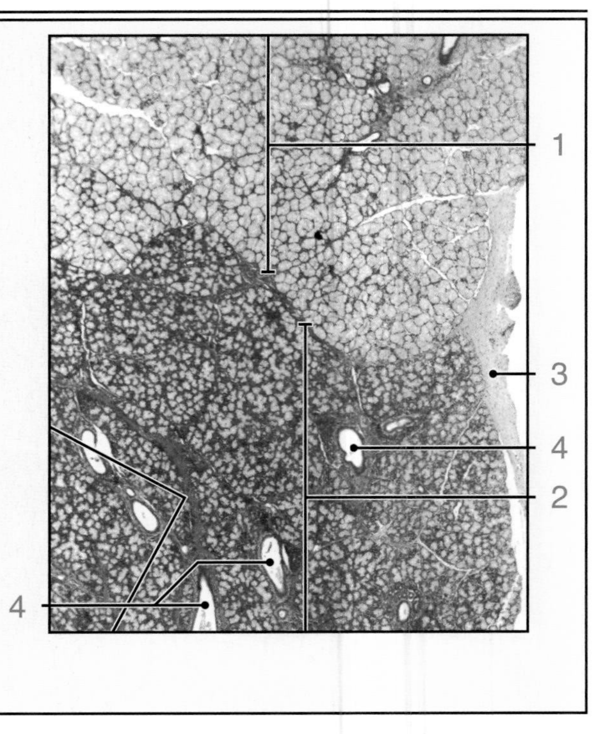

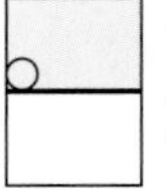

Sublingual gland, human, H&E, x175; inset x355.

This micrograph is a higher magnification of the boxed area on the lower left of the orientation micrograph. It reveals only seromucous glands. **Connective tissue septa** (1) subdivide the gland into lobules. The ducts seen in the septa are **intralobular ducts** (2). They are the equivalent of the striated duct of the submandibular and parotid glands. They lack the extensive basal infoldings and mitochondrial array that create striations. The lack of striations in these ducts is characteristic of mucus-secreting glands in which there is no need to resorb water, as is the case with the serous-type gland. The **inset** is a higher magnification of the circled area in the lower right corner of the micrograph. It represents an area in which several **intercalated ducts** (3) are present. Careful examination of the mucous acini at this relatively low magnification reveals that they are elongate or tubular structures with branching outpockets. Thus, each acinus is rather large, and much of it is usually not seen in most sections. One of the **glands** (4) has been cut in a manner that best reveals the complexity of its organization.

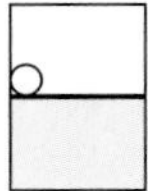

Sublingual gland, human, H&E, x725.

This higher magnification micrograph shows a number of acini with their serous demilumes. The **serous cells** (5) appear to form a cap around the mucous cells. Several of the acini have been cut in a manner that reveals the **lumen** (6) of the acinus. A small **group of serous cells** (7) that appear unrelated to the mucous cells is visible. Most likely, the group of cells reflects a tangential section of the serous demilume cells of one of the mucous acini. Also included in the micrograph is an **intercalated duct** (8). Note the small lumen. In contrast, the wall of a larger, **intralobular duct** (9) is seen in the lower right of the micrograph. An important feature of the connective tissue stroma of the salivary glands is the presence of numerous **plasma cells** (10). They are responsible for the production of salivary IgA.

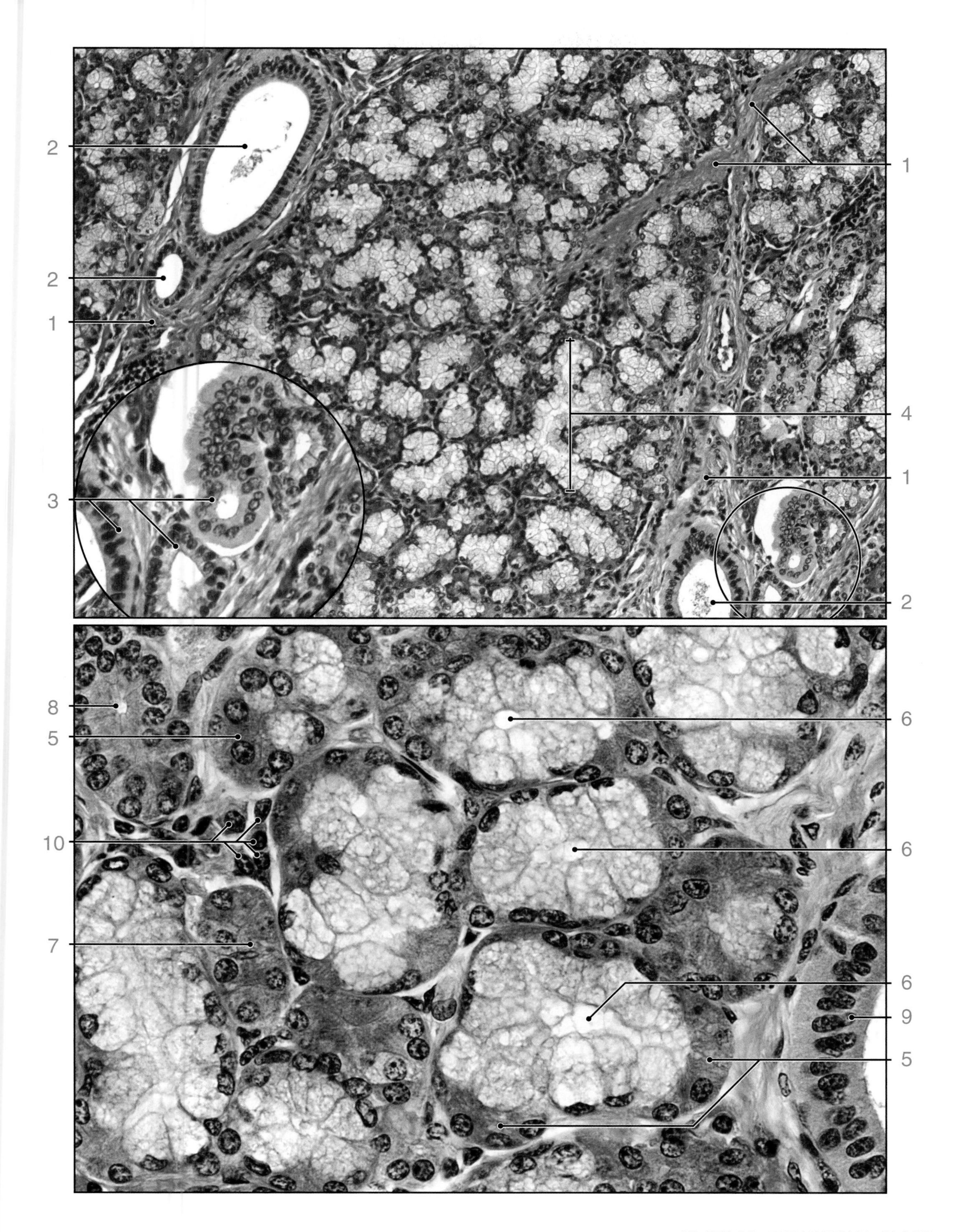

PLATE 81. SUBLINGUAL GLAND

PLATE 83. **DEVELOPING TOOTH**

Initiation of tooth development is signified by the proliferation and growth of oral cavity epithelium into the underlying mesenchyme of the jaw. This proliferation of the epithelium, known as a dental lamina, leads first to a bud stage structure that gives origin to the enamel organ (primordium of enamel). Mesenchymal cells beneath the tooth bud begin to differentiate, forming the precursor of the dental papilla, which will give rise to the pulp and dentin of the tooth. Further development leads to the cap stage where the cells located in the concavity of the cap differentiate into columnar cells (ameloblasts), which form the inner enamel epithelium. Next, at the bell stage, the dental lamina, which forms a connection between the developing tooth and the oral epithelium, separates. The enamel organ consists of a layer of epithelial cells, forming the outer enamel epithelium, and an inner enamel epithelium, formed by ameloblasts and a condensation of cells that form the stratum intermedium. More widely spaced cells form the stellate reticulum, which lies above the stratum intermedium. The dental papilla is deeply invaginated against the inner enamel organ.

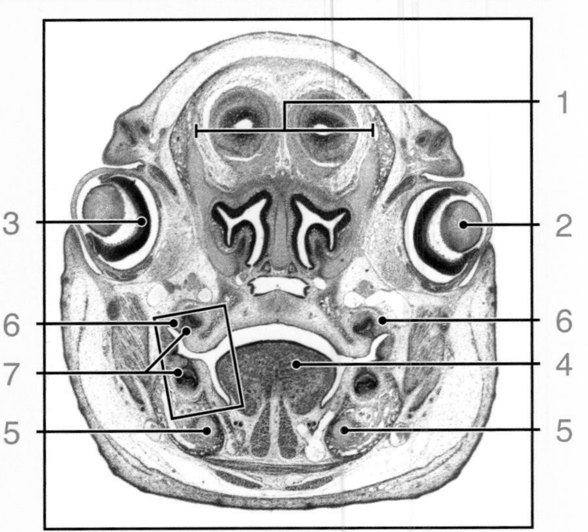

ORIENTATION MICROGRAPH: This micrograph is a frontal section through the head of a fetus. It shows the developing **brain** (1), the **lens** (2), the **optic cup** (3) of the developing eye, the **tongue** (4), the **mandible** (5), and the **maxilla** (6). Developing **teeth** (7) are also seen in the maxilla and mandible.

Developing tooth, human, H&E, x125.
This micrograph is a higher magnification of the boxed area on the orientation micrograph. The clear space is the developing **oral cavity** (1). Part of the **tongue** (2) is also evident. Two **developing teeth** (3) are seen in this plane of section. The upper developing tooth is in the presumptive maxilla. A very small **spicule of bone** (4) of the developing maxilla is present. Similarly, several **spicules of bone** (5) are present in the developing mandible. All of the developing teeth are in the cap stage.

Developing tooth, human, H&E, x250.
This micrograph shows the tooth bud seen in the box on the top left micrograph at higher magnification. It shows the **oral epithelium** (6). The enamel organ consists of a single layer of cuboidal cells that form the **outer enamel epithelium** (7) and the **inner enamel epithelium** (8), which has differentiated into columnar ameloblasts. The **stratum intermedium** (9), which appears as a condensation of cells, lies just above the inner enamel epithelium. Mesenchyme of the **dental papilla** (10) has formed another condensation of cells and pushed into the enamel organ. The cellular component surrounding the tooth bud, referred to as the **dental sac** (11), gives rise to periodontal structures (e.g., the cementum and periosteal fibers).

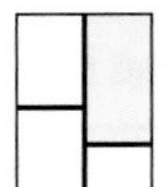

Developing tooth, human, H&E, x25.
The developing tooth shown here is at the bell stage, in which early crown formation takes place. The dental lamina that connect the developing tooth with the oral epithelium has degenerated. The **outer enamel epithelium** (12) is seen as a layer covering the **ameloblasts** (13). The bright red stained material is the **enamel** (14). Beneath the enamel is the forming **dentin** (15). The inner surface of the dentin is the site of the **odontoblasts** (16) that are producing the dentin. The tissue in the center of the developing tooth is the **dental papilla** (17).

Developing tooth, human, H&E, x50.
The boxed area on the top right micrograph is shown here at higher magnification. It shows to advantage the **outer enamel epithelium** (18), the tall, columnar cells known as **ameloblasts** (19), the **enamel** (20), and the **dentin** (21). A small portion of the cellular **dental papilla** (22) is also seen.

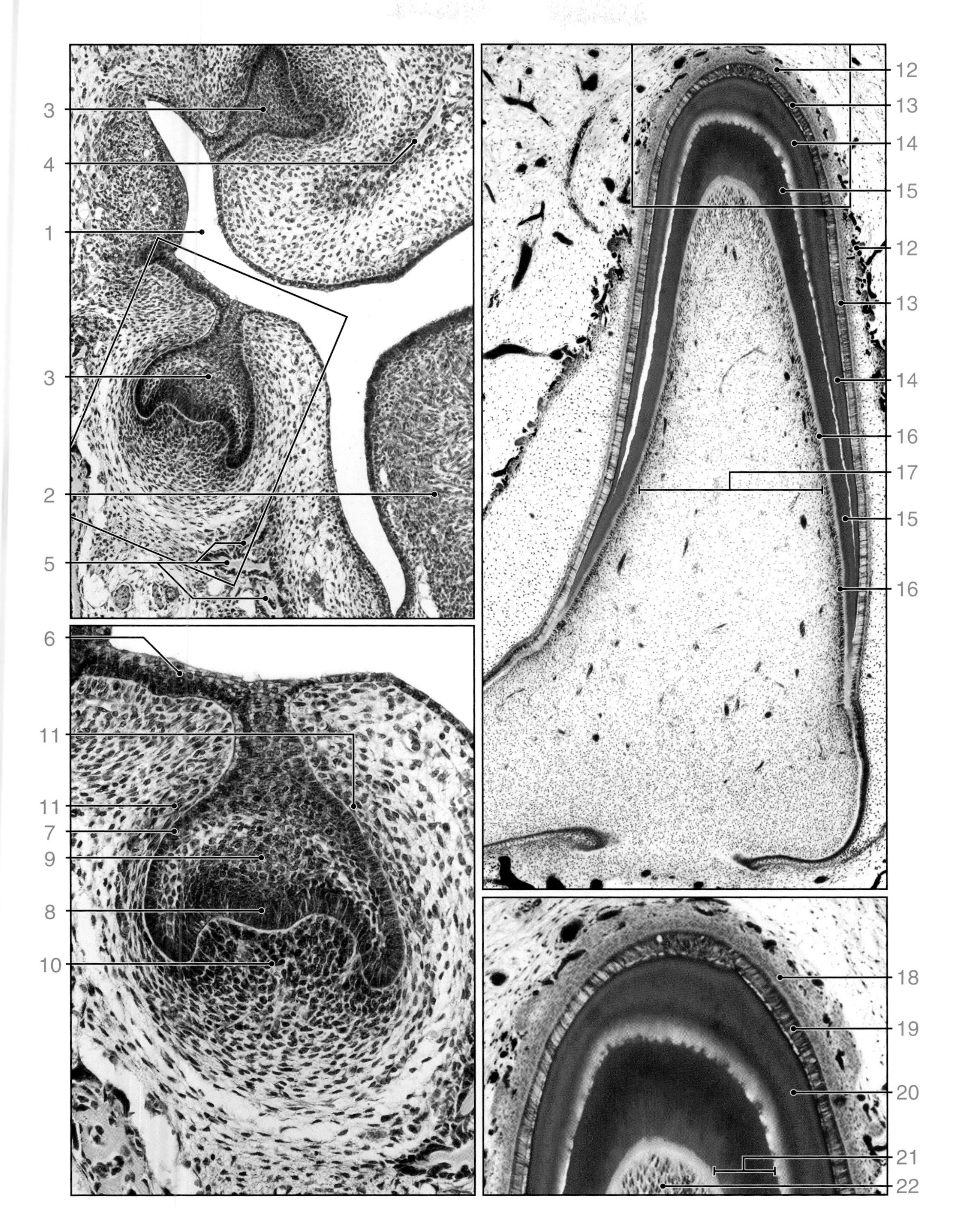

PLATE 83. DEVELOPING TOOTH

CHAPTER 13

Digestive System II: Esophagus and Gastrointestinal Tract

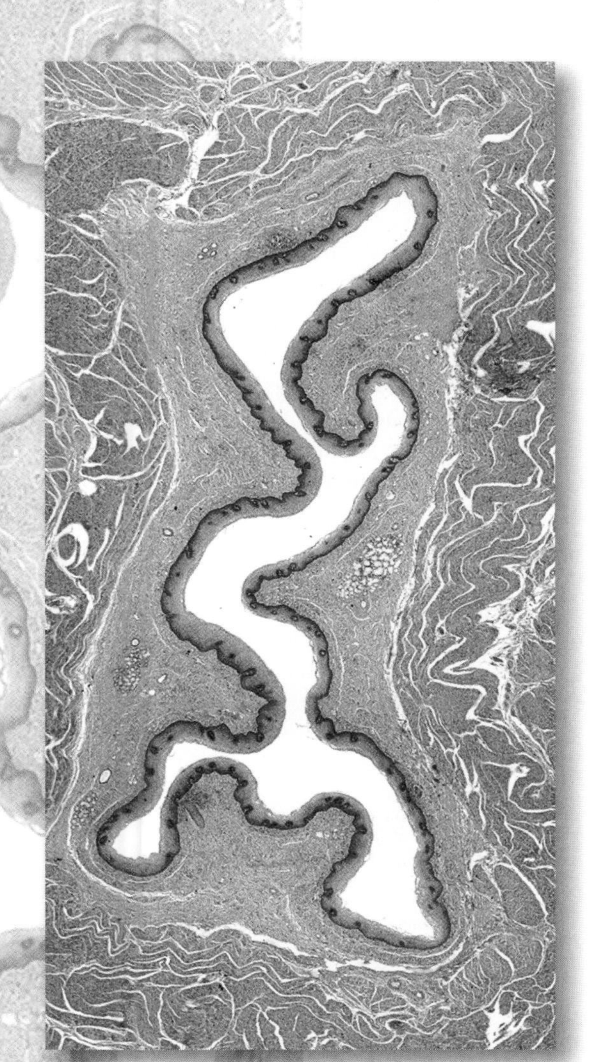

PLATE 84. ESOPHAGUS

The esophagus, the first segment of the alimentary canal, extends from the oropharynx to the stomach. Other than serving as a conduit, it has no other major function. Throughout its length, it is attached to adjacent structures by its outermost layer, the adventia. However, after the esophagus passes through the diaphragm, it is unattached and covered by a serosa. It possesses a muscular wall (the muscularis externa) whose peristaltic activity facilitates passage of a solid food bolus to the stomach. The muscularis externa of the initial portion of the tube consists of striated muscle. This gradually gives way in the middle third of the tube to smooth muscle, which is then found throughout the remainder of the alimentary canal.

ORIENTATION MICROGRAPH: The orientation micrograph, showing a cross section through the lower esophagus, reveals its partially collapsed **lumen** (1) and the **mucosal epithelium** (2). (The muscularis mucosa and lamina propria are not evident at this low magnification.) The next recognizable layer is the **submucosa** (3), which exhibits several **mucous glands** (4). The **muscularis externa** (5), the next layer, is actually two layers, an internal, circular layer of **smooth muscle** (6) and an external **longitudinal layer** (7). Some of the outermost layer, the **adventitia** (8), is also evident. It consists of a fibroelastic connective tissue that binds the esophagus to adjoining structures. The short portion of the esophagus below the diaphragm is not attached to surrounding structures and possesses a serosa instead of an adventia.

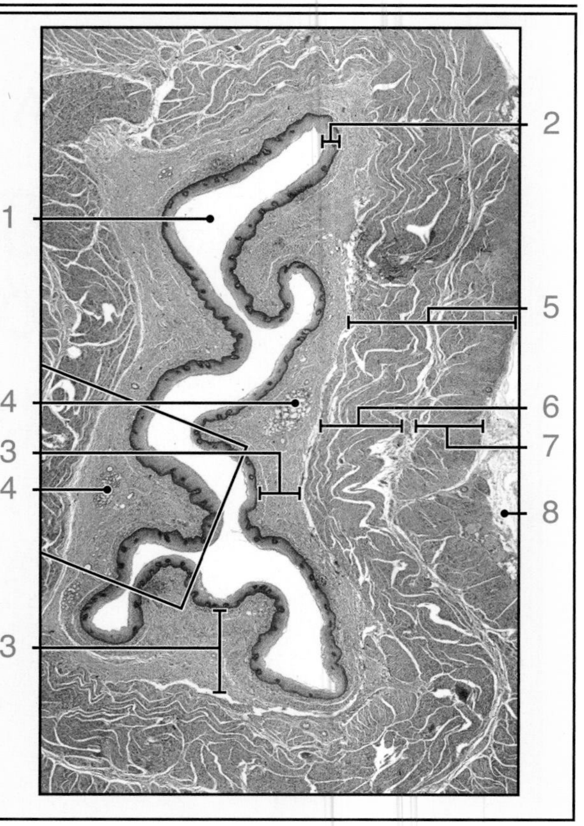

Esophagus, human, H&E, x50; inset x200.
This micrograph shows the region of the esophagus within the box on the orientation micrograph. The **epithelium** (1) presents a smooth surface contour, but its basal surface is marked by irregularly spaced **connective tissue papillae** (2) from the underlying **lamina propria** (3). The **muscularis mucosa** (4) is the next layer, but is difficult to identify at this magnification. The **submucosa** (5) consists of dense connective tissue and contains **submucousal glands** (6) that secrete mucus to help lubricate the epithelial lining surface of the esophagus. The **inset** shows part of the gland, from the smaller boxed area at higher magnification, with its typical **alveoli** (7) and two ducts. The **smaller duct** (8) has a simple cuboidal epithelium. The **larger duct** (9) has a stratified cuboidal epithelium; note the two layers of nuclei. A small portion of the muscularis externa is seen in the upper left of the main micrograph. The **circular layer** (10) presents itself with the smooth muscle cells cut longitudinally, whereas the **longitudinal layer** (11) consists of smooth muscle cells cut in cross-section. This pattern is consistent with a cross-sectional cut of the esophagus.

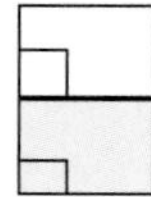

Esophagus, human, H&E, x175; inset x250.
This higher magnification of the area in the larger box on the above micrograph shows the esophageal mucosa. The **epithelium** (12) exhibits cuboidal cells at its basal surface that become squamous by the time they reach the free surface. The **inset** shows the transition more clearly; also note the elongate, desquamated epithelial cell at the surface. These cells are typically disc-shaped. The **lamina propria** (13) is a relatively cellular connective tissue, as evidenced by the numerous small round nuclei. Most of these nuclei belong to lymphocytes. Also within this layer are many small **blood vessels** (14) and a **lymphatic vessel** (15). Several round nuclei are evident in the lumen of the lymphatic vessel; they belong to lymphocytes. The **muscularis mucosa** (16) is identified by the presence of clusters of smooth muscle cells, which stain more intensely than the surrounding connective tissue.

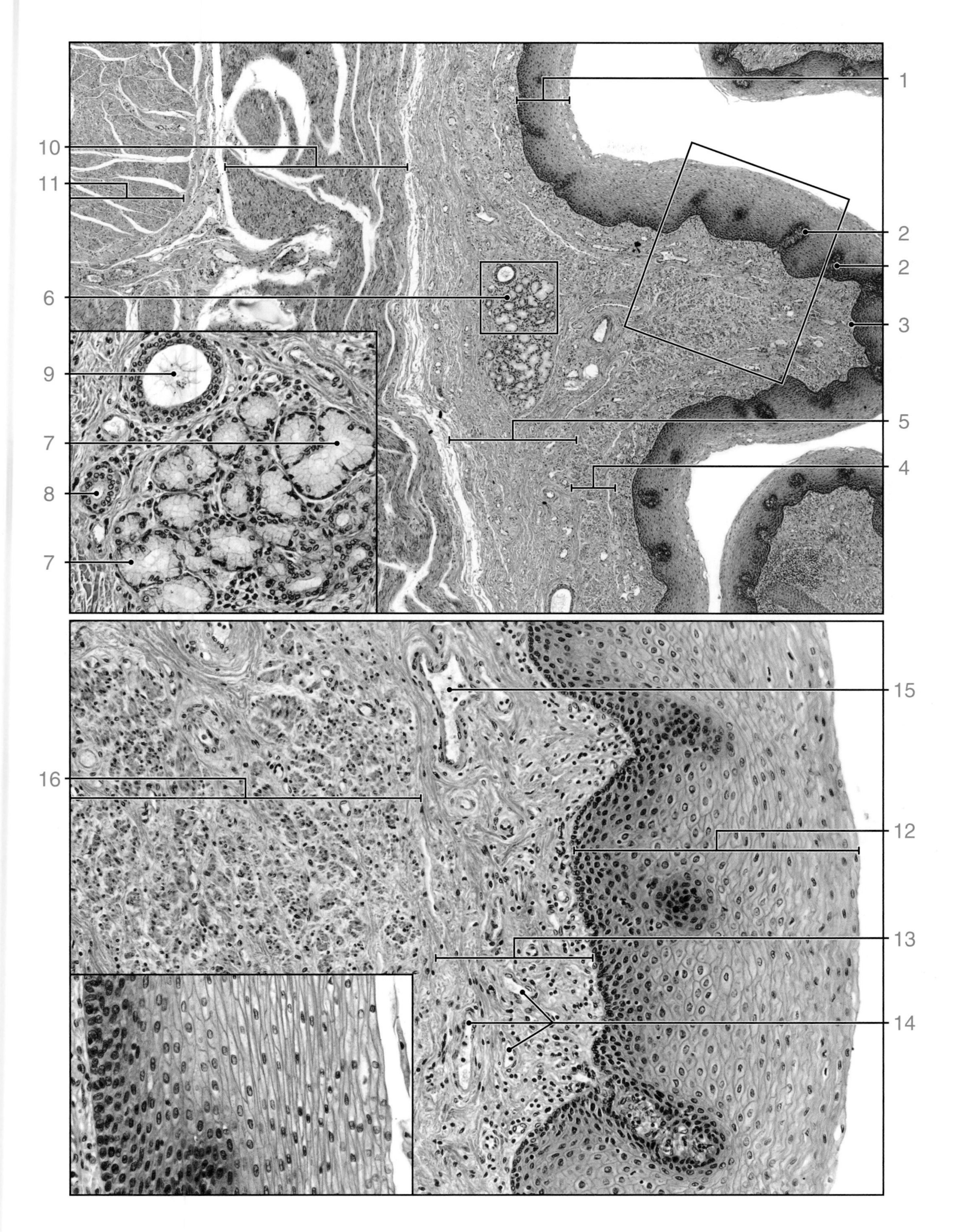
1
10
11
2
2
6
3
9
7
5
4
8
7
15
16
12
13
14

PLATE 85. ESOPHAGOGASTRIC JUNCTION

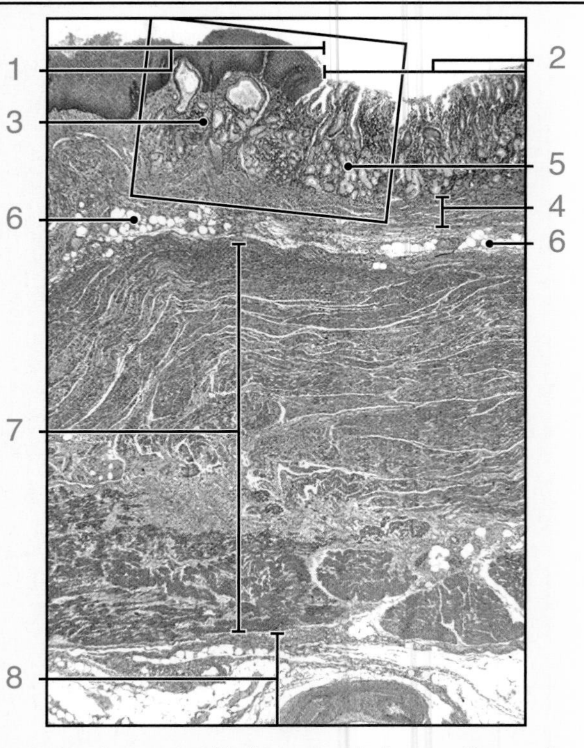

The transition from esophagus to stomach, the esophagogastric junction, is marked by an abrupt change from the stratified squamous epithelium lining the esophagus to the simple columnar, mucus-secreting epithelium lining the stomach. Another prominent feature of the esophagogastric junction is the presence of mucosal esophageal glands in the terminal esophagus. These glands lie within the mucosa in contrast to the esophageal submucosal glands (see Plate 84), which are located in the submucosa at various sites along the length of the esophagus. Because the mucosal esophageal glands are similar in appearance to the adjacent cardiac glands in the cardiac part of the stomach, they are also called cardiac glands. The esophageal cardiac glands and the cardiac glands of the stomach, both mucus-secreting, help protect the esophageal lining from regurgitated gastric contents.

ORIENTATION MICROGRAPH: This micrograph shows the point of transition from the **esophagus** (1) to the cardiac part of the **stomach** (2). Note the abrupt surface epithelial change between the two structures; also note the **mucosal esophageal glands** (3) in the terminal portion of the esophagus. The relatively thick **muscularis mucosa** (4) lies beneath the **cardiac glands** (5) and below this layer is the submucosa, which contains **adipose tissue** (6). The **muscularis externa** (7) at the junctional site, as well as in the remainder of the stomach, appears less ordered in that its smooth muscle forms incomplete layers: Some layers have an oblique array and others have longitudinal or circular arrays. This is characteristic of hollow, spheroidal organs. The outermost layer, the **serosa** (8), is seen only in part. It contains considerable adipose tissue, and thus it is relatively thick.

Esophagogastric junction, human, H&E, x90.
The mucosa of the esophagogastric junction from the boxed area on the orientation micrograph is shown here at higher magnification. The right side of the micrograph shows the beginning of the cardiac stomach. Note the **stratified squamous epithelium** (1) of the esophagus and its abrupt termination at the cardiac portion of the stomach. In the underlying lamina propria of the esophagus, several **ducts of the mucosal esophageal glands** (2) are evident. Both are somewhat dilated, a common feature. Deeper in the lamina propria are the **secretory components** (3) of the mucosal esophageal glands. Beneath the lamina propria is the **muscularis mucosa** (4). The surface of the stomach is lined by a simple columnar epithelium consisting of mucus-secreting cells (stripped off by physical damage). Extending downward from the surface are **gastric pits** (5) lined by the same type of mucus-secreting epithelial cells. Deeper in the lamina propria are the **cardiac glands** (6), which empty into the gastric pits. They are made up entirely of mucus-secreting cells.

Stomach, cardiac region, glands, human, H&E, x225.
This micrograph is a higher magnification of the boxed area in the above micrograph. At the top of the micrograph are the bottoms of the **gastric pits** (7). The **secretory portions of the gastric glands** (8) are seen in a variety of profiles. The nuclei of these cells tend to be flat compared to the nuclei of the gastric pits, which have a round to slightly elongate profile. Note the nature of the **lamina propria** (9), a highly cellular loose connective tissue, which contains numerous lymphocytes and plasma cells.

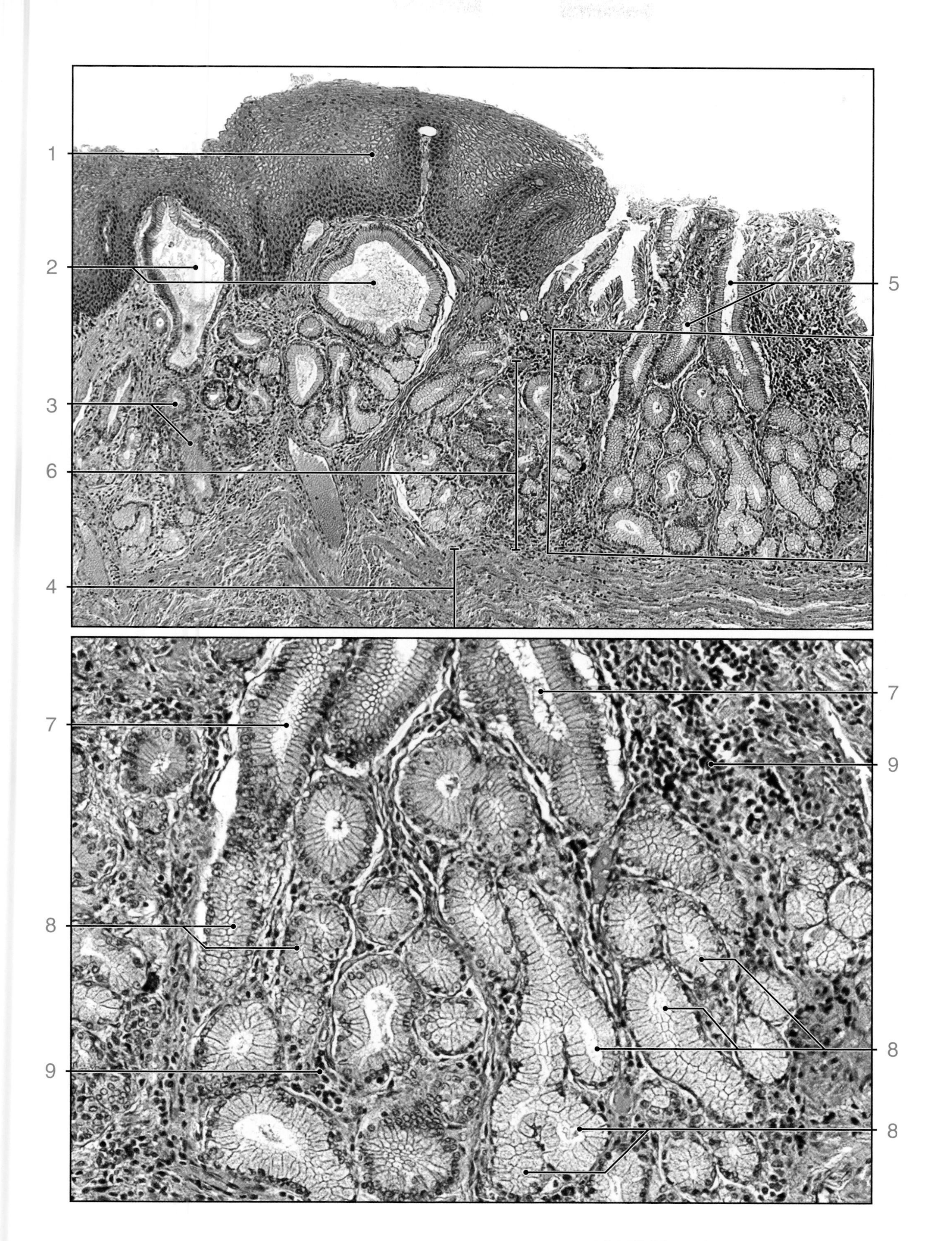

PLATE 85. ESOPHAGOGASTRIC JUNCTION

PLATE 86. STOMACH, FUNDUS

The fundic region of the stomach lies between the narrow cardiac stomach, adjacent to the esophagus, and the distal pyloric region of the stomach. Thus, it comprises the bulk of the stomach. The interior of the empty stomach reveals a series of longitudinal folds or ridges except in the upper expanded portion, which is relatively smooth. These ridges, called rugae, are composed of the mucosa and submucosa. They disappear upon filling and consequent distension of the stomach.

Examination of the mucosa with a simple lens reveals grooves that subdivide the surface into slightly bulging gastric areas 1 to 6 mm in diameter. These are called mammillated areas. Further observation of the mucosal surface reveals the gastric pits. These are surface openings of the underlying gastric glands, which in this area of the stomach are also called fundic glands.

ORIENTATION MICROGRAPH: This micrograph is of a section through the fundic stomach. It shows a **ruga** (1), some **mammillated areas** (2), the full thickness of the **mucosa** (3), the **submucosa** (4), the **muscularis externa** (5), and some of the larger **blood vessels** (6) present in the submucosa.

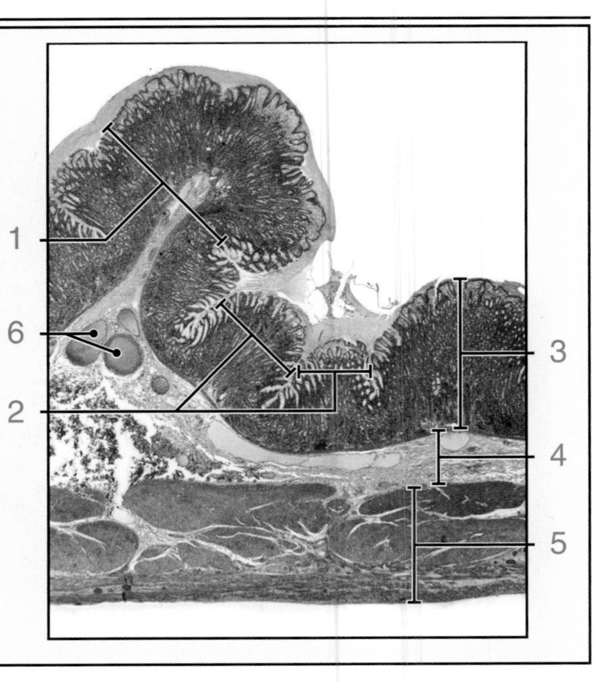

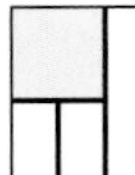

Stomach, fundic region, mammillated area, human, H&E, x80.
This micrograph shows one of the mammillated areas in the orientation micrograph. The stomach **lumen** (1) is at the top of the micrograph. The surrounding **groove** (2), which is continuous with the stomach lumen, enhances the total surface area of the stomach. Note the **gastric pits** (3). They are lined by surface mucous cells. Where the gastric pits appear to end (4), they are actually continuous with one to several **fundic glands** (5). The fundic glands occupy most of the mucosa and extend down to the **muscularis mucosa** (6).

Stomach, fundic region, gastric pits and fundic glands, human, H&E, x175.
The full thickness of the gastric mucosa from lumen to muscularis mucosa is seen here at higher magnification. Note the **gastric pits** (7) with their lining of mucous surface cells. One of the **fundic glands** (8) can be seen where it empties into a **gastric pit** (9). As many as 3 to 5 fundic glands may empty into one gastric pit.

Stomach, fundic region, gastric pits and fundic glands, human, H&E, x325.
This micrograph is a higher magnification of the bottom of a **gastric pit** (10) and the upper portion, or neck region, of a **fundic gland** (11). At this site, cells replicate. Those that migrate in an upward direction become mucous surface cells. Those that migrate downward maintain the cell population of the **fundic gland** (12), becoming mostly **parietal cells** (13) and **chief cells** (14).

Stomach, fundic region, glands, human, H&E, x160.
The basal portions of the fundic glands are shown here. The **chief cells** (15) are readily identified by the basophilic staining of their cytoplasm due to the extensive ergastoplasm (rough endoplasmic reticulum) that they contain, whereas **parietal cells** (16) stain intensely with eosin due to their numerous mitochondria and extensive membrane content. Also, parietal cell nuclei appear in the center of the cell. In contrast, the nucleus of the chief cell is near the base of the cell. The base of one of the glands (17) is seen in cross-section, a result of acute and typical bending of the gland at its base.

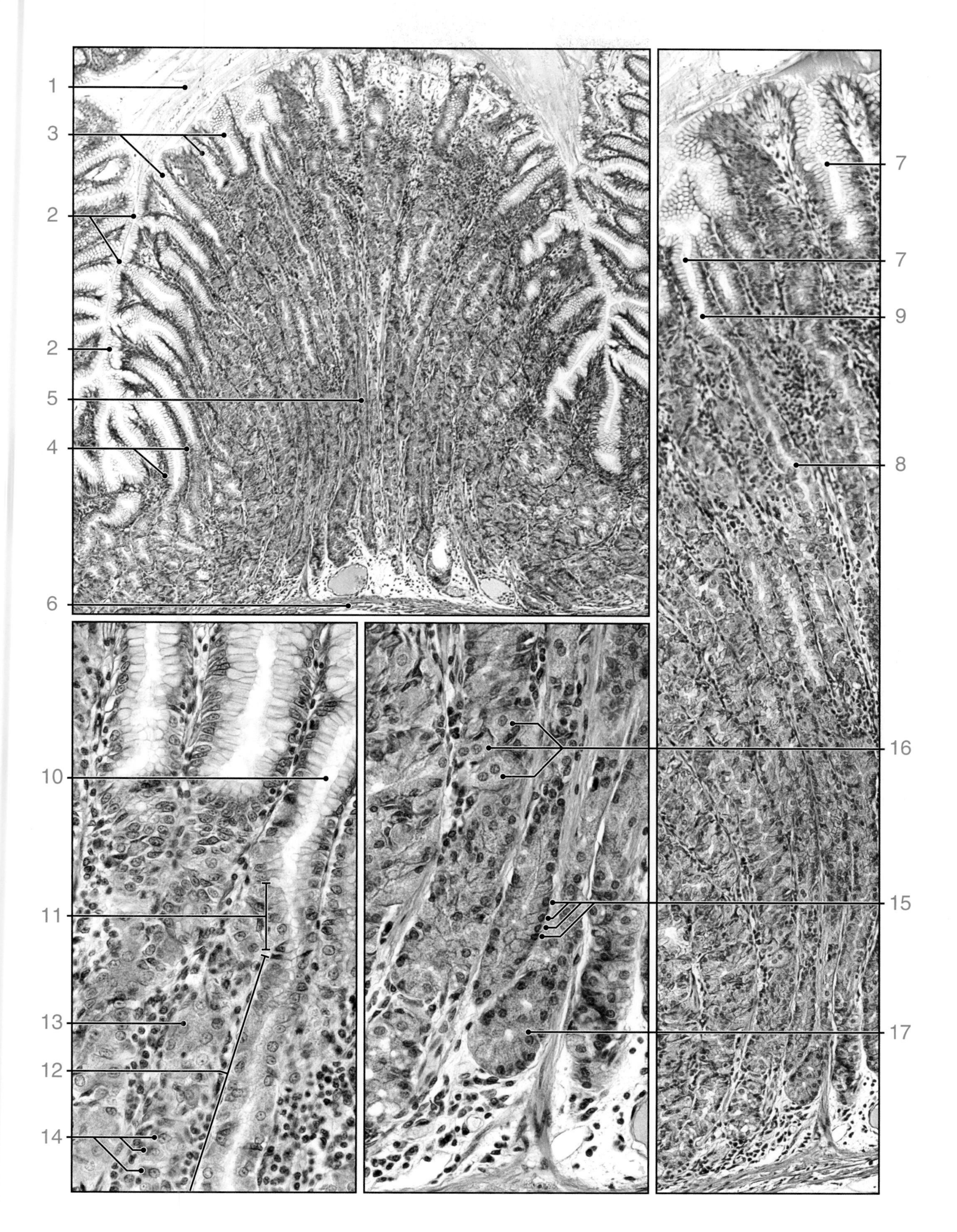

PLATE 86. STOMACH, FUNDUS

PLATE 87. GASTRODUODENAL JUNCTION

ORIENTATION MICROGRAPH: The gastroduodenal (pyloric) junction is the site where the stomach ends and the small intestine, namely, the duodenum, begins. The most prominent feature at this site is the thickening of the circular layer of the muscularis externa, which forms the **gastroduodenal (pyloric) sphincter** (1). It controls the passage of chyme from the stomach into the intestine. Another significant feature at this site is the abrupt change from the surface mucous cells lining the stomach to the columnar enterocytes (absorptive cells) that are present on the surface of the **villus protrusions** (2) of the small intestine. A third important feature is the presence of submucosal mucus-secreting **glands of Brunner** (3) that occur in the initial portion of the duodenum. Their secretions help neutralize the chyme as it enters the intestine.

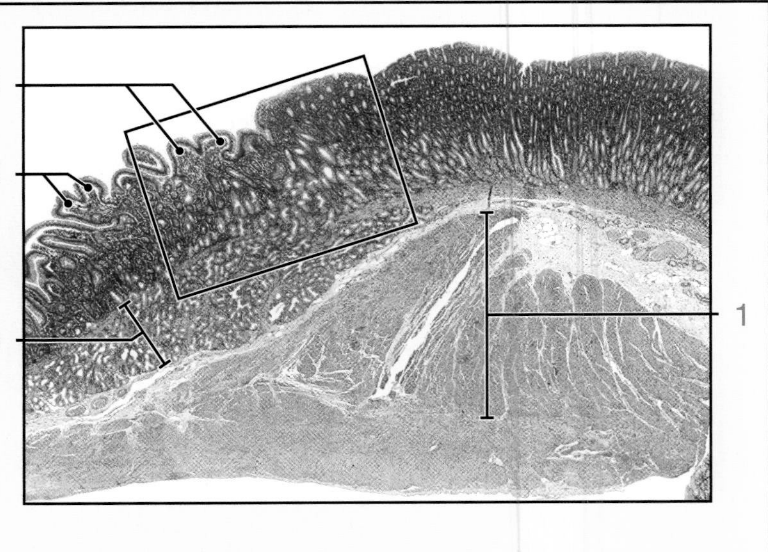

Gastroduodenal junction, human, H&E, x90. This micrograph is a higher magnification of the boxed area on the orientation micrograph. The transition of surface mucous cells of the stomach to the enterocytes of the intestine is marked by the *arrowhead*. The stomach mucosa is to the right of the arrowhead and the duodenal mucosa is to the left, where the **villi** (1) begin to appear. The **pyloric glands** (2) that occur in the pyloric region of the stomach secrete mucus. They extend upward from the **muscularis mucosa** (3) to the luminal surface of the pyloric stomach. The duodenum exhibits similar-looking mucous glands, which lie below the muscularis mucosa. These are the **glands of Brunner** (4), some of which have extended laterally to lie under the muscularis mucosa of the pyloric stomach. The glands of Brunner empty into the **intestinal crypts** (5) (of Lieberkuhn). The secretion of the submucosal glands reaches the intestinal surface via the intestinal crypts.

Gastroduodenal junction, human, H&E, x700. This high magnification micrograph shows the smaller boxed area on the above micrograph. The *arrowhead* is at the site of the transition from the surface mucous cells of the stomach (upper portion of the micrograph) to the enterocytes that line the surface of the duodenum. Note the apical mucus-containing cup of the **surface mucous cells** (6) in contrast to the absorptive **enterocytes** (7) in the lower left of the micrograph. The enterocytes have an apical border of microvilli, which are barely visible at this magnification.

Gastroduodenal junction, human, H&E, x160. This micrograph shows a higher magnification of the larger boxed area on the above figure. It shows the **Brunner's glands** (8) and a portion of the **muscularis mucosa** (9). As noted, the Brunner's glands lie below the level of the muscularis mucosa. However, it can be seen here that they interrupt this thin layer of smooth muscle as the ducts enter the mucosa and empty into the intestinal crypts.

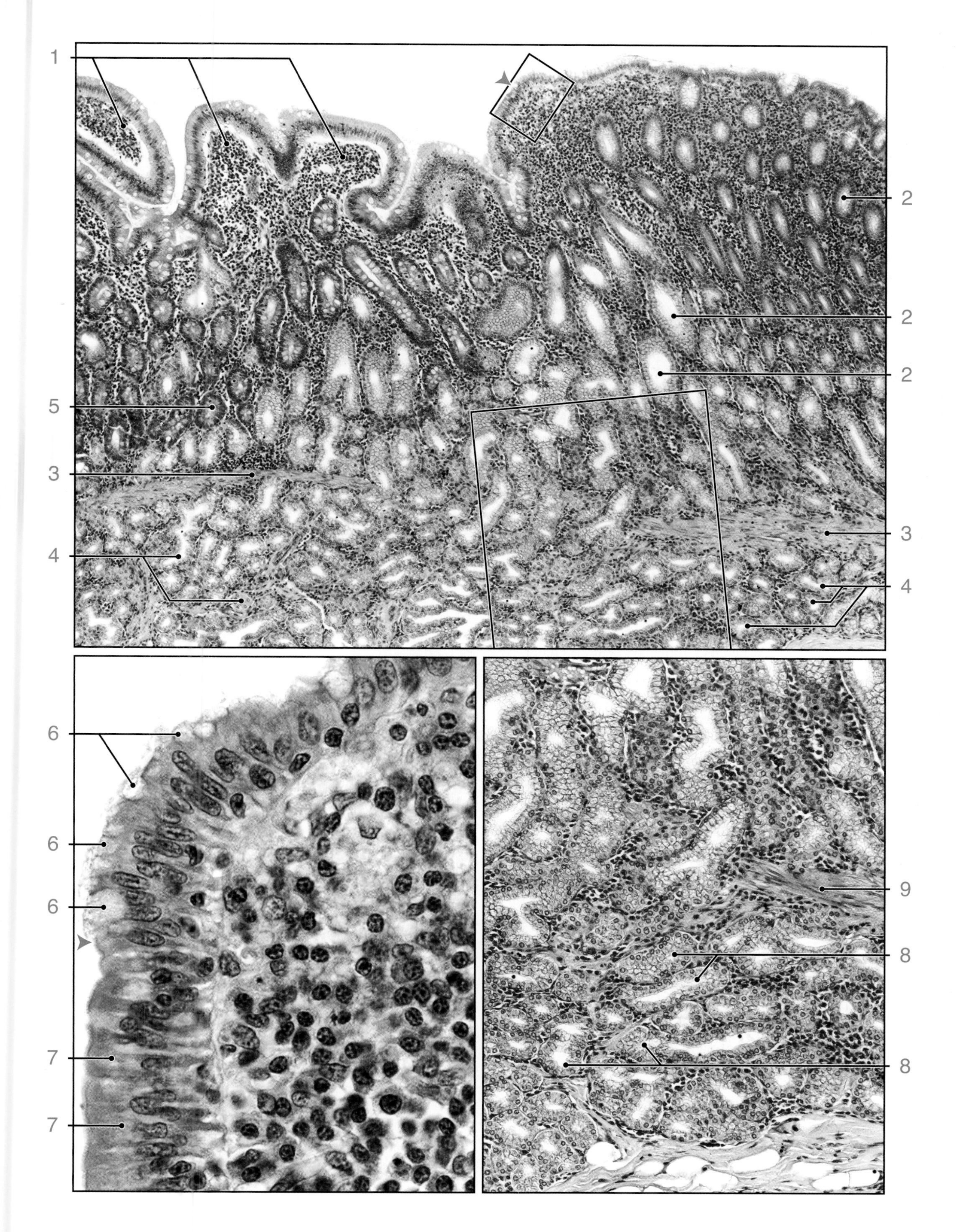

PLATE 87. GASTRODUODENAL JUNCTION

PLATE 88. DUODENUM

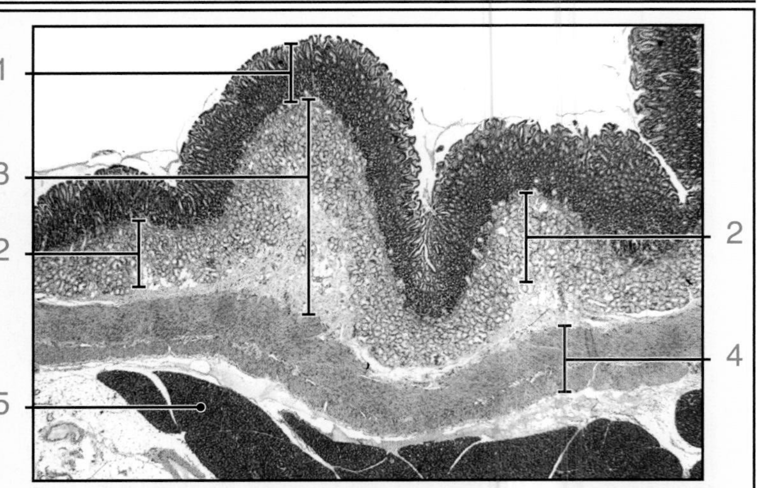

The duodenum is the first segment of the small intestine and measures approximately 25 cm in length. It receives the partially digested bolus of food and chyme from the stomach. It also receives digestive enzymes from the stomach, pancreas, and also from the liver via the gallbladder. Because of the high acidity of the gastric contents, much of the length of the duodenum possesses prominent submucosal glands (Brunner's glands) that secrete an alkaline mucus to help neutralize the acidic chyme.

In order to maximize the absorptive surface of the small intestine, its luminal surface possesses plicae circulares, which are fixed transverse folds whose core is made up of submucosa, and villi, the leaflike and fingerlike projections of the mucosa. The epithelial surface of the villi consists mainly of enterocytes, the cells that absorb the digested metabolites from the intestinal lumen. These cells further increase the absorptive surface by their numerous microvilli. The enterocyte is not only an absorptive cell; it also synthesizes enzymes that are integrated into the membrane of the microvilli for digestion of disaccharides and dipeptides. Other cells present on the villi are the mucus-secreting goblet cells. The cells that line the intestinal crypts are Paneth cells. At the crypt basestem cells that give rise to the enterocytes, goblet cells, and enteroendocrine cells.

ORIENTATION MICROGRAPH: This micrograph presents a longitudinal section of the duodenum. Note the **mucosa** (1), which contains the villi and intestinal crypts (of Lieberkühn) that lie at the base of the villi. Beneath the crypts is the muscularis mucosa (not identifiable at this magnification). The **Brunner's glands** (2) occupy most of the **submucosa** (3). Beneath the submucosa is the **muscularis externa** (4). The segment of the duodenum shown here lies next to the **pancreas** (5), thus there is no serosa; connective tissue binds the two structures together.

Duodenum, human, H&E, x180.
The histological features of the duodenal mucosa are seen in this micrograph. The **villi** (1) are closely packed and exhibit a small amount of intervillus space. In the duodenum, the villi are mostly broad, leaflike structures rather than more uniform, fingerlike projections. Because of their shape, they tend to exhibit variations in terms of their width and general shape. Thus, some of the villi appear relatively broad and others appear narrow, a reflection of the plane in which the villi were sectioned. The **intestinal crypts** (2) begin just above the **muscularis mucosa** (3) and extend upwards where they end at the base of the villi. The crypts consist of stem cells and developing cells that migrate from the base of the crypts to become mature goblet cells and enterocytes that line the surface of the villi.

Duodenum, human, H&E, x650.
This micrograph shows at high magnification the epithelial cells present on the surface near the tip of two adjacent villi. Several **goblet cells** (4) are readily identified by their mucus cups. The remaining cells are the absorptive enterocytes. They exhibit a **striated border** (5), which consists of numerous, closely packed, fine, cylindrical processes known as microvilli. Beneath the epithelial cells is the core of the villus, the lamina propria. It is made up of a collagenous stroma in which there is a population of closely packed **connective tissue cells** (6). Most of the cells present in the lamina propria are lymphocytes and plasma cells. Neutrophils and smooth muscle cells are present in smaller numbers. Also present are capillaries, both vascular and lymphatic.

Duodenum, human, H&E, x310.
This micrograph reveals the base of the **crypts of Lieberkühn** (7), the underlying **muscularis mucosa** (8), and the **glands of Brunner** (9) that lie in the submucosa. The micrograph shows the merging of several Brunner's glands (*asterisk*) into a very short, duct-like structure. The area within the *circle* shows where the mucous secreting cells of the gland are emptying into the base of a crypt. The crypt cells are notably smaller and more closely packed than the mucus-secreting cells of the Brunner's gland. These very small cells of the crypt are the stem cells, which will give rise to the goblet cells and enterocytes.

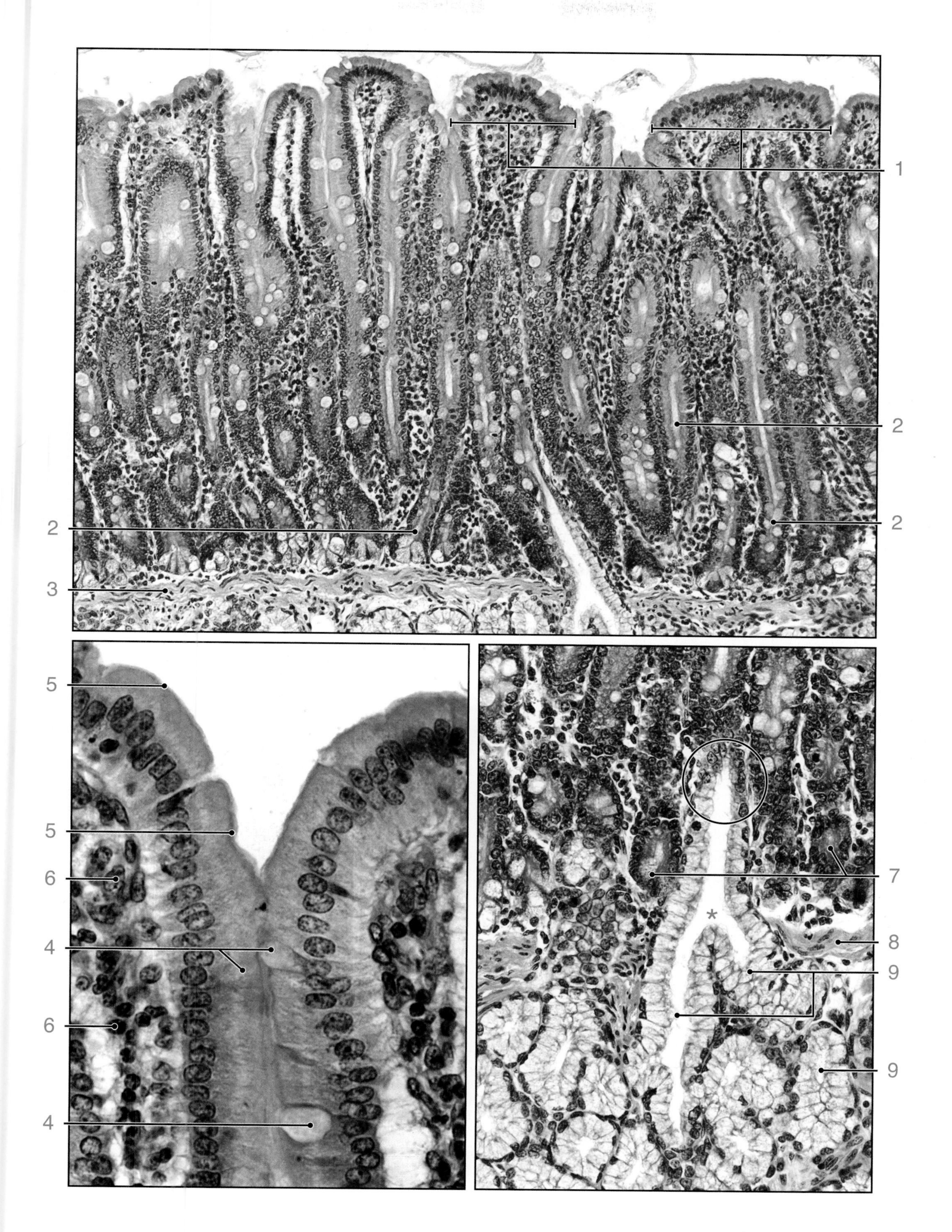

PLATE 88. DUODENUM

PLATE 89. JEJUNUM

The jejunum, the second segment of the small intestine, measures approximately 2.5m in length. Structurally, the jejunum and the ileum are very similar to the duodenum. The plicae circulares are most numerous in the distal half of the duodenum and the proximal part of the jejunum. In the distal half of the jejunum, they become smaller and less numerous and virtually disappear in the middle of the ileum. Although the villi usually have a leaflike configuration, they become more fingerlike in shape as the ileum is approached. In the ileum, most of the villi have the fingerlike shape. Brunner's glands are absent in the jejunum and ileum. They gradually decrease in number and finally disappear in the last third of the duodenum. The epithelial cell types that line the crypts and villi in the jejunum and ileum are, for the most part, the same as those found in the duodenum.

ORIENTATION MICROGRAPH: This micrograph shows a longitudinal section of the jejunum, obtained from a site near the midpoint of its length, at low magnification. Four of the **plicae circulares** (1) are evident. Extending from the plicae are the villi. The core of each plica consists of dense connective tissue. The plicae form most of the submucosa. Below the submucosa is the **circular layer of the muscularis externa** (2), which is relatively thick. The **longitudinal layer of the muscularis externa** (3) is thinner. Most of the remainder of the specimen consists of a thick layer of dense connective tissue and an epithelial surface (mesothelium), which together are known as the **serosa** (4).

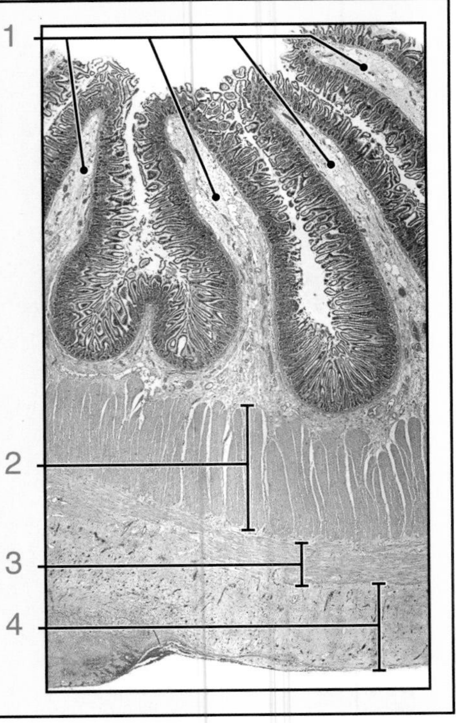

Jejunum, human, H&E, x90.
This low power micrograph shows the **mucosa** (1) and the underlying **submucosa** (2). The **villi** (3) are relatively tall and occupy more than half of the thickness of the mucosa. The **intestinal crypts** (4) empty at the base of the villi. The lamina propria, which constitutes the core of the villi and the tissue surrounding the crypts, is a loose connective tissue, like that of the duodenum, which contains numerous lymphocytes and plasma cells, and, in lesser numbers, neutrophils and smooth muscle cells, as well as both vascular and lymphatic vessels. Barely identifiable in this micrograph is the **muscularis mucosa** (5), consisting of smooth muscle cells. The dense connective tissue that makes up the submucosa contains **blood vessels** (6) that supply the mucosa and muscularis externa (not shown).

Jejunum, human, H&E, x255.
The upper portion of several villi from the boxed area of the top micrograph are shown here at higher magnification. Note the very cellular **lamina propria** (7). The epithelial cells lining the villi are mostly **enterocytes** (8) with occasional, interspersed **goblet cells** (9).

Jejunum, human, H&E, x320; inset x700.
Crypts are shown in this micrograph at higher magnification. **Paneth cells** (10) are found at the extreme base of the crypts. The **replicating cells** (11) occupy most of the deep portion of the crypts where occasional **mitotic figures** (12) are seen. As the epithelial cells move toward the upper region of the crypts, they stop dividing. They continue to migrate upwards on the villi until they reach the villus tip where they undergo apoptosis. The **inset**, an even higher magnification of the base of a crypt, reveals the Paneth cells. Note the eosinophilic secretory granules in the apical portion of the cell.

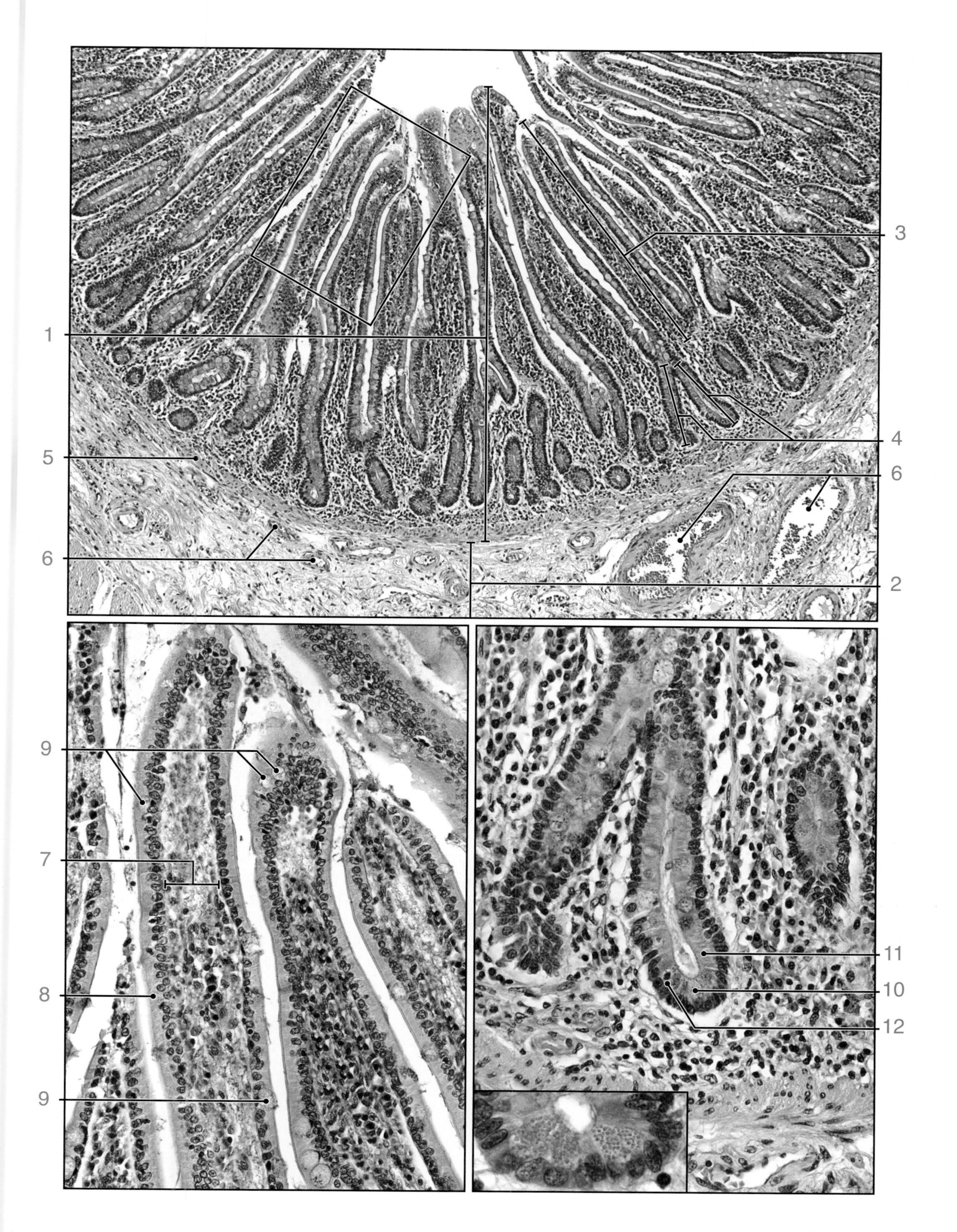

PLATE 89. JEJUNUM

PLATE 90. ILEUM

The ileum, the distal segment of the small intestine, is the longest, measuring approximately 3.5 m in length. In most respects, the ileum is structurally similar to the other parts of the small intestine with the exception already noted: that the plicae circulares are smaller and fewer. They disappear midway along its length. Also as noted, the villi become more fingerlike and are easily recognized as villi in most sections. Furthermore, a unique feature of the ileum is the presence of numerous, closely packed lymphatic nodules that lie in the submucosa and penetrate the mucosa. These nodules, known as aggregated nodules or Peyer's patches, are preferentially located along the ileum opposite to the site where it is attached to the mesentery. The number of nodules changes with age. They become maximal around puberty, but then become fewer during later life. In histological specimens, aggregated nodules present as an extremely basophilic region, in contrast to a fresh, gross specimen where they appear as aggregates of white specks.

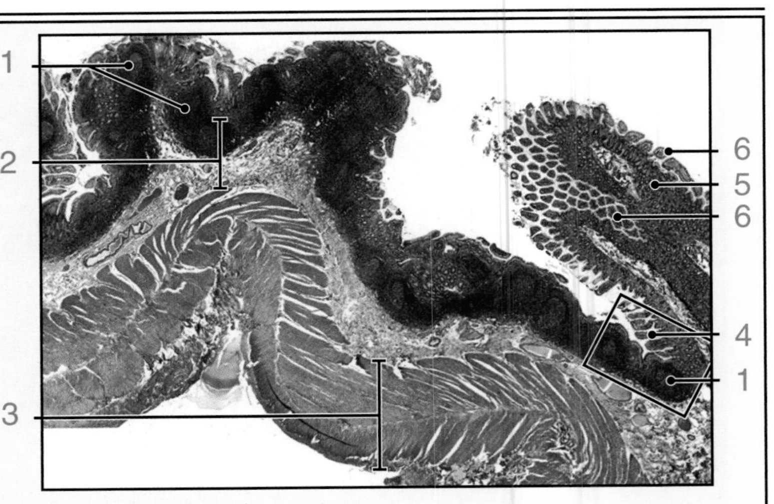

ORIENTATION MICROGRAPH: This micrograph shows a longitudinal section of the ileum that includes the **aggregated nodules** (1). The specimen also shows the **submucosa** (2) consisting of dense connective tissue and lymphatic nodules. Beneath the submucosa is the **muscularis externa** (3). Note that where the section has obliquely passed to the side of the nodules, the fingerlike **villi** (4) are present. The remaining portion of the intestinal wall shows a tangentially cut **plica circularis** (5). Most of the **villi** (6) in this region are seen in cross-section.

Ileum, human, H&E, x90.

The boxed area on the orientation micrograph is shown here at higher magnification. Two **lymphatic nodules** (1) are evident. Note the absence of villi in this region. Where present, the nodules typically intrude into most of the mucosa leaving, at best, very rudimentary villi or simply an irregular surface. There is also a paucity of crypts of Lieberkühn in the area of Peyer's patch. Only a few **crypts** (2) are evident here. The muscularis mucosa is typically disrupted by the nodules and, as a consequence, one sees eosinophilic clusters of tissue. These sites are groups of **smooth muscle cells** (3) of the disrupted muscularis mucosa. Where the nodules are absent, typical, **fingerlike villi** (4) and associated **intestinal crypts** (5) are present. Many of the crypts here are seen in cross-section, a reflection of the oblique angle at which this part of the intestinal wall was sectioned.

Ileum, dog, H&E, x320.

This micrograph is from the same specimen at a site where the section has passed in a true cross-sectional plane of the intestinal wall, thus showing the basal portion of several **crypts of Lieberkühn** (6). At this site, the lymphatic nodules did not intrude; therefore the muscularis mucosa appears intact as an uninterrupted layer. Careful examination of the muscularis mucosa reveals two layers: an **inner circular layer** (7) and an **outer longitudinal layer** (8). Note that the smooth muscle nuclei appear as elongate profiles in the inner circular layer and more or less round profiles in the outer longitudinal layer. At the bottom of the micrograph is the **submucosa** (9), consisting of dense connective tissue. It should also be noted that the Paneth cells, which are located at the very base of the crypts of Lieberkühn, are not distinctive in this specimen, presumably due to the loss of the granules during tissue preparation. This is not an uncommon occurrence.

Ileum, dog, H&E, x320.

This specimen was cut in a similar plane to that seen in the micrograph on the left, but it reveals an area that contains a lymphatic nodule. The basal portion of several crypts of Lieberkühn are evident at this site. Again, note that the granules characteristic of Paneth cells are not evident. The periphery of the nodule has penetrated into the mucosa. Note the extensive mass of lymphocytes. Also note that the **muscularis mucosa** (10) is discontinuous. Aggregates of smooth muscle are seen dispersed within the lymphatic tissue.

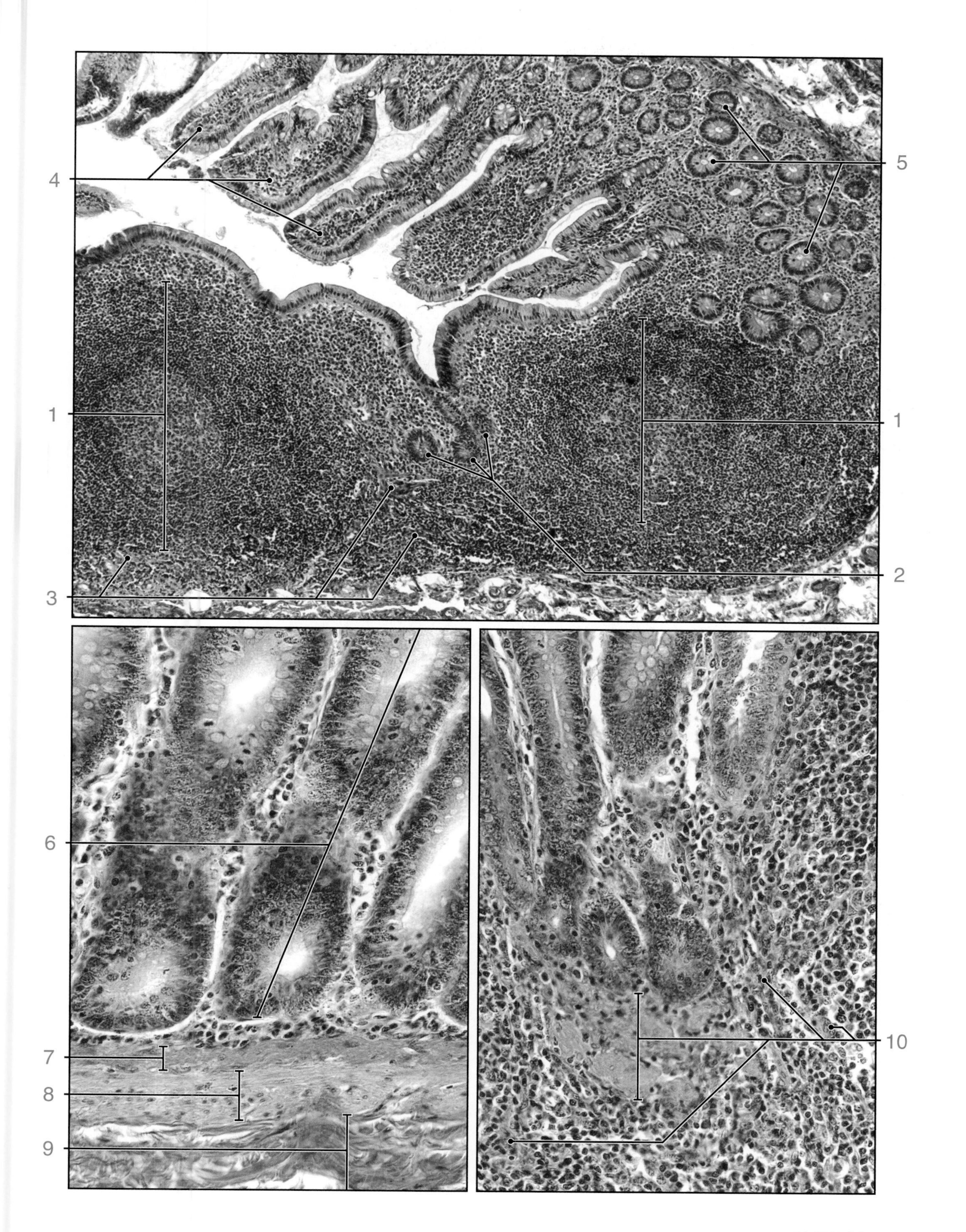
4
5
1
1
2
3
6
7
8
9
10

PLATE 91. ILEOCECAL JUNCTION

The ileum terminates at the site where its contents empty into the cecum through the ileocecal valve. The cecum, a blind ending, pendulus sack from which the appendix extends as a fingerlike structure, is the initial part of the large intestine. The valve consists of cecal mucosa overlying circular muscle fibers of the ileum. It helps prevent retrograde movement of fecal contents back into the ileum. Structurally, the major change seen in the region of this junctional site is a reduction in the size and number of villi in the ileum. The cecal portion of the large intestine is marked by a smooth luminal surface and the absence of villi. The intestinal glands of the cecum appear the same as those in the ileum except that they are somewhat deeper. The muscularis mucosa of the small intestine continues uninterrupted into the large intestine. The other layers, namely the submucosa and muscularis externa, also extend from small to large intestine without interruption.

ORIENTATION MICROGRAPH: This micrograph shows a longitudinal section of the ileocecal junction. The upper portion of the micrograph shows the **villi** (1) of the ileum whereas the **cecum** (2) lacks villi and exhibits a smooth luminal surface. Note that the **submucosa** (3) and the **muscularis externa** (4) appear as continuous layers with no distinction in appearance from ileum to cecum. The micrograph does not adequately reveal the structural organization of the ileocecal valve.

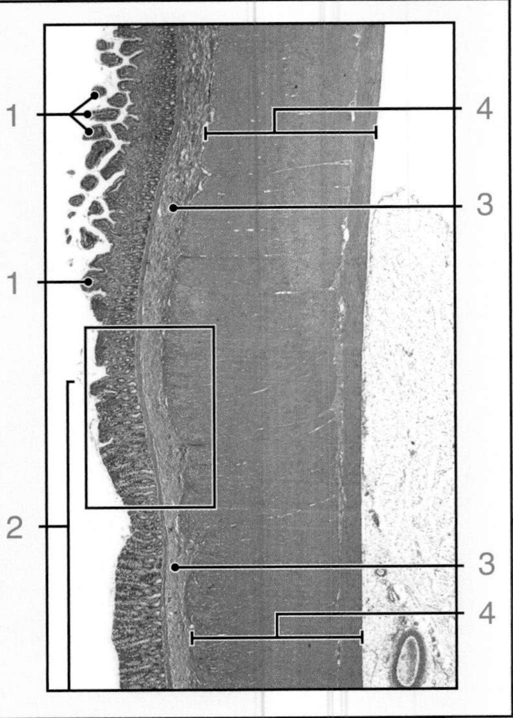

Ileocecal junction, human, H&E, x180.
This micrograph shows at higher magnification the boxed area on the orientation micrograph, which includes the junction between ileum and cecum. Rudimentary **villi** (1) indicate that the right half of the micrograph is ileum. A transition from ileum to cecum is seen at the *arrow*. Note the uniform luminal surface of the **cecum** (2). The **muscularis mucosa** (3), the underlying **submucosa** (4), and **muscularis externa** (5) appear as a continuum from ileum to cecum.

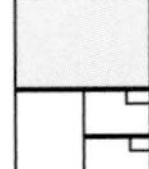

Ileocecal junction, human, H&E, x255.
A higher magnification of the mucosa of the cecum and a small portion of the terminal ileum from the boxed area on the top micrograph is shown here. Note the **surface absorptive cells** (6), belonging to the ileum, and their termination at the cecum. The mucosa of the cecum contains straight, tubular, intestinal glands (crypts of Lieberkühn) that extend through its full thickness. The epithelium of the glands is simple columnar, like that of the ileum, and consists mostly of mucus-secreting **goblet cells** (7) and groups of intervening **absorptive cells** (8). Also, as in the small intestine, occasional **enteroendocrine cells** (9) are found interspersed among the absorptive and goblet cells. The **lamina propria** (10) occupies the space between the glands and, as in the small intestine, is highly cellular.

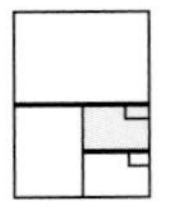

Ileocecal junction, human, H&E, x180; inset x440.
The micrograph shown here reveals the **muscularis mucosa** (11), the dense irregular connective tissue of the **submucosa** (12), and a small portion of the inner circular layer of the **muscularis externa** (13). Within the submucosa, part of the network of unmyelinated nerve fibers and ganglion cells that constitute the **submucosal plexus** (14) (Meissner's plexus) is evident. One of the nerve bundles is shown in the **inset** at higher magnification. The clear circles represent the unstained, unmyelinated axons. The nuclei belong to Schwann cells. Their cytoplasm outlines the axons that they surround.

Ileocecal junction, human, H&E, x180; inset x440.
This micrograph reveals the outer portion of the cross-sectioned inner circular layer of the **muscularis externa** (15) as well as a portion of the longitudinally sectioned outer **longitudinal layer of smooth muscle** (16). Between the two muscle layers is a thin connective tissue layer that contains the myenteric plexus (Auerbach's plexus). Several large **ganglion cells** (17) belonging to the plexus can be seen here. The **inset** shows the unmyelinated nerve fibers and Schwann cells that surround the axons.

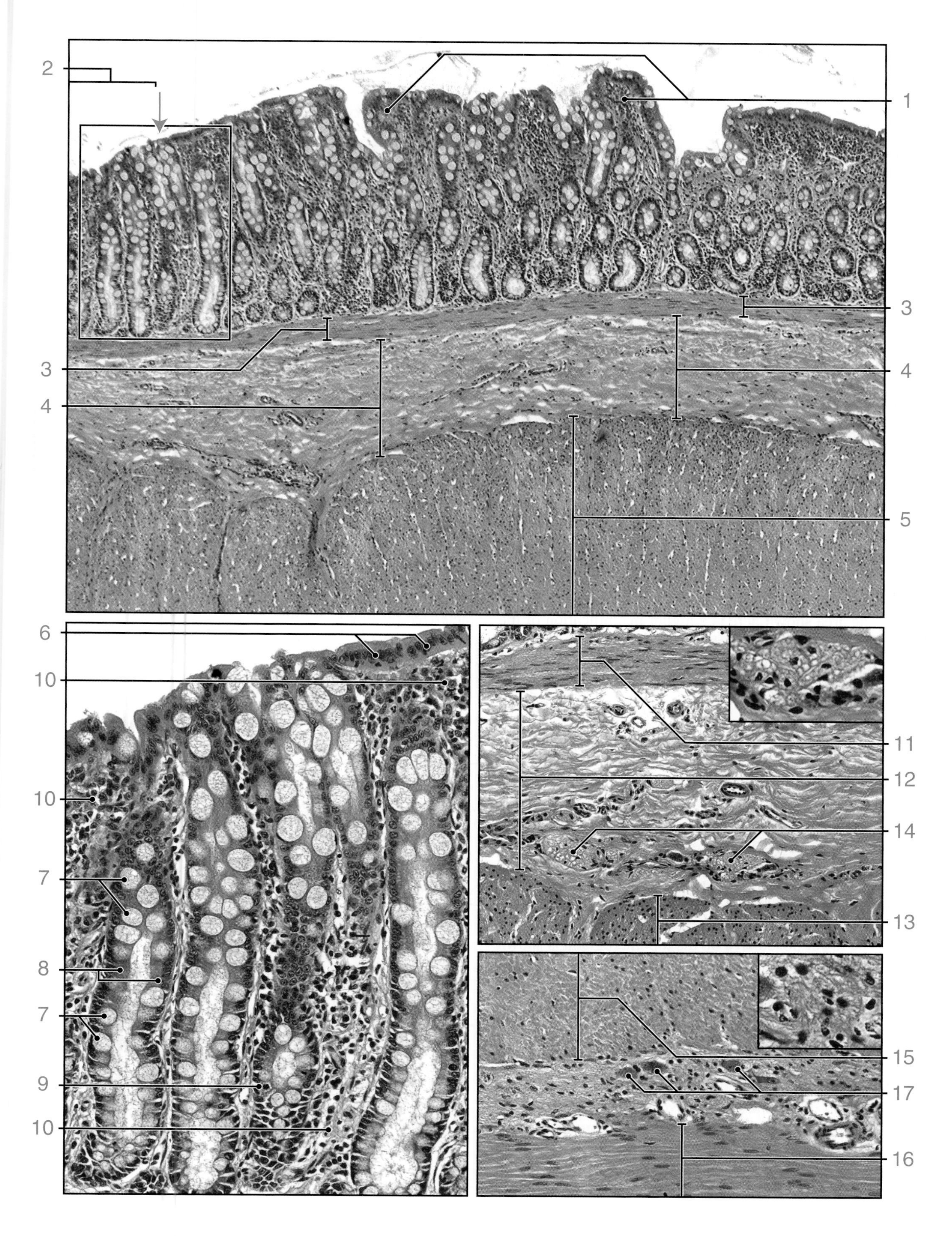

PLATE 91. ILEOCECAL JUNCTION

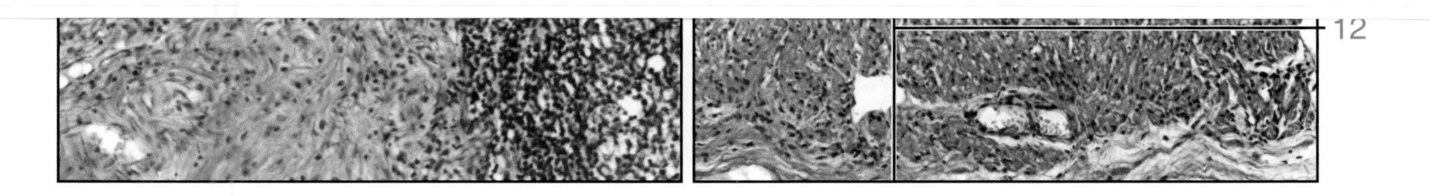

PLATE 93. APPENDIX

PLATE 93. APPENDIX

The vermiform appendix, or simply the appendix, is a fingerlike structure that arises from the cecum (the cecum is a pouch just distal to the ileocecal valve, the first segment of the large intestine). The appendix is a narrow, blind-ending tube that ranges in length from 2.5 cm to as much as 13 cm. Because it is a blind-ended pouch, intestinal contents may be trapped or sequestered in the appendix, often leading to inflammation and infection. In infants and young children, it contains numerous lymphatic

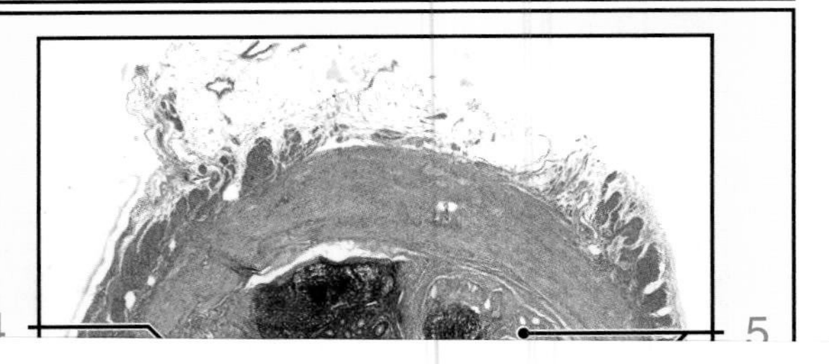

PLATE 94. ANAL CANAL AND ANOCUTANEOUS JUNCTION

The structure of the mucosa of the rectum is essentially the same as that of the colon. As the anal canal is approached, a transformation occurs in which the simple columnar epithelium undergoes a change to a stratified epithelium. The epithelium in the anal canal is variable in character. It may be stratified columnar or cuboidal, changing to a stratified squamous with patches of stratified columnar or cuboidal, and ultimately becoming a keratinized stratified squamous epithelium at the anus—the transition zone. Associated with the keratinized stratified squamous epithelium are sebaceous glands and hair follicles. At the level of the anocutaneous junction, the muscularis mucosa ends. At the same level, the smooth muscle of the circular level of the muscularis externa thickens to become the internal anal sphincter. The external anal sphincter is formed by the striated muscles of the pelvic floor and perineum.

ORIENTATION MICROGRAPH: This micrograph shows a section through the wall of the rectum and anal canal. It reveals the **crypts of Lieberkühn** (1) in the rectal region and below this, the anal canal. The portion of the micrograph below the *arrow* is the anus, which is characterized by skin. Other recognizable features include the termination of the circular layer of the **muscularis externa** (2), which is expanded, forming the internal anal sphincter; the **external anal sphincter** (3) (*within the dashed line*), consisting of bundles of striated muscle; and the transition zone (*arrow*).

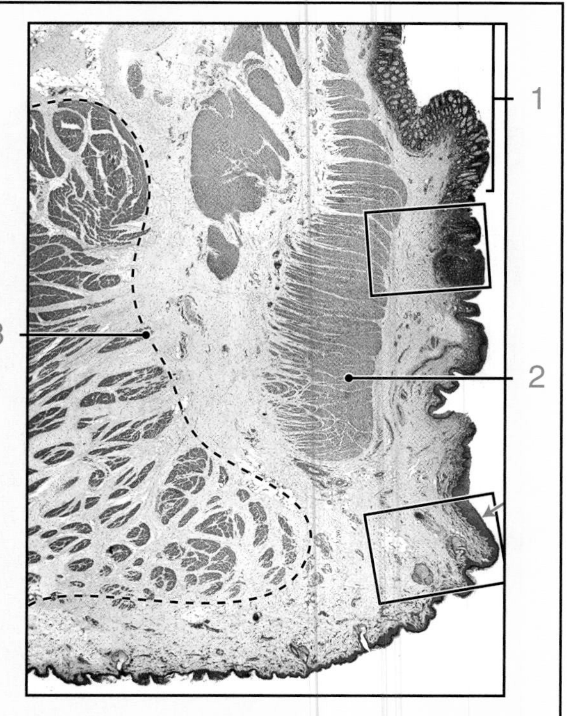

Anal canal, anorectal junction, human, H&E, x65. This micrograph shows the upper boxed area on the orientation micrograph at higher magnification. It includes mucosa, submucosa, and a portion of the muscularis externa. The lining epithelium shows the transition from **simple columnar epithelium** (1) to **stratified cuboidal epithelium** (2). Also present are the last of the **crypts of Lieberkühn** (3). A **lymph nodule** (4) is present below the stratified cuboidal epithelium. At the bottom of the micrograph is a portion of the inner circular layer of the **muscularis externa** (5). The **submucosa** (6), in the absence of a muscularis mucosa at this site, directly adjoins the mucosa.

Anal canal, anorectal junction, human, H&E, x255. The area shown here is from the boxed area on the top left micrograph. Note the **simple columnar epithelium** (7) and its transition to **simple cuboidal epithelium** (8) at this site. In association with the simple columnar epithelium is one of the last **crypts of Lieberkühn** (9). The underlying lamina propria is extremely cellular, containing large numbers of lymphocytes.

Anal canal, anorectal junction, human, H&E, x65. This micrograph shows at higher magnification the lower boxed area on the orientation micrograph. The epithelium changes from stratified squamous to keratinized stratified squamous at the *arrow*. This the transition zone between mucous membrane and skin. Note that associated with the skin are **sebaceous glands** (10) and **hair follicles** (11).

Anal canal, anorectal junction, human, H&E, x255. The area shown here is from the boxed area on the bottom left micrograph. The **epithelium** (12) appears as typical epidermis. The various layers of epidermis are evident. Note the **stratum granulosum** (13), a characteristic feature of keratinized epithelium. The underlying **connective tissue** (14) is considerably less cellular, but typical dermal **melanocytes** (15) are present, contributing to the pigmentation of the anal region.

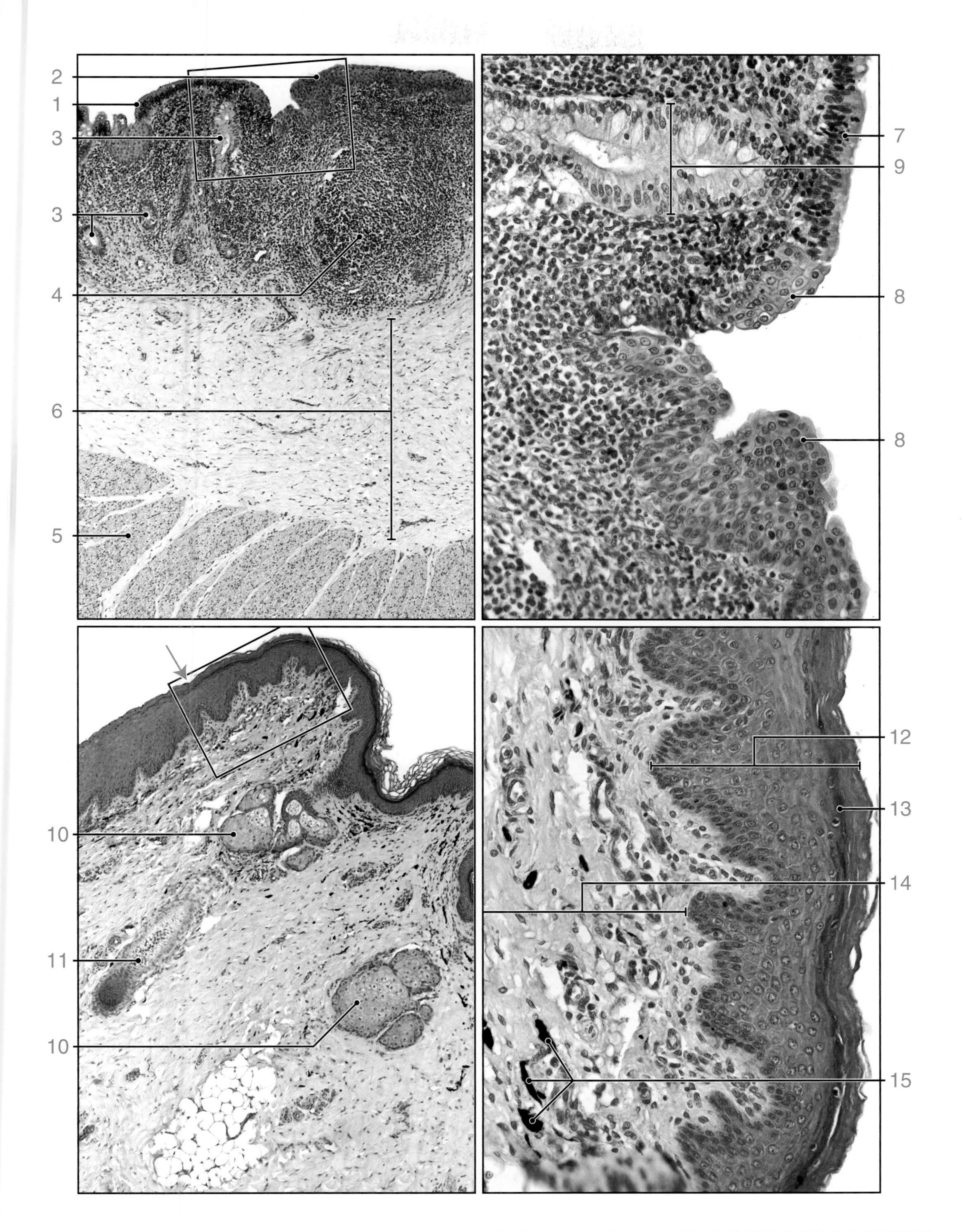

PLATE 94. ANAL CANAL AND ANOCUTANEOUS JUNCTION

CHAPTER 14

Digestive System III: Liver and Pancreas

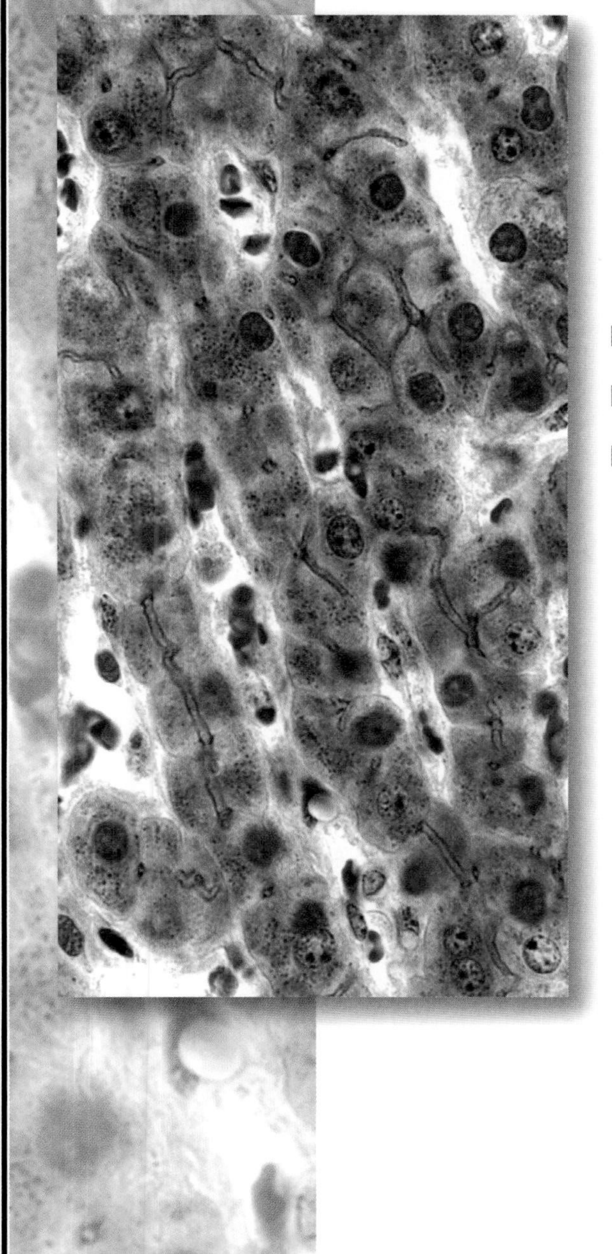

PLATE 95. LIVER I

The liver is the largest mass of glandular tissue in the body and the largest internal organ. It has an unusual blood supply, in that it receives most of its blood from the hepatic portal vein, which carries venous blood from the gastrointestinal tract, pancreas, and spleen. Thus, the liver is directly in the pathway that conveys materials, absorbed in the intestine, to the rest of the body. This gives the liver the first exposure to metabolic substrates and nutrients; it also makes the liver the first organ exposed to noxious and toxic substances absorbed from the intestine. One of the major roles of the liver is to degrade or conjugate toxic substances to render them harmless. It can, however, be seriously damaged by an excess of such substances.

Each liver cell, or hepatocyte, has both exocrine and endocrine functions. The exocrine secretion of the liver is called bile; bile contains conjugated and degraded waste products that are delivered back to the intestine for disposal. It also contains substances that bind to metabolites in the intestine to aid absorption. A series of ducts of increasing diameter, beginning with canaliculi between individual hepatocytes and ending with the common bile duct, deliver bile from the liver and gallbladder to the duodenum.

The endocrine secretions of the liver are released directly into the blood that supplies the liver cells and include albumin, nonimmune α- and β-globulins, prothrombin, and glycoproteins, including fibronectin. Glucose, released from stored glycogen, and triiodothyronine (T_3), the more active deiodination product of thyroxine, are also released directly into the blood.

Functional units of the liver, described as lobules or acini, are made up of regular interconnecting sheets of hepatocytes, separated from one another by the blood sinusoids (see Plate 96 regarding parenchymal organization and functional units).

Liver, human, H&E, x75.

At the low magnification shown here, large numbers of hepatocytes appear to be uniformly distributed throughout the specimen. The hepatocytes are arranged in one cell thick plates, but when sectioned, they appear as interconnecting cords, one or more cells thick, depending on the plane of section. The sinusoids appear as light areas between the cords of cells. Blood flows along the sinusoids to the **central vein** (1). The sinusoids and central vein are more clearly shown at higher magnification in the micrograph below.

Also present in this figure is a portal canal. A portal canal is a connective tissue septum that contains the branches of the **hepatic artery** (2) and **portal vein** (3); **bile ducts** (4); and lymphatic vessels and nerves. The artery and vein, along with the bile duct, are collectively referred to as a portal triad.

The hepatic artery and the portal vein are easy to identify because they are found in relation to one another within the surrounding connective tissue of the portal canal. The vein is typically thin walled; the artery is smaller in diameter and has a thicker wall. The bile ducts are composed of a simple cuboidal or columnar epithelium, depending on the size of the duct. Multiple profiles of the blood vessels and bile ducts may be evident in the canal either because of their branching or their passage out of the plane of section and then back in again.

Liver, human, H&E, x75; inset x 25.

The **central veins** (5) are the most distal radicals of the hepatic vein, and like the hepatic vein, they travel alone. Their distinguishing features are the **sinusoids** (6), which penetrate the wall of the vein, and the paucity of surrounding connective tissue. These characteristics are shown to advantage in Plate 96. Both the sinusoids and the vein are lined by a discontinuous sheet of **endothelial cells** (7) interspersed with **Kupffer cells** (8).

It is best to examine low-magnification views of the liver to define the boundaries of a lobule. A lobule is best identified when it is cut in cross-section. The central vein then appears as a circular profile, and the hepatocytes appear as cords radiating from it. Such a lobule is outlined by the *dashed line* in the top micrograph.

The limits of the lobule are defined, in part, by the portal canal. In other directions, the plates of the lobule do not appear to have a limit; that is, they have become contiguous with plates of an adjacent lobule. One can estimate the dimensions of the lobule, however, by approximating a circle with the central vein as its center and incorporating those plates that exhibit a radial arrangement up to the point where a portal canal is present. If the lobule has been cross-sectioned, the radial limit is set by the location of one or more of the portal canals.

The **inset** shows the vessel through which blood leaves the liver, the **hepatic vein** (9). It is readily identified because it travels alone and is surrounded by an appreciable amount of **connective tissue** (10). A small **lymphatic nodule** (11) is seen in the connective tissue surrounding the vein.

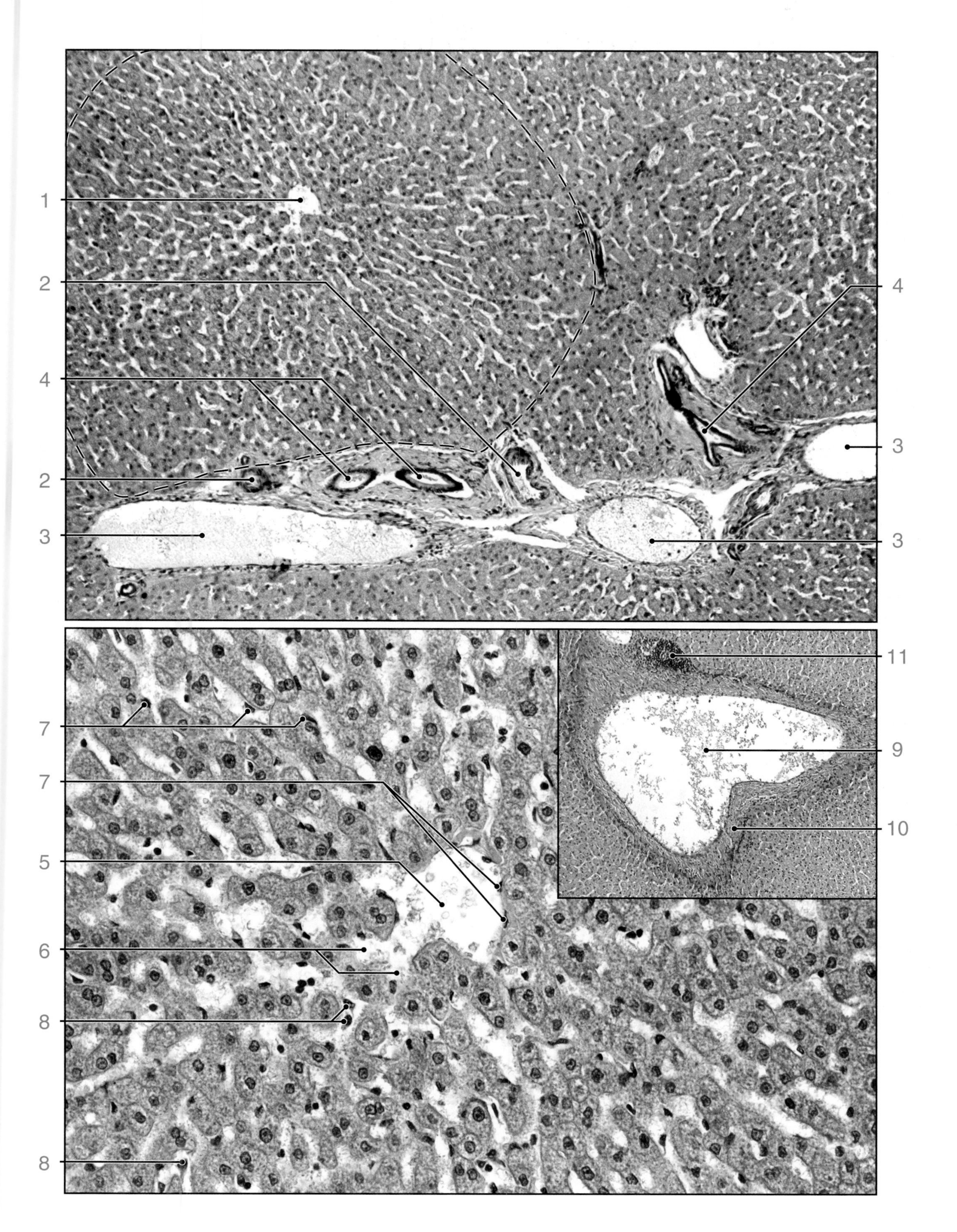
1
2
4
2
3
4
3
3
7
7
5
6
8
8
11
9
10

PLATE 96. **LIVER II**

The structure of the liver in terms of a functional unit is described in three ways: the classic lobule, the portal lobule, and the liver acinus. The traditional way to describe the organization of the liver parenchyma is the classic lobule, which is a roughly hexagonal mass of tissue. It is based on the distribution of the branches of the portal vein and hepatic artery and the pathway that blood flows as it perfuses the liver cells (see diagram). It can be visualized as plates of hepatocytes, one cell thick, separated by a system of sinusoids that perfuse the cells with mixed portal and arterial blood. At the center of this lobule is a relatively large vein, the central vein, into which the sinusoids drain.

The portal lobule emphasizes the parenchyma in terms of its exocrine functions. It defines a roughly triangular block of tissue that includes portions of three classic lobules (see diagram). The concept of the portal lobule emphasizes hepatic parenchymal structure comparable to that of other exocrine glands.

The liver acinus is a lozenge-shaped mass of tissue that represents the smallest functional unit of the hepatic parenchyma. It provides the best correlation between blood perfusion, metabolic activity, and liver pathology. The short axis of the acinus is defined by the terminal branches of the portal canal that lie along the border between two classic lobules (see diagram). The long axis is a line drawn between the two central veins closest to the short axis. The liver acinus is further described as three concentric elliptical zones surrounding the short axis. The zones are numbered 1 to 3 based on their distance from the short axis (see zones of liver acinus in enlarged diagram). Cells in zone 1 are the first to receive oxygen, nutrients, and toxins from the sinusoidal blood and the first to show morphologic changes. They are the last to die if circulation is impaired and the first to regenerate. Conversely, cells in zone 3 are the first to show ischemic necrosis in situations of reduced perfusion and the first to show fat accumulation. They are the last to respond to toxic substances. Cells in zone 2 have functional and morphological characteristic and responses intermediate to those of zones 1 and 3.

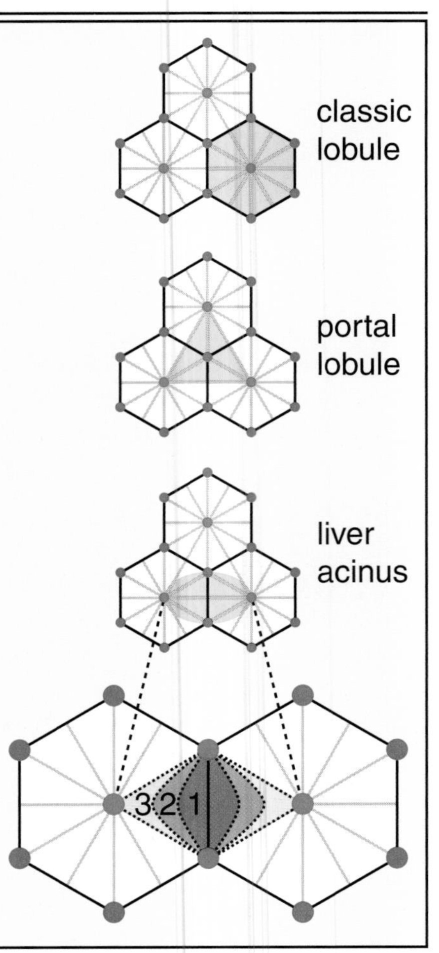

Liver, human, H&E, x870.

This micrograph is a higher magnification of the bottom micrograph seen in Plate 95. The center space is the **central vein** (1). It is collecting blood from the **sinusoids** (2). At two sites, the sinusoids (*arrows*) can be seen emptying into the central vein. The central vein is lined by endothelial cells whose **nuclei** (3) are readily apparent. Between the endothelium and the liver parenchymal cells is a thin layer of **connective tissue** (4). The **endothelial cells of the sinusoids** (5) are identified by their flattened nuclei. Their cytoplasm, as noted previously, is not recognizable. The **Kupffer cells** (6) are distinguished by their relatively large nuclei, their apparent location within the sinusoid, and often the cytoplasm surrounding the nucleus. It should also be noted that some hepatic parenchymal cells are **binucleate** (7), as evidenced by several examples in this micrograph.

Liver, bile canaliculi, human, iron hematoxylin, x1140.

This specimen was stained to demonstrate the canalicular bile system. The micrograph shows a number of hepatic plates. The anastomosing character of the plates is seen in the upper part of the micrograph. The lower portion of the micrograph shows a **bile canaliculus** (8) in one of the plates that was longitudinally sectioned. It reveals the linear direction of the canaliculus as it lies between opposing cells in the plate. At several points, circular profiles (9) are seen that represent canaliculi arranged at right angles to and joining the longitudinally sectioned canaliculus. In another site, where the hepatocytes within a plate have been sectioned in an anastomosing area, it can be seen that the **bile canaliculi** (10) completely surround the hepatocyte. In effect, the bile canaliculi are formed by the indentation of adjoining hepatocytes. Other features worth noting in this micrograph are **endothelial cells** (11), lining the sinusoids, and occasional **Kupffer cells** (12), located in the sinusoids.

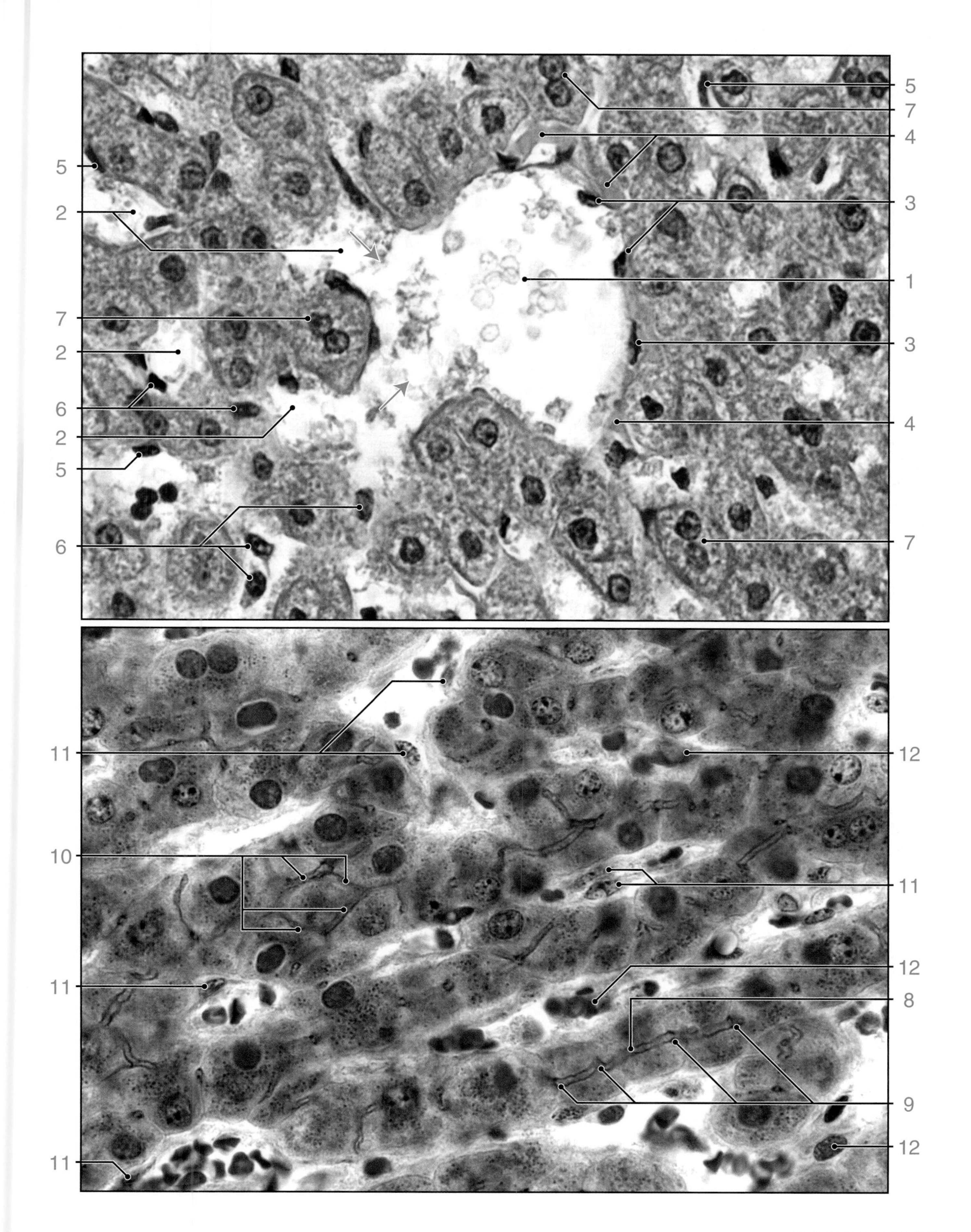
5
7
4
3
1
3
4
7
5
2
7
2
6
2
5
6
11
12
10
11
11
12
8
9
11
12

PLATE 97. LIVER III AND FETAL LIVER

Liver, rat, glutaraldehyde-osmium fixation, toluidine blue, x900.

This micrograph shows a plastic-embedded liver specimen that was prepared by a method used for electron microscopy. In contrast to a typical H&E-stained preparation, it demonstrates to advantage the cytologic detail of the hepatocyte and the sinusoidal structure. The cytoplasm of the hepatocytes is colored blue by the toluidine blue stain. However, careful examination reveals small, irregular, magenta-colored masses. Glutaraldehyde fixation has retained the **glycogen** (1) stored in the cell and the toluidine blue stain has colored the glycogen magenta. In contrast, the glycogen is lost in typical H&E preparations, creating empty-appearing spaces within the cell. Also evident in the cytoplasm are **lipid droplets** (2), which are seen in varying sizes and numbers in many of the cells. The lipid has been preserved and stained black by the osmium used as the secondary fixative. The quantities of lipid and glycogen are variable and under normal conditions reflect dietary intake. Careful examination of the hepatocyte cytoplasm also reveals small, punctate, dark blue bodies in contrast to the lighter blue background of the cell. These are mitochondria. Another feature of this specimen is the clear representation of the **bile canaliculi** (3) located between adjacent liver cells. Most of the bile canaliculi appear as circular profiles, which indicate cross sections of the canaliculi. At several sites (4), the bile canaliculus is seen as a somewhat elongate structure, indicating that the canaliculus has been sectioned in an oblique or longitudinal plane. A number of very prominent cells are seen in the sinusoids. These are the **Kupffer cells** (5). They exhibit a relatively large nucleus and extensive cytoplasm. The surface of the Kupffer cell exhibits a very irregular or jagged contour because of the numerous processes that provide the cell with an extensive surface area. The **endothelial cells** (6) have a considerably smaller nucleus, their cytoplasm is much more attenuated, and they present a relatively smooth contour. A third cell type, less frequently observed, the perisinusoidal lipocyte (Ito cell), is not seen in this micrograph. These cells contain numerous lipid droplets. The droplets contain stored vitamin A. In this specimen, some of the sinusoids contain **red blood cells** (7), which have stained blue.

Liver, human, H&E, x175; inset x700.

The liver is the principal site of hemopoiesis beginning in the third month of gestation. The process continues until shortly before birth, but its potential is preserved even after birth. If there is a loss of the bone marrow's ability to produce blood cells, caused, for example, by malignancy or fibrous tissue formation, hemopoiesis can resume to some degree. The specimen shown in this micrograph reveals hemopoiesis in a late fetal liver. The structural organization of the liver parenchyma at this stage is essentially the same as that found during post-natal life. The parenchyma is arranged in typical chords of cells, between which are the sinuses that empty into the **central vein** (8). Careful examination of the organ, even at this low magnification, reveals clumps of small, round, densely stained nuclei. These are mostly nuclei of developing **erythrocytes** (9). The **inset** reveals these cells at higher magnification. Although it is difficult to discern, these cells are located between the liver parenchymal cells and the endothelium lining the vascular sinuses.

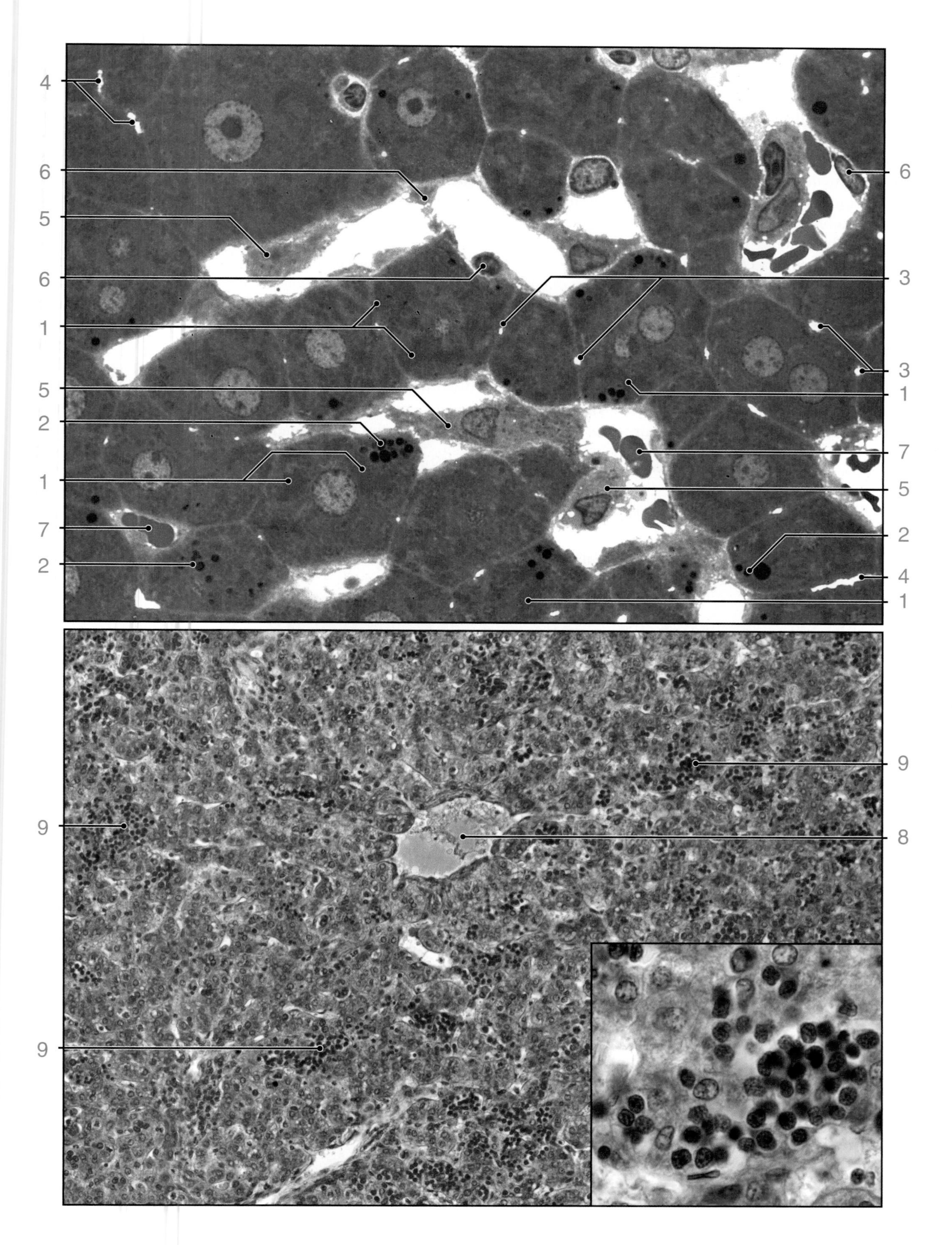

PLATE 97. LIVER III AND FETAL LIVER

CHAPTER 15

Respiratory System

PLATE 103. OLFACTORY MUCOSA

PLATE 109. TERMINAL AND RESPIRATORY BRONCHIOLES, ALVEOLUS

Alveoli, whether present in the respiratory bronchioles, alveolar ducts, or alveolar sacs, lead into surrounding alveolar spaces. They represent the terminal air spaces of the respiratory system and thus serve as the actual sites of gas exchange between the air and the blood. While alveoli often appear as closed spaces, small openings in the alveolar wall or septum, referred to as alveolar pores, permit circulation between alveoli. The septum is made up of an epithelial lining on both of its surfaces, and interposed between epithelial cells are the pulmonary capillaries. In regions where gas exchange occurs, the septum is very thin. At such sites, the type I pneumocyte is immediately adjacent to the pulmonary capillary. Electron microscopy reveals that the basement membrane (lamina) is fused with that of the capillary, leaving no intervening space. In regions where the septum is relatively thick, there are varying amounts of connective tissue containing collagen fibrils, thin elastic fibers, fibroblasts, macrophages, and occasional eosinophils. The macrophages function in the connective tissue of the septa as well as in the alveoli. They enter the alveoli, where they scavenge the surface of the septa to remove inhaled particulate matter (e.g., dust and pollen).

The lining of the alveolus contains three types of epithelial cells: type I pneumocytes, type II pneumocytes, and brush cells. Type I pneumocytes are exceedingly thin squamous cells that line most (95%) of the surface of the alveoli. Type II pneumocytes, also called septal cells, are secretory cells. They are cuboidal in shape and are interspersed among the type I cells. They are as numerous as type I cells, but because of their shape, they cover only about 5% of the alveolar air surface. They secrete surfactant, a surface-active agent that coats the pneumocytes. The surfactant layer reduces the surface tension at the air-epithelium interface and prevents collapse of the alveoli upon exhalation. Unlike the type I pneumocyte, which is not capable of cell division, type II pneumocytes also serve as progenitor cells for type I pneumocytes. In response to lung injury, they are able to proliferate and restore both types of pneumocyte. The brush cells are few in number and are best identified by their short, blunt microvilli, recognizable in the electron microscope. It is thought that they act as receptors that monitor air quality in the lung.

Lung, terminal bronchiole, human, H&E, x515.

This micrograph shows the wall of a terminal bronchiole at a higher magnification than that shown in Plate 108. The epithelium is reduced in height and changes from a **ciliated pseudostratified epithelium** (1) to a **nonciliated simple cuboidal epithelium** (2), consisting of Clara cells. Careful examination of the region containing the ciliated cells reveals occasional, interspersed **Clara cells** (3). They are identified by their lack of cilia. Bundles of circumferentially arranged **smooth muscle cells** (4) are typically present in the underlying connective tissue.

Lung, Terminal bronchiole and alveolar duct, human, H&E, x35.

This very low power micrograph shows lung tissue and part of the **lung surface** (5). The *asterisk* denotes an area in which the lung tissue has been damaged and the surface is missing. The upper portion of the micrograph reveals a **respiratory bronchiole** (6). One branch of the respiratory bronchiole continues in the plane of section as an **alveolar duct** (7). They are the terminal segments of the airway, having almost no walls, only alveoli surrounding the duct space. They open into a space referred to as an **alveolar sac** (8).

Lung, alveolar sac, human, H&E, x255.

The **alveolar sacs** (9) shown here are characterized by an irregular space surrounded by **alveoli** (10). Included in this micrograph is the surface of the lung, showing its simple squamous **mesothelial cells** (11) and underlying **connective tissue** (12).

Lung, alveoli, human, H&E, x515; inset x1550.

This micrograph shows several **alveoli** (13) at higher magnification. The wall of the alveolus is made up mostly of thin, type I pneumocytes with occasional **type II pneumocytes** (14), identified by their large, round nuclei. Capillaries are present between the walls of adjacent alveoli. The **inset** is a higher magnification of the boxed area, showing the septum with a capillary between two adjacent alveoli. Three blood cells are seen within the capillary. The cytoplasmic components consist of type I pnumocyte cytoplasm and the endothelium of the capillary. It is at these very thin-walled sites that gas exchange occurs.

Lung, alveolar macrophages, human, H&E, x515; inset x1550.

Several **alveolar macrophages** (15) are shown in this micrograph. These cells exhibit a large nucleus and a variable amount of surrounding cytoplasm. They occur in the connective tissue of the alveolar interseptal wall, but are best recognized when they enter the alveolar space. The **inset** shows the alveolar macrophage from the boxed area of the micrograph at higher magnification. The nucleus exhibits a small indentation in its surface. The cytoplasm reveals several small, round inclusions.

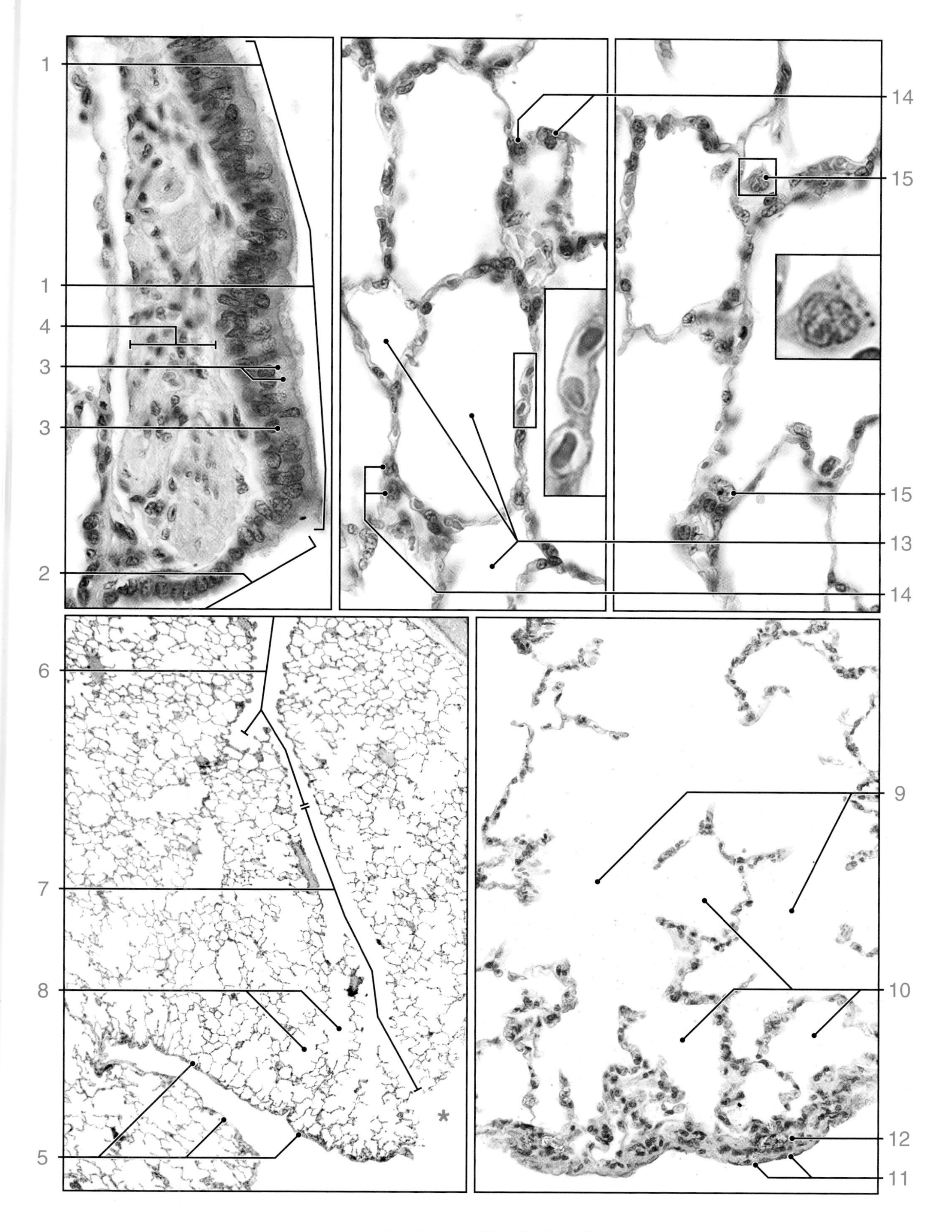

PLATE 109. TERMINAL AND RESPIRATORY BRONCHIOLES, ALVEOLUS

CHAPTER 16

Urinary System

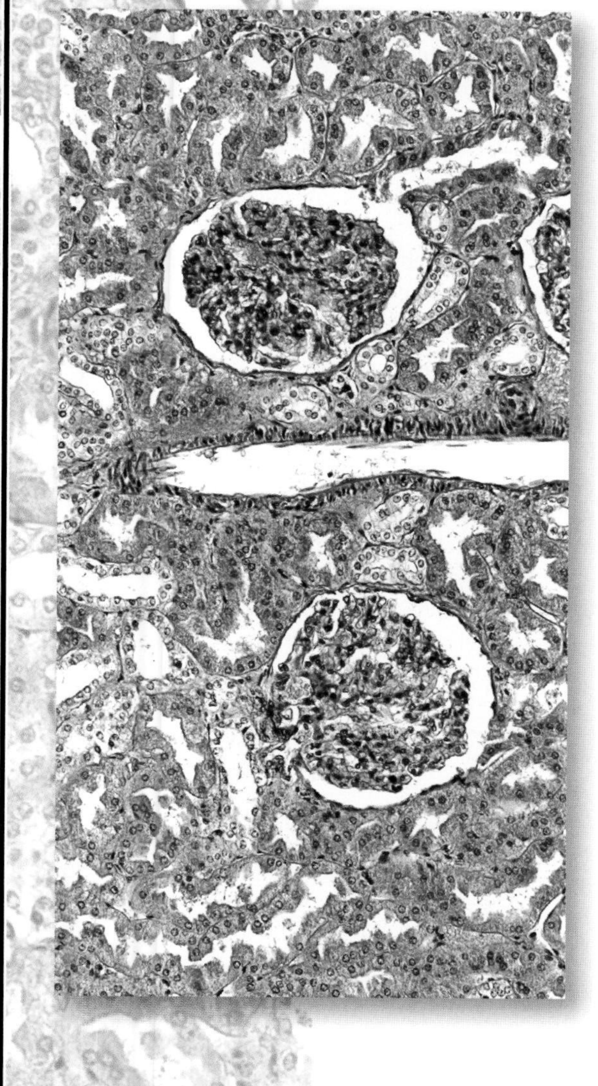

PLATE 110. KIDNEY I

The kidneys are bean-shaped, paired organs whose functions include the conservation of body fluid and electrolytes; the removal of metabolic wastes through filtration of the body's blood; and modification of the filtrate through selective resorption and specific secretion by a system of tubules that constitutes the bulk of the kidney. The kidneys also function as an endocrine organ. They synthesize and secrete erythropoietin, a regulator of red blood cell production; and renin, an enzyme involved in the control of blood volume and pressure. The kidneys are also responsible for hydroxylation of an inactive form of vitamin D_3 to the highly active form, 1,25-$(OH)_2$. Histologically, the kidney exhibits a complex array of tubules, an extensive vascular system and ducts, that collect, modify, and deliver urine into the ureter for discharge from the body.

ORIENTATION PHOTOGRAPH: The specimen shown here is a fresh, hemisected human kidney. It reveals two distinctive regions: a reddish-brown, outer region, the **cortex** (1), and an inner portion, the **medulla** (2), that changes to a lighter color in its deeper part. The medulla can be further subdivided into an **outer medulla** (3), which in the fresh state has a deep red color, and an **inner medulla** (4), which is lighter in color. In a histologic preparation, this distinction of outer and inner medulla is reflected in the arrangement of the tubules and vascular supply. At the boundary between cortex and medulla there is a system of blood vessels, the **arcuate arteries** and **veins** (5), that parallel the surface of the kidney and give off ascending and descending branches to supply the cortex and the medulla.

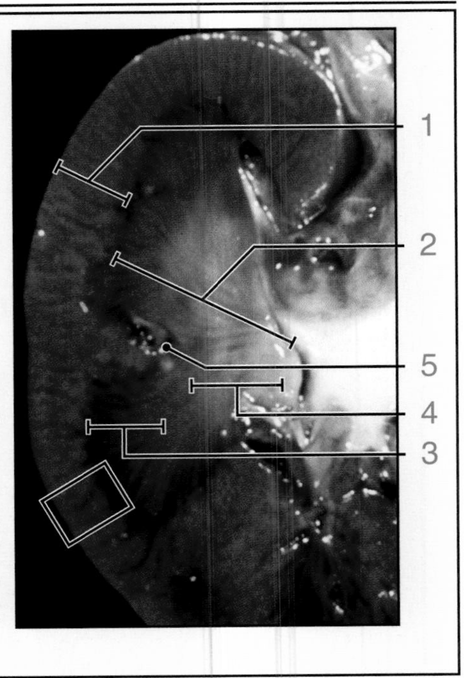

Kidney, human, H&E, x25; inset x350.

This low magnification micrograph is comparable to the boxed area on the orientation photograph. The essential features shown here include a relatively thin **capsule** (1) that surrounds the organ, the **cortex** (2) and the **outer medulla** (3). The cortex displays several structural or organizational features that characterize this part of the kidney. Note the linear arrayed structures referred to as **medullary rays** (4). They are so named because they appear as raylike structures that emanate from the medulla. The medullary rays consist of straight tubules. Because the plane of section in a sagittal section such as this may not be precisely aligned with a medullary ray, this structural unit may not appear to extend all the way to the periphery of the cortex. Also, the medullary ray narrows as it extends towards the cortex, a reflection of the number of straight tubules within the medullary ray that are present at any given level through the cortex. Between the medullary rays are the **cortical labyrinths** (5). The cortical labyrinths consist of numerous tortuous tubules that are continuous with the straight tubules of the medullary rays. The spherical structures, known as **renal corpuscles** (6), are located in the cortical labyrinth and are also continuous with the tortuous tubules. Each corpuscle contains a capillary network that is responsible for filtration of the blood. At the cortical medullary boundary an **arcuate vein** (7) and an **arcuate artery** (8) can be seen. The ascending branches from these vessels consist of the **interlobular veins** (9) and the **interlobular arteries** (10). The **inset** shows the capsule at higher magnification. Of significance is the nature of the capsule, namely, it consists of an outer layer containing collagen fibers and **fibroblasts** (11) typical of capsules seen in other organs. However, the inner half of the capsule is more cellular and its cells have the character of **smooth muscle cells** (12). Its functional role, however, is not known.

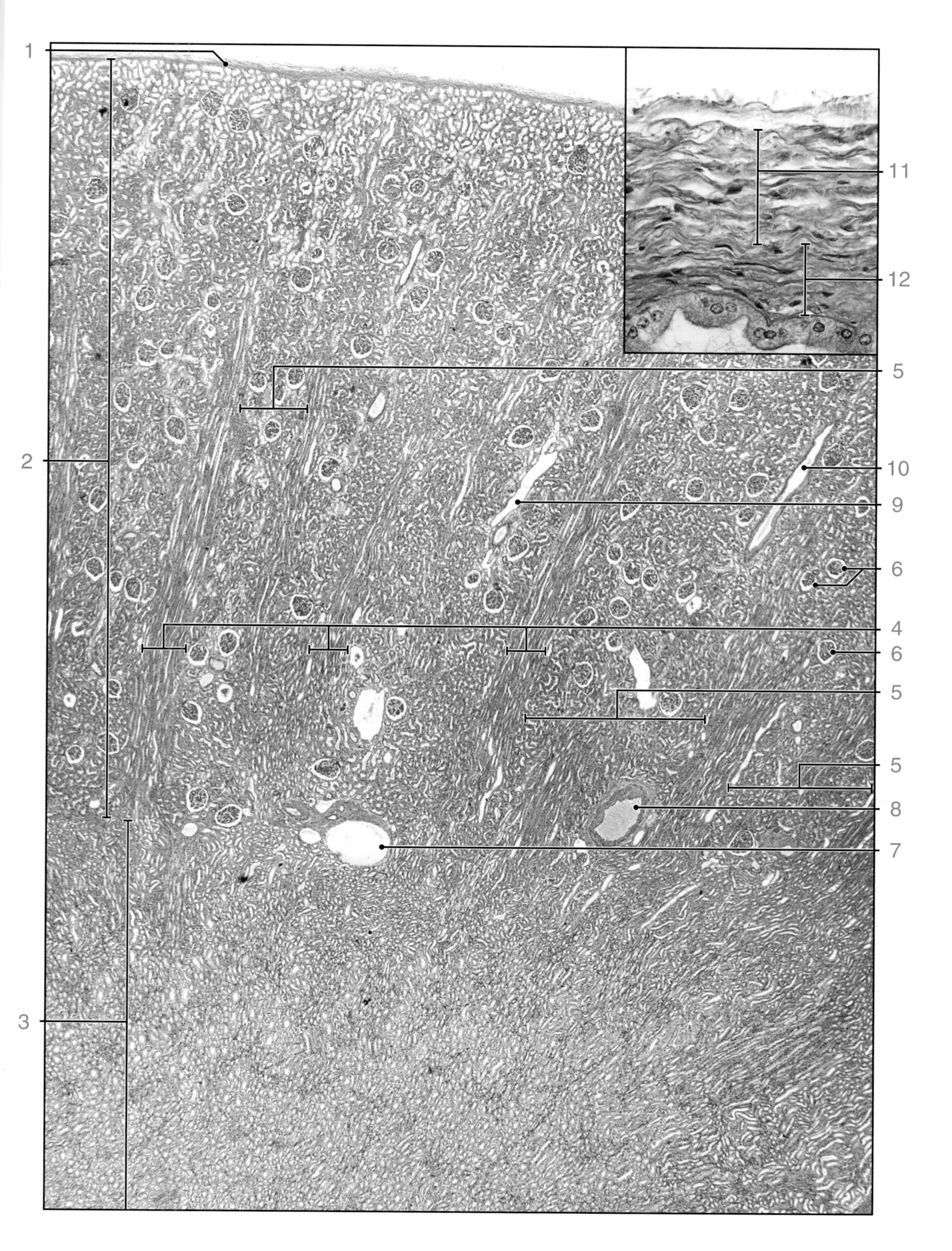
1
2
3
11
12
5
10
9
6
4
6
5
5
8
7

PLATE 111. **KIDNEY II**

The nephron is the structural and functional unit of the kidney. Nephrons are responsible for the production of urine and can be compared to the secretory components of other glands. The collecting ducts, which drain the nephrons, are responsible for the final concentration of the urine and are comparable to the ducts of exocrine glands that modify the concentration of the secretory product. Each of the approximately 2 million nephrons in a kidney begins with a glomerulus, a complex tuft of capillaries surrounded by a ball-shaped layer of simple squamous epithelial cells, referred to as Bowman's capsule, that becomes continuous with the epithelial cells that make up the proximal convoluted tubule. The proximal convoluted tubule is a highly tortuous structure and, because of its length and tortuosity, makes up the greater proportion of the cortical labyrinth. After an extensive series of repetitive tight turns, the tubule follows a straight course as it enters the medullary ray and descends into the medulla as the descending limb of the loop of Henle. It then makes a sharp turn upon itself and ascends as a straight tubule within the medullary ray, the ascending limb of the loop of Henle. The tubule then leaves the medullary ray to return to the renal corpuscle from which the nephron originated and undergoes another series of convoluted turns within the cortical labyrinth. This part of the nephron, the distal convoluted tubule, is less tortuous than the proximal. The distal convoluted tubule then joins with a collecting tubule, which then connects to a collecting duct located in the medullary ray.

ORIENTATION MICROGRAPH: This micrograph shows the outer half of the cortex from the same specimen as shown in Plate 110.

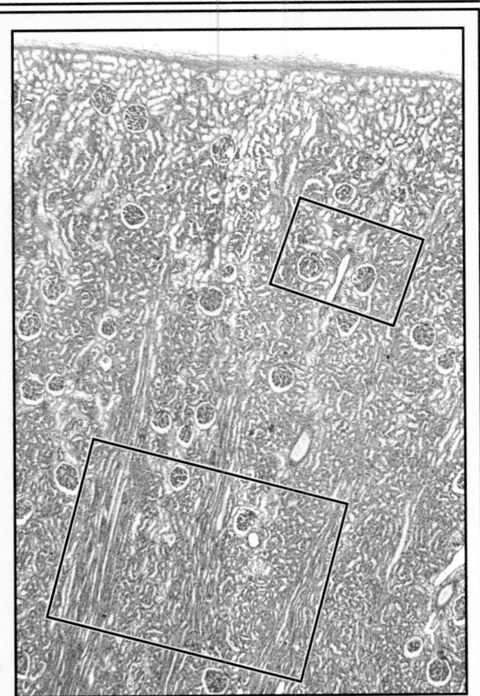

Kidney, human, H&E, x90.
The lower boxed area on the orientation micrograph is shown here at higher magnification. Included in the micrograph are three profiles of **medullary rays** (1). Between them are the **cortical labyrinths** (2). In comparing the glandular nature of the kidney, it should be pointed out that each medullary ray, along with half of the cortical labyrinth that surrounds that medullary ray, constitutes a **renal lobule** (3). The artery and vein that supply the structures within the cortical labyrinth, namely the interlobular artery and vein, course through the center of the cortical labyrinth. A single profile of an **interlobular artery** (4) and an **interlobular vein** (5) are seen in the cortical labyrinth on the right.

Kidney, human, H&E, x180.
This micrograph shows the cortical labyrinth from the upper boxed area on the orientation micrograph. At the center of the micrograph is an **interlobular artery** (6). Above it is a profile of the accompanying **interlobular vein** (7). Both are branches of the arcuate artery and vein that lie at the boundary between cortex and medulla. These vessels are located midway between adjacent medullary rays. Although the boundaries between lobules are not distinct, the vessels course between lobules, as their name indicates. Thus, the **renal corpuscles** (8), as seen here, lie in adjacent kidney lobules. Their blood supply comes from afferent arterioles, which are branches of the interlobular artery. The remaining components of the cortical labyrinth consist of proximal convoluted tubules and distal convoluted tubules. The **proximal convoluted tubules** (9) are readily recognized in this micrograph by their dark, eosinophilic-staining cytoplasm as opposed to the lighter staining cytoplasm of the **distal convoluted tubules** (10). Because the proximal convoluted tubule follows a more convoluted course than the distal convoluted tubule, a significantly greater number of sectioned profiles of the proximal convoluted tubule are seen compared to the distal convoluted tubule. These tubules are examined in greater detail in Plate 112.

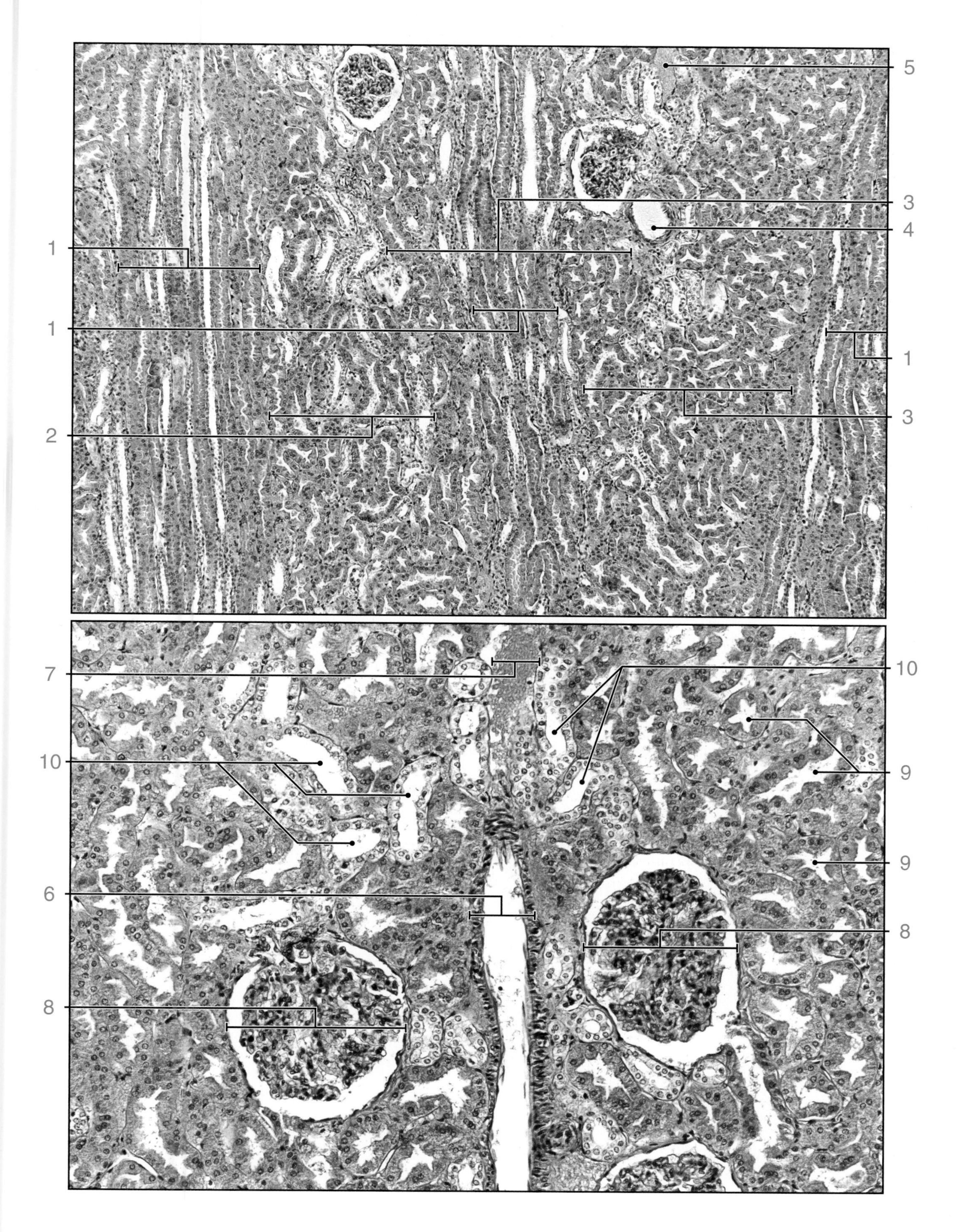
5
3
4
1
1
1
3
2
10
7
10
9
9
6
8
8

PLATE 112. **KIDNEY III**

Kidney, proximal and distal convoluted tubules, human, H&E, x240, inset x450.

This micrograph reveals a section through the cortical labyrinth cut parallel to the surface of the kidney. Thus, this section is perpendicular to the section shown in the micrographs on Plate 111. The tubules observed in this micrograph consist of **proximal convoluted tubules** (1) and **distal convoluted tubules** (2). Note that the proximal convoluted tubules appear to be greater in number than the distal convoluted tubules. Although there are an equal number of both types of tubules, the discrepancy seen in a section of cortical labyrinth is due to the fact that the proximal convoluted tubule is considerably more tortuous than the distal, and thus there are more cross-sectional profiles of the former. This, in part, helps to distinguish between the two types of tubules. Other differences include the generally larger diameter of the proximal convoluted tubule and the larger size of its cells. Because of the larger size of the proximal tubule cells, the nucleus of each cell making up the tubule wall may not be included in the section through the cell, thus the nuclei appear to have a very irregular distribution. In comparison, the cells of the distal convoluted tubule, being considerably smaller, have most of their nuclei included within the section, and thus they appear more uniformly distributed. Also, the lumen of the proximal convoluted tubule, when observed in cross-section, has an irregular star-shaped lumen, whereas the distal convoluted tubule has a more round lumen. Another feature characteristic of the proximal convoluted tubule that can be observed in very well preserved specimens is the presence of the **brush border** (3) (**inset**) at the apical surface of the tubule cells. This is a reflection of the microvilli at the apical surface of these cells. Lastly, the cytoplasm of the proximal convoluted tubule cells stain more intensely with eosin than do distal convoluted tubule cells.

Kidney, proximal and distal straight tubules, human, H&E, x240.

This micrograph is a section cut parallel to the surface of the kidney through a medullary ray. The distinction between **proximal straight tubules** (4) and **distal straight tubules** (5) is readily apparent. Compared to proximal and distal convoluted tubules, the only significant difference is that the proximal and distal straight tubules have a slightly smaller diameters. The more significant feature in this micrograph is that the number of proximal straight tubule profiles is equal to the number of distal profiles. Two of the tubules in the micrograph are **collecting tubules** (6). Careful examination of these tubules reveals that **cell boundaries** (7) are evident between the collecting tubule cells. In contrast, proximal and distal tubules, whether straight or convoluted, do not show a boundary between cells in the light microscope. Also present in this micrograph is a **proximal convoluted tubule** (8). Note the turn that it makes. It is part of the cortical labyrinth.

Kidney, renal corpuscle, human, H&E, x350.

The renal corpuscle shown here has been cut in a plane that includes the **vascular pole** (9) and shows one of the blood vessels, either an afferent or efferent arteriole that has entered the corpuscle. Another feature that can be identified is the simple squamous epithelium of Bowman's capsule. Between **Bowman's capsule** (10) and the **glomerulus** (11) is the **urinary space** (12). The nuclei that are seen associated with the glomerulus belong to three cell types. They consist of endothelial cells of the glomerular capillaries; podocytes that lie against the outer wall of the glomerular capillaries and are enclosed by the basal lamina of the capillary endothelial cells; and mesangial cells that lie along the vascular stalk of the glomerulus and at the interstices of adjoining glomerular capillaries. Adjacent to the entry and exit of the afferent and efferent arterioles is the terminal portion of the distal straight tubule. At this site, the wall of the tubule contains cells of the **macula densa** (13). The nuclei of these cells are tightly packed and typically appear partially superimposed over one another.

Kidney, renal corpuscle, human, H&E, x350.

The renal corpuscle shown here has been cut in a plane that includes the urinary pole. At the urinary pole, the wall of Bowman's capsule becomes continuous with a **proximal convoluted tubule** (14). Thus, the filtrate that empties into the urinary space is able to flow into the proximal convoluted tubule. Like the previous micrograph, this glomerulus reveals many nuclei, most of which are difficult to identify in terms of their cell type. Occasional nuclei are seen on the outer wall of the capillary facing the urinary space. These are most likely **podocyte nuclei** (15).

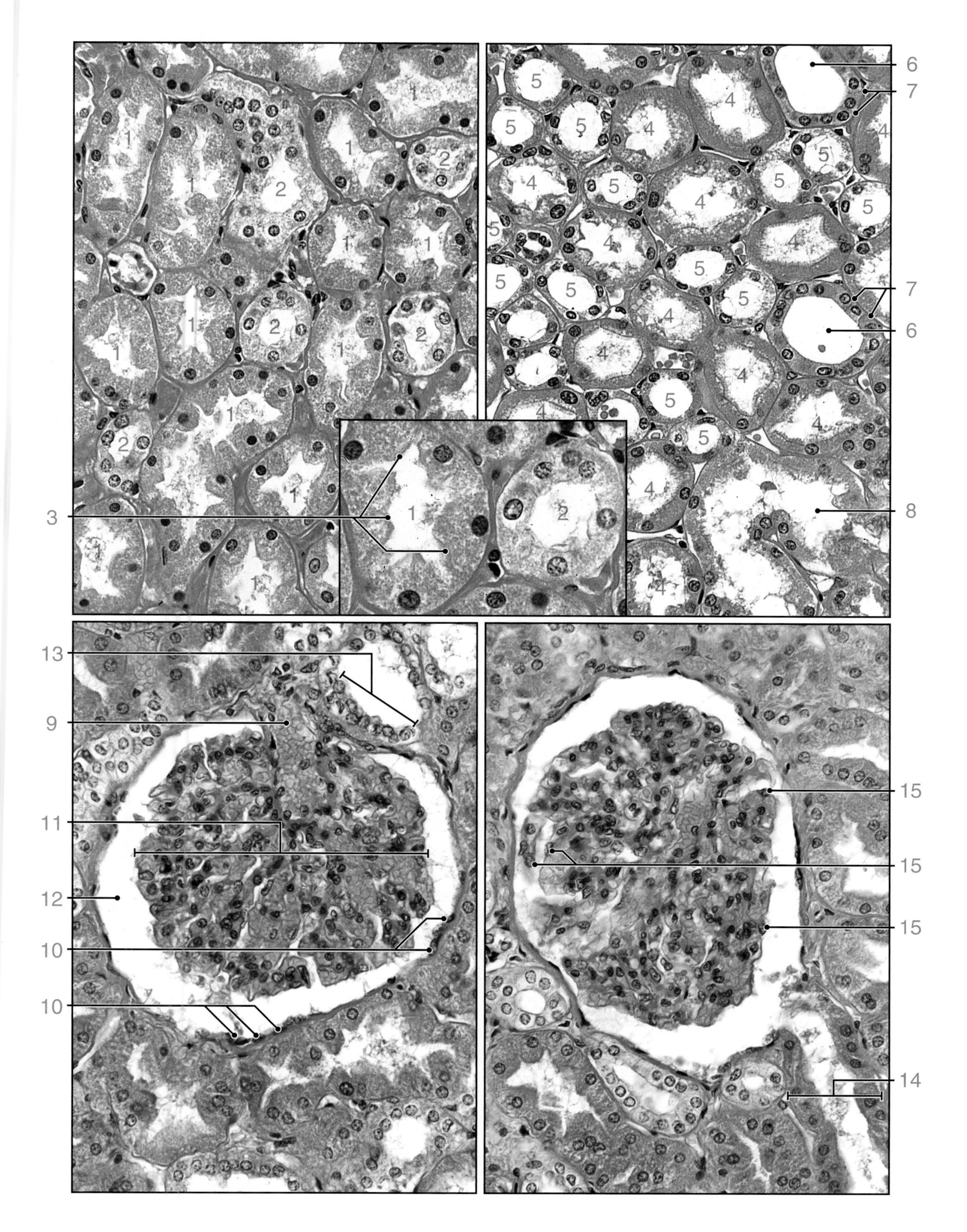

1
2
3
4
5
6
7
8
9
10
11
12
13
14
15

PLATE 114. **FETAL KIDNEY**

As indicated in Plate 113, the human kidney usually has 8 to 12 pyramids, but as many as 18 pyramids may be present. Each pyramid and the associated cortical tissue at its base and sides, namely, one half of each adjacent renal column, constitute a lobe of the kidney. This lobar organization of the kidney is conspicuous during fetal development. Each lobe appears as a convexity on the surface of the kidney, but usually disappears after birth. In some cases, the convexities may persist to some degree for a period of time after birth, occasionally into early adulthood. Examination of the fetal kidney reveals a sequential development pattern of the individual nephrons. Those nephrons in the deep portion of the cortex undergo initial development followed by nephrons located in progressively more superficial portions of the cortex. This is readily seen in the developmental state of the renal corpuscles associated with these nephrons. In late fetal life, the renal corpuscles have a relatively simple form. Some renal corpuscles maintain this form, particularly those in the subcapsular area of the cortex, up to a year following birth. By six years of age, the majority, if not all, of the renal corpuscles have a mature structure.

ORIENTATION MICROGRAPH: This micrograph shows a section of a **fetal kidney** (1) and the **adrenal gland** (2) adjacent to the superior pole of the kidney. The slice through this kidney reveals two hemisected **lobes** (3) that extend from the kidney surface to the papilla of the lobe.

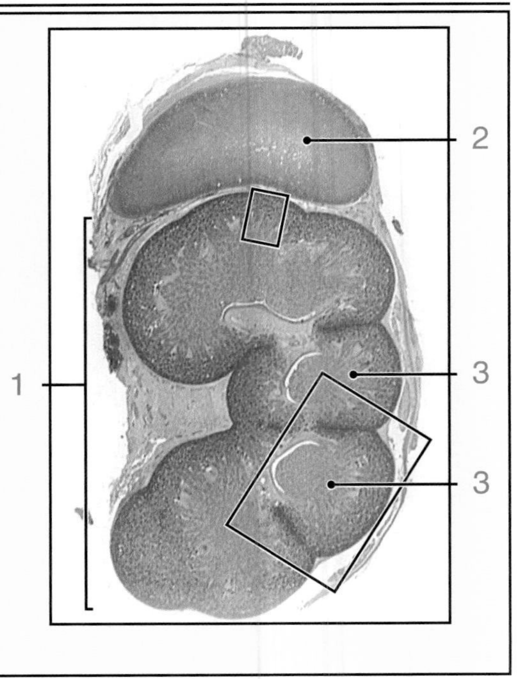

Kidney, fetal human, H&E, x90; inset x515.
This micrograph shows the lower boxed area on the orientation micrograph. It includes an entire lobe. The **cortex** (1) exhibits numerous small, round bodies, the developing renal corpuscles. Beneath this is the **medulla** (2), containing the developing ascending and descending limbs of the loops of Henle along with collecting ducts. The deepest portion represents a **medullary pyramid** (3). The collecting ducts within the pyramid are seen here in cross-section, due to the angle of the cut. They empty into the space below, the **minor calyx** (4). Lastly, the **renal columns** (5) can be seen on the lateral side of the lobule. The **inset** is a higher magnification of a cross section of a **proximal convoluted tubule** (6), a longitudinally sectioned **collecting tubule** (7), as well as a longitudinal section of what will become the thin segment of the **loop of Henle** (8).

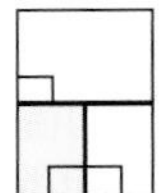

Kidney, fetal human, H&E, x65; inset x415.
The area in the upper boxed area on the orientation micrograph is shown here at higher magnification. It includes the **capsule** (9) of the kidney, the entire thickness of the **cortex** (10), and the upper portion of the **medulla** (11). The cortex exhibits a number of renal corpuscles in varying stages of development. Those in the deeper portion of the cortex are more advanced in their developmental stage than those in the upper portion of the cortex. The **inset** shows one of the renal corpuscles in a fairly advanced stage of development. The **urinary space** (12) is present, and the glomerulus has already formed, providing a functional structure.

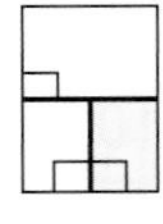

Kidney, fetal human, H&E, x255; inset x545.
The outer part of the cortex is shown at higher magnification in this micrograph. Note the **capsule** (13). The renal corpuscles at this level of the cortex, compared to those in the deeper portion of the cortex, are in a much earlier stage of development. The **inset** shows at higher magnification the boxed area on this micrograph. It clearly shows the beginning of the **proximal convoluted tubule** (14) of the renal corpuscle. Also evident at this stage of development is the parietal layer of **Bowman's capsule** (15), consisting of squamous cells, as well as the visceral layer of Bowman's capsule, which at this stage of development consists of **cuboidal cells** (16). Between the visceral layer of Bowman's capsule and the initial portion of the proximal convoluted tubule is the developing vascular part of the renal corpuscle, namely, the **glomerulus** (17). Compare this developing renal corpuscle with the renal corpuscle in inset of the bottom left micrograph.

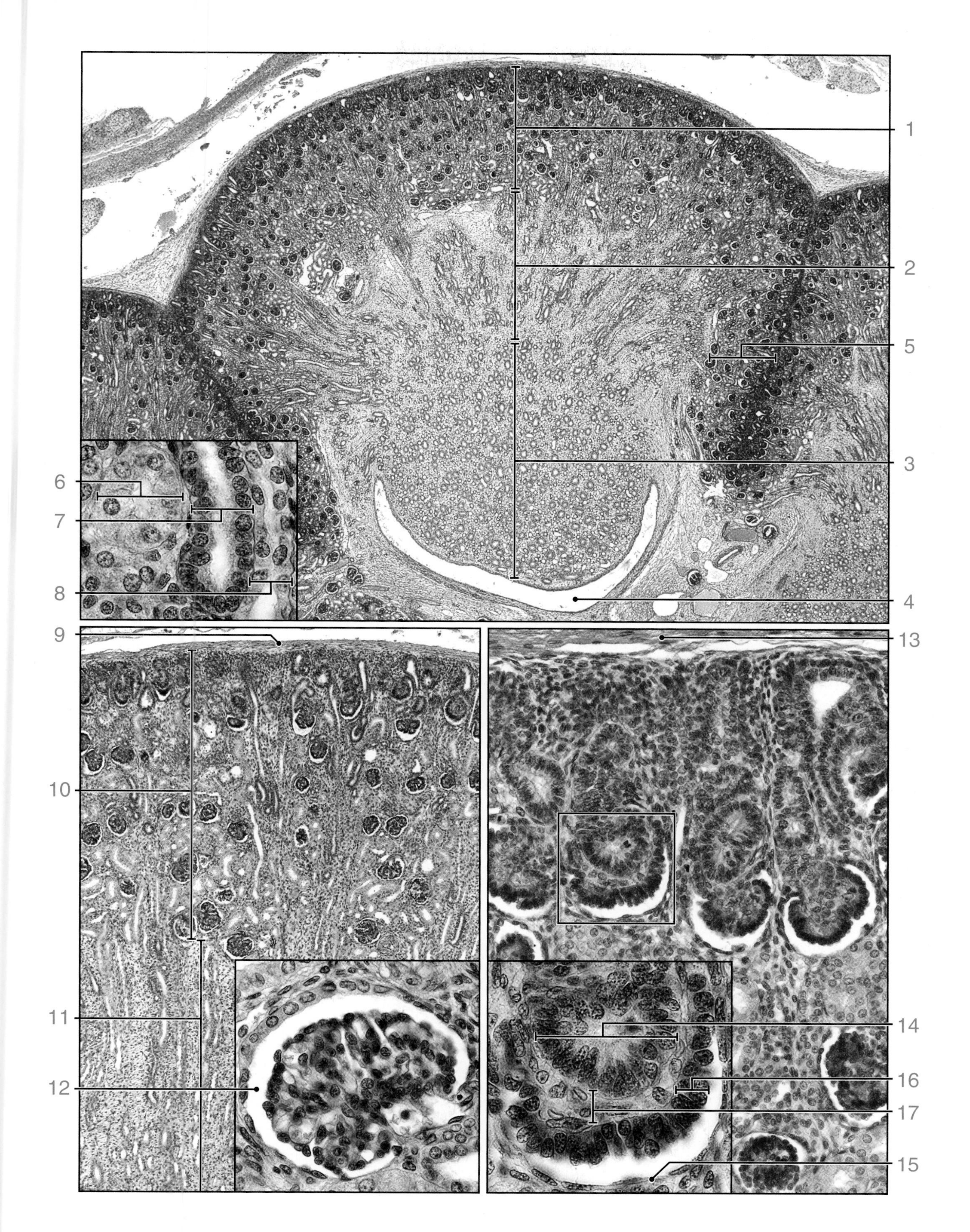

PLATE 114. FETAL KIDNEY

PLATE 115. URETER

The urine that leaves the renal papillae is collected in the minor calyces and passes from there to the renal pelvis where it enters a ureter. The paired ureters, one from each kidney, conduct the urine to the urinary bladder. Passage of the urine is assisted by peristaltic action of smooth muscle in the ureter wall. Each ureter lies in a retroperitoneal position, like the kidney, and is surrounded by adipose and connective tissue, which form the adventitia of the ureter. However, where the ureter is immediately adjacent to the abdominal cavity, a serosa may be identified as part of the ureter wall. The lining of the ureter consists of transitional epithelium (urothelium), which lies on a layer of dense connective tissue. The urothelium and underlying connective tissue constitute the mucosa. Between the mucosa and the outer component of the ureter (the adventitia) are bundles of smooth muscle that make up the muscularis. The smooth muscle is arranged throughout most of the length of the ureter as an inner longitudinal layer and a surrounding circular layer. The bundles of smooth muscle that make up these two layers tend to have an imperfect arrangement and sometimes it is difficult to determine a layers' orientation. This is due to a slight helical direction that the two layers follow. Near the bladder, a third, outer longitudinal layer of smooth muscle is added, which continues into the bladder to form a principle component of the bladder wall.

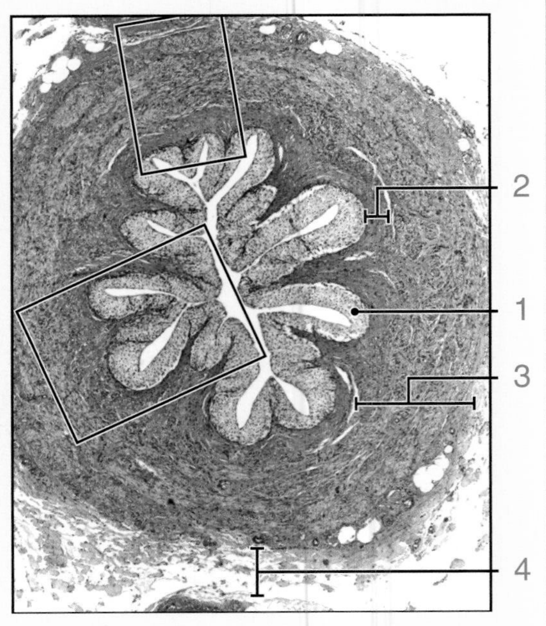

ORIENTATION MICROGRAPH: This micrograph reveals a cross section of a ureter near the bladder. The cytoplasm of the **urothelial cells** (1) appears lightly stained with eosin in this preparation. The adjacent **dense connective tissue** (2), in contrast, is heavily stained with eosin. The smooth muscle component of the ureter that makes up the **muscularis** (3) is seen between the dense connective tissue and the **adventitia** (4). Typically, the luminal surface of a cross-sectioned ureter exhibits multiple folds due to contraction of the smooth muscle in its wall.

Ureter, human, H&E, x180; inset x565.

This micrograph is a higher magnification of the lower boxed area on the orientation micrograph. Note the **transitional epithelium** (1) lining the lumen and the multilayering of its cells. The epithelial cytoplasm in this preparation is very lightly stained; the nuclei appear as the more prominent element. The underlying **connective tissue** (2) is very densely stained with eosin. It contains small **blood vessels** (3) and nerves. The epithelium and its underlying connective tissue constitute the **mucosa** (4). Note, there is no muscularis mucosa. Beneath the mucosa is the **muscularis** (5). Only a portion of the longitudinal layer of the smooth muscle is evident. The **inset** reveals a high power micrograph of the transitional epithelium. A feature characteristic of transitional epithelium is the presence of **binucleate cells** (6) in the surface cells. Another feature is a characteristic curvature, often described as a dome-shaped contour, of the apical surface of these cells. Also note the presence of several degenerating or **apoptotic cells** (7). Beneath the surface cells and extending down to the connective tissue, the remaining epithelial cells appear randomly organized, a feature characteristic of the contracted bladder. In a fully distended bladder, the multilayering is reduced to three to four layers of cells.

Ureter, human, H&E, x225.

This micrograph is a higher magnification of the upper boxed area on the orientation micrograph. It includes the **mucosa** (8), the underlying **muscularis** (9), and a small portion of the **adventitia** (10). As noted, the muscularis consists of several layers. The innermost layer is the **longitudinal layer** (11). Note that its smooth muscle cell nuclei appear as round profiles since these cells have been cross-sectioned. The next layer is the **circular layer** (12). The smooth muscle cell nuclei appear here in longitudinal profile. The third and outermost layer is an additional **longitudinal layer** (13) that is found only in the distal end of the ureter. Compare the nuclear profiles here with those of the inner longitudinal layer. The small amount of adventitia seen in the micrograph reveals a small **artery** (14) and **vein** (15).

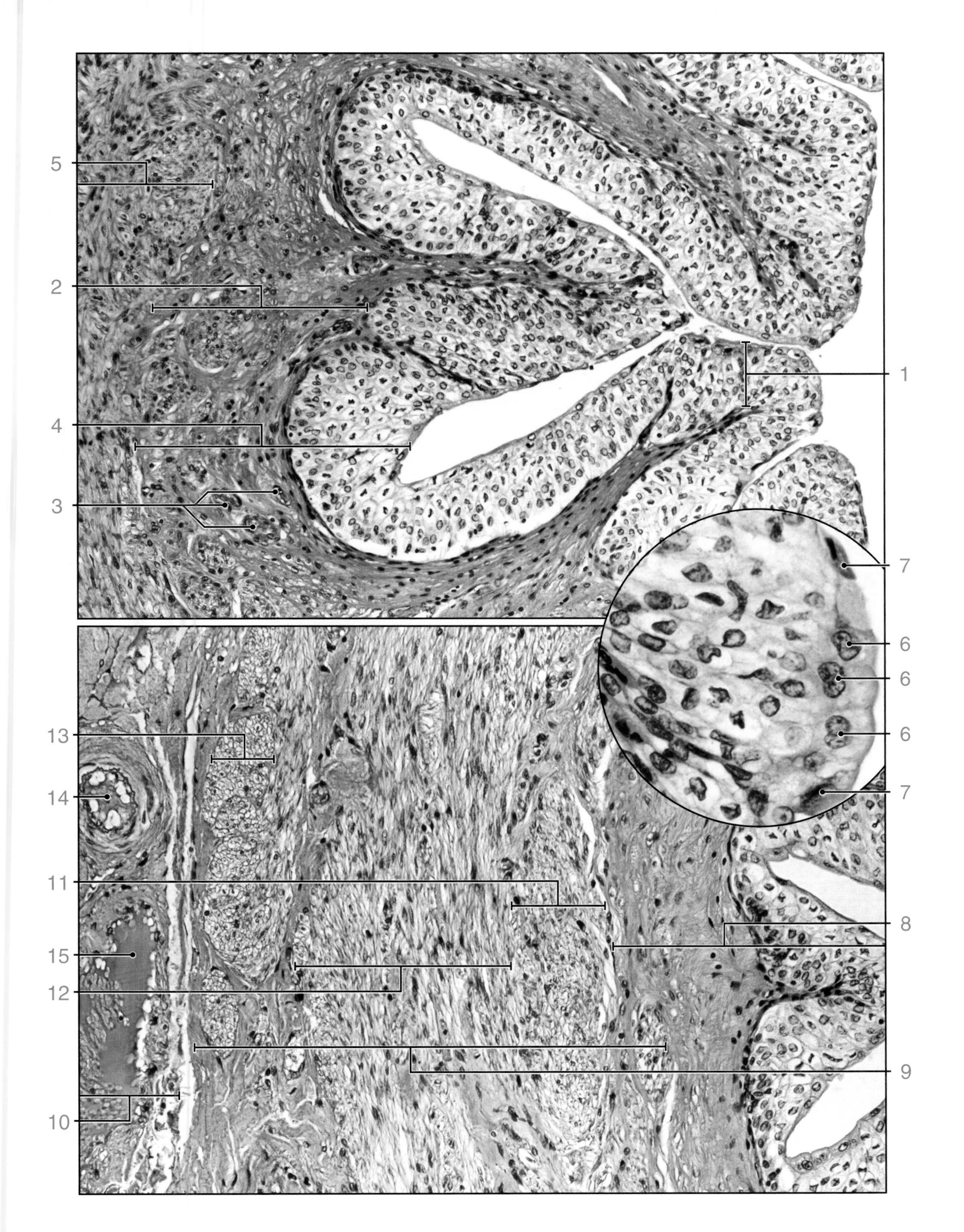

PLATE 115. URETER

PLATE 117. **URETHRA**

The urethra is a fibromuscular tube that allows urine to flow from the bladder to the exterior. In the male, the urethra can be divided into three segments. The proximal, or prostatic, urethra extends from the neck of the bladder through the prostate gland and is lined with transitional epithelium. The membranous urethra extends from the prostate gland to the bulb of the penis. This portion is lined with a stratified or pseudostratified columnar epithelium. The longest portion is the penile "spongy" urethra, which extends the length of the penis and opens on the body surface at the glans penis. It is lined by pseudostratified columnar epithelium, but at its distal end, it becomes stratified squamous epithelium that is continuous with the skin of the penis. The bulbourethral glands (Cowper's glands) and mucus-secreting urethral glands (glands of Littré) empty into the penile urethra.

The female urethra corresponds to the prostatic urethra in the male; it is short, extending for only 3 to 5 cm from the bladder to the vestibule of the vagina. The epithelial lining is initially transitional, but changes to stratified squamous just before its termination. Numerous urethral glands open into the urethra lumen. Other glands, the paraurethral glands, which are homologous to the prostate gland, secrete into the common paraurethral ducts, which open on each side of the external urethral orifice. The lamina propria contains numerous venous vessels.

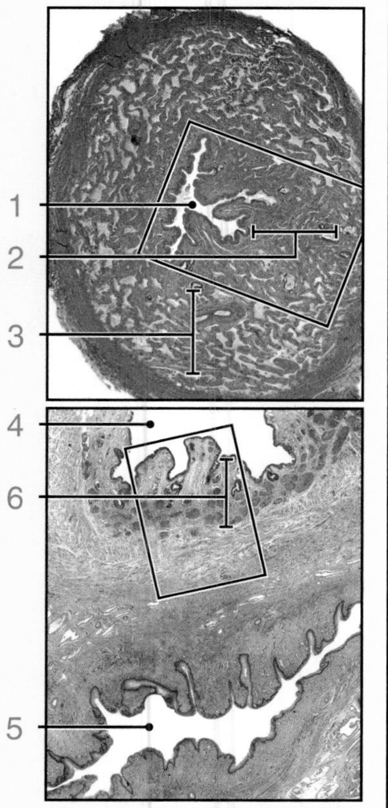

ORIENTATION MICROGRAPHS: The *upper micrograph* is a section through the glans penis. The **urethra** (1) in this segment when viewed in cross-section reveals a stellate profile; sometimes it appears star-shaped or sometimes a T-shape, as is seen here. The mucosal **lamina propria** (2) consists of fibroelastic tissue with scattered bundles of smooth muscle, mainly longitudinally oriented but with some circular bundles in the outer part. Peripheral to the mucosa is the **vascular erectile layer** (3).

The *lower micrograph* is a section showing part of the lumen of the **female urethra** (4) and the **vagina** (5) below. The urethral **lamina propria** (6) is a loose connective tissue with abundant elastic fibers and a venus plexus that is homologous to the cavernous vasculature of the corpus spongiosum in the male. The mucosa is surrounded by smooth muscle; the inner layers are longitudinally arranged fibers and the outer layer is circularly arranged.

Urethra, male, human, H&E, x50; inset x300.

The region shown here is from the boxed area in the upper orientation micrograph. It shows the stellate **lumen** (1) of the urethra, the **lining epithelium** (2), and the **lamina propria** (3), which merges with the **cavernous region** (4) of the corpus spongiosum. Careful examination of this micrograph reveals intra-epithelial groups of **mucus-secreting cells** (5), which are readily recognized by their light staining. In many areas, these cells form recesses or outpockets, the **lacunae of Morgagni** (6). Some of these outpockets extend deeper as branching glandular structures, the **glands of Littré** (7). The **cavernous sinuses** (8) appear as the empty spaces in the corpus spongiosum. The **inset**, a higher magnification of the boxed area, reveals the **pseudostratified columnar epithelium** (9) that lines the urethra and a region containing the **intra-epithelial mucus-secreting cells** (10). The mucous cells, in contrast, are arranged in a simple columnar epithelium. Note that the nuclei of the mucus-secreting cells are at the bases of the cells and are characteristically flattened, whereas the nuclei of the typical urethral epithelium are ovoid, conforming to the columnar shape of the surface cells.

Urethra, female, human, H&E, x360.

The wall of the female urethra from the boxed area on the lower orientation micrograph is shown here. Its mucous membrane tends to form **longitudinal folds** (11). The epithelium, as in the male urethra, forms invaginations with **mucus-secreting glands** (12) in the lamina propria. Also present in the lamina propria is a well-developed system of **venous vessels** (13), which are very prominent in this micrograph due to engorgement with red blood cells. These vessels are homologous to the vasculature of the corpus spongiosum in the male. The mucosa is surrounded by a thick layer of **smooth muscle** (14), consisting of an inner layer of longitudinally arranged fibers and an outer layer of circular arranged fibers.

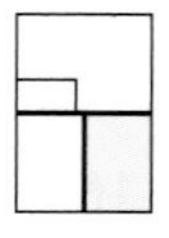

Urethra, female, human, H&E, x360.

This micrograph shows at higher magnification the mucosa of the female urethra from the boxed area on the bottom left micrograph. The epithelium is **pseudostratified columnar** (15), but at the very distal end of the urethra, it becomes stratified squamous. The underlying lamina propria tends to be very cellular with lymphocytes being the predominant cell type present. The **venous vessels** (16) that form the plexus have a structure more characteristic of typical veins than the complex anastomosing venous vessels in the male.

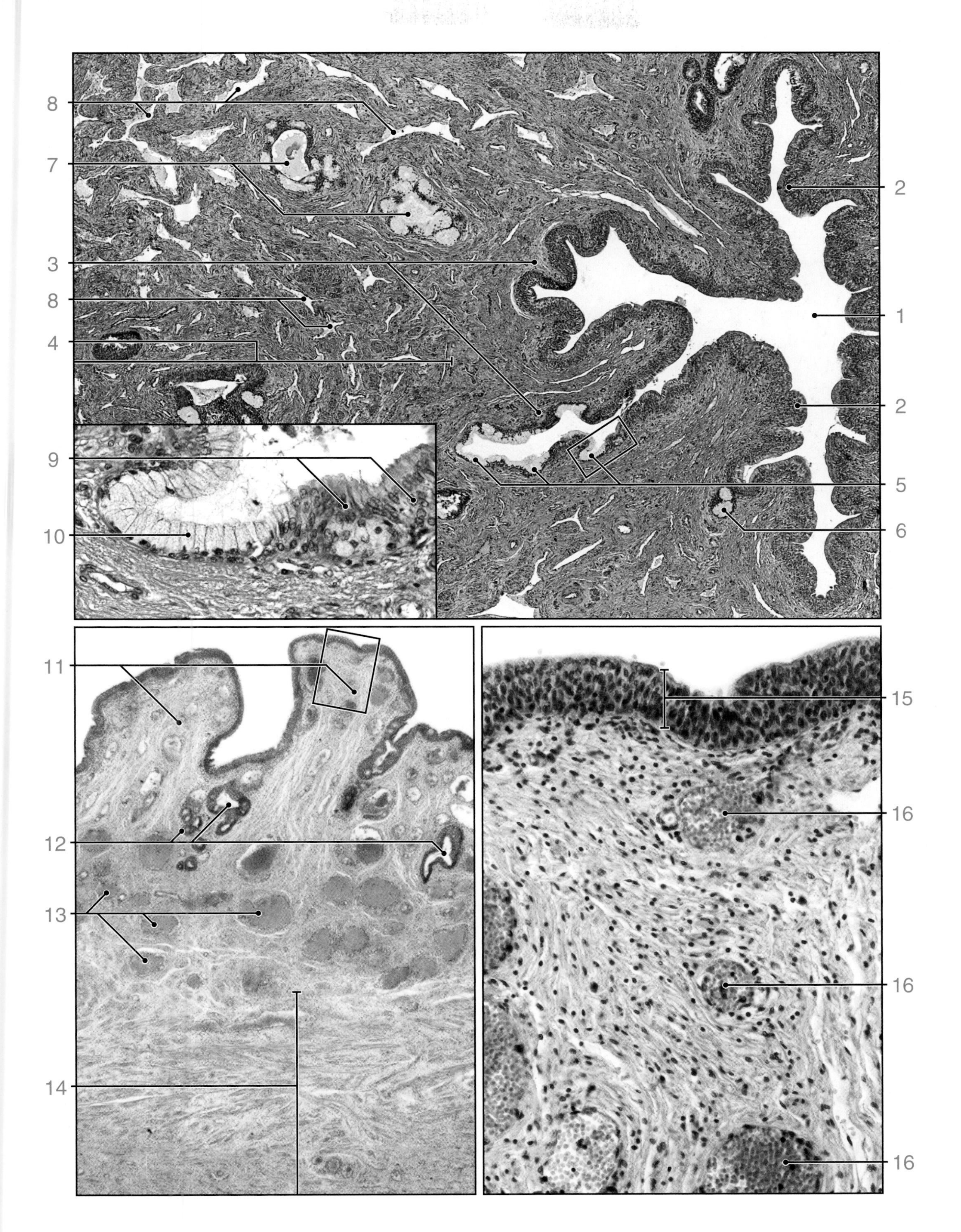

8
7
3
8
4
9
10
2
1
2
5
6
11
12
13
14
15
16
16
16

CHAPTER 17

Endocrine System

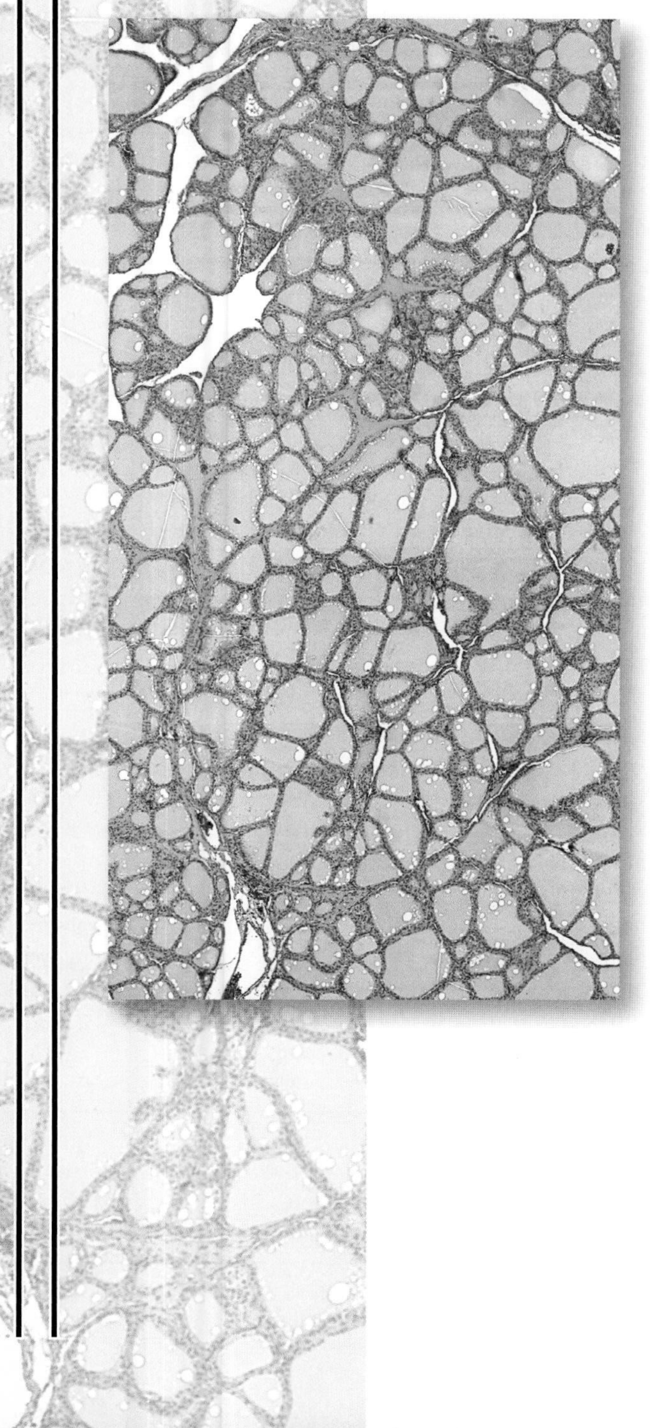

PLATE 118. **PITUITARY GLAND I**

The pituitary gland is a pea-sized endocrine gland located at the base of the brain and contained in a depression of the sphenoid bone, the sella turcica. A short stalk, the infundibulum, connects it to the hypothalamus. The gland has two distinct components: an anterior lobe (adenohypophysis), which is a glandular epithelial tissue, and a posterior lobe (neurohypophysis), which is a neural secretory tissue. The posterior lobe consists of the pars nervosa and the infundibulum. The anterior lobe is made up of three parts: the pars distalis, comprising the bulk of the anterior lobe; the pars intermedia, which lies between the pars distalis and the pars nervosa; and the pars tuberalis, a sheath around the infundibulum.

The pars distalis consists of cells arranged in cords and groups of cells with a rich capillary network. Three cell types can be identified based on the staining reactions of their cytoplasm with acidic and basic dyes. They are acidophils, which stain with acidic dyes, basophils, which stain with basic dyes, and chromophobes, which take up little or no dye. The chomophobes are thought to be a heterogeneous group of cells; many are considered depleted acidophils or basophils. With the use of immunocytochemical reactions, the acidophils and basophils are functionally classified into five cell types, according to their secretory product:

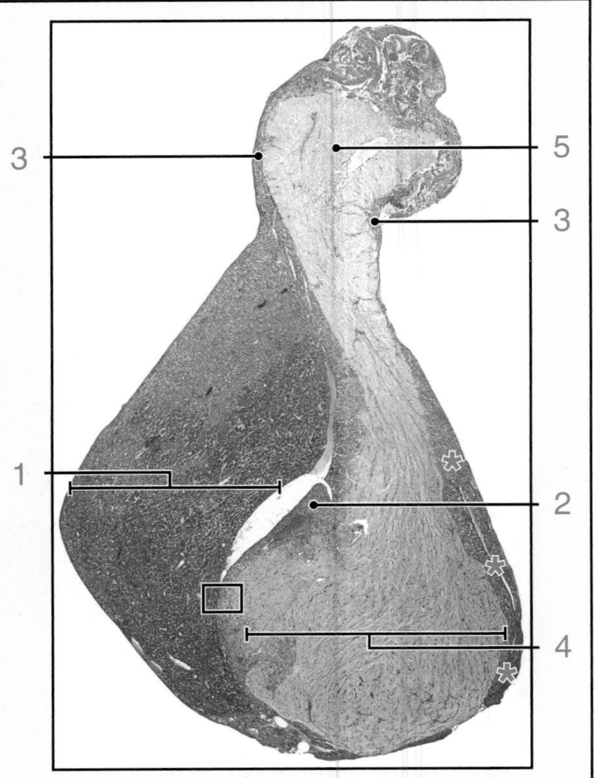

- Somatotrope, an acidophil that produces somatotropin, a growth hormone.
- Lactotrope, an acidophil that produces prolactin, which promotes mammary gland development and initiates milk production.
- Corticotrope, a basophil that produces adrenocorticotropic hormone (ACTH), which stimulates secretion of glucocorticoids and gonadocorticoids by the adrenal gland.
- Gonadotrope, a basophil that produces follicle-stimulating hormone (FSH), which stimulates follicular development in the ovary and spermatogenesis in the testis, and luteinizing hormone (LH), which regulates final maturation of ovarian follicles, ovulation, and corpus luteum formation in the female and is essential for maintenance of and androgen secretion by Leydig cells of the testis in the male.
- Thyrotrope, a basophil that produces thyrotropic hormone (TSH), which stimulates production and release of thyroglobulin and thyroid hormones.

ORIENTATION MICROGRAPH: This micrograph shows the pituitary gland from a monkey. The anterior lobe (adenohypophysis) consists of the **pars distalis** (1), the **pars intermedia** (2), and the **pars tuberalis** (3). A portion of the pars distalis is seen on the right of the micrograph (*asterisks*). This unusual appearance is due to the plane of section. The posterior lobe (neurohypophysis) consists of the **pars nervosa** (4) and the **infundibulum** (5); both contain neural secretory axons.

Pituitary gland, monkey, H&E, x360.

The boxed area on the orientation micrograph is shown here at higher magnification. It includes a portion of the **pars distalis** (1) on the left, which consists of cords and clumps of parenchymal cells, and a portion of the **pars nervosa** (2) on the right. In the center is the **pars intermedia** (3). In the pars distalis, at this relatively low magnification, it is possible to distinguish between the **acidophils** (4), which are stained a bright red with eosin, from the **basophils** (5), which, in contrast, exhibit a bluish color from the hematoxylin stain. The **chromophobes** (6) are in the minority. They can be distinguished by their small nuclei and unstained cytoplasm. The pars intermedia exhibits numerous basophils, which are lightly stained with hematoxylin. Also present are chromophobes, which have the same appearance as those in the pars distalis. (In humans, the pars intermedia is not as extensive as in the species shown here. Also, the pars intermedia in humans consists mostly of parenchymal cells that surround colloid-filled follicles and of basophils and chromophobes, as in the species shown here.) The pars nervosa consists of unmyelinated axons. The majority of nuclei in the pars nervosa belong to specialized glial cells called **pituicytes** (7). These cells have numerous processes, some of which terminate in the perivascular spaces.

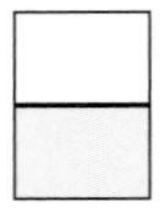

Pituitary gland, monkey, H&E, x500.

This higher magnification of an area of the pars distalis shows to advantage the eosin-stained **acidophils** (8) and the contrasting bluish-stained **basophils** (9). The **chromophobes** (10) are relatively few in number in the area shown here. They have a small nucleus and little surrounding cytoplasm that does not stain. Between the cords and clumps of parenchymal cells are **capillaries** (11).

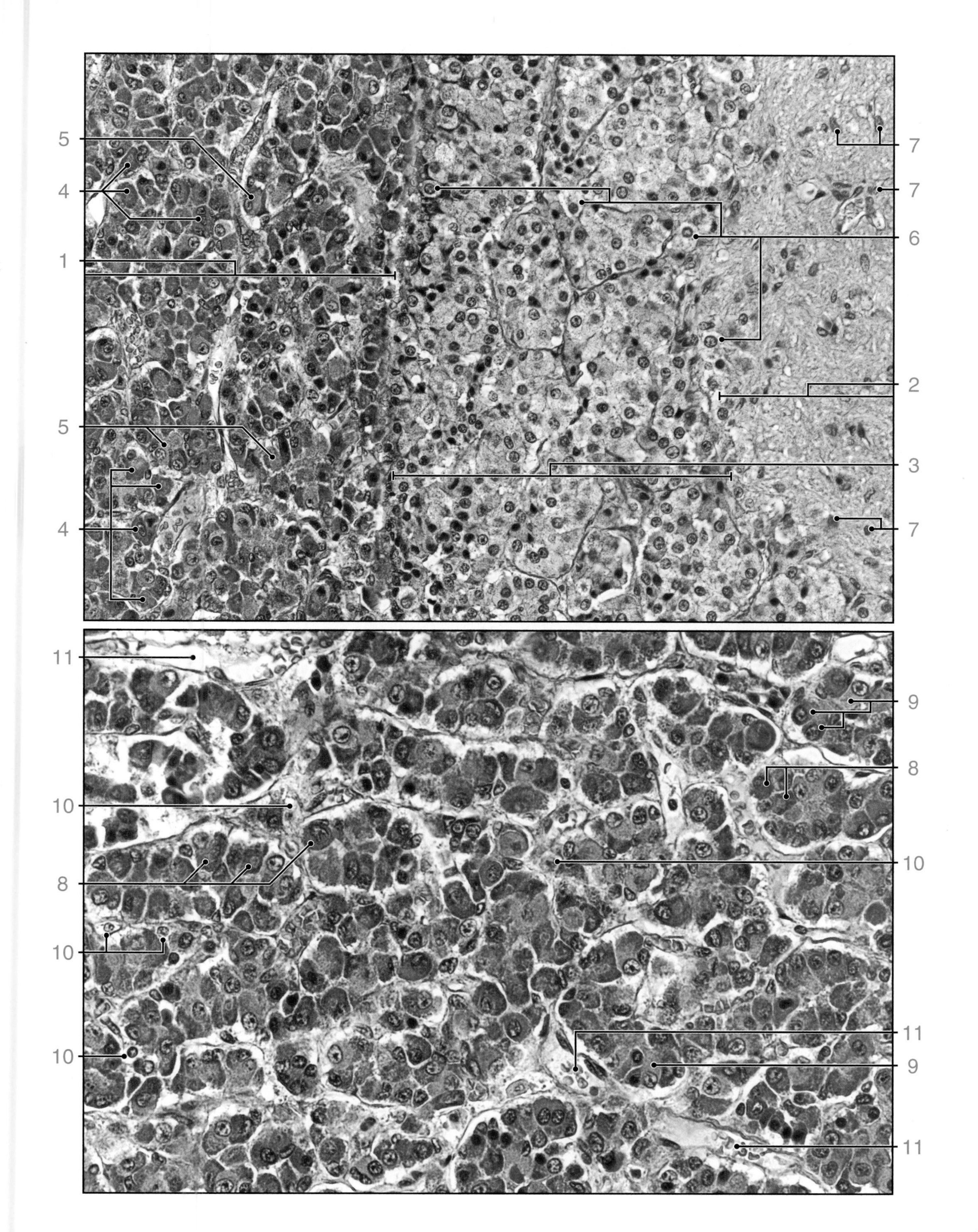
5
4
1
5
4
7
7
6
2
3
7
11
10
8
10
10
9
8
10
11
9
11

PLATE 119. **PITUITARY GLAND II**

The posterior lobe of the pituitary gland is an extension of the central nervous system. It consists of the pars nervosa and the infundibulum. The infundibulum, the stalklike component, connects to the hypothalamus. The pars nervosa, the neural lobe of the pituitary, contains unmyelinated axons and their endings. The cell bodies of these axons lie in the supraoptic and paraventricular nuclei of the hypothalamus. The posterior lobe is not an endocrine gland but a storage site for neurosecretions of the neurons of the supraoptic and paraventricular nuclei. The unmyelinated axons that extend from the cell bodies in these nuclei convey the neurosecretory products from the cell body. Thus, the bulk of the infundibulum and pars nervosa consists of axons. At the light microscopic level, the terminals of these axons are seen as dilations, called Herring bodies, where the secretory product is stored. The secretions contain either oxytocin or vasopressin (antidiuretic hormone, ADH). Other neurons from the hypothalamus release secretions into the fenestrated capillaries of the infundibulum, the first capillary bed of the hypophyseal portal system that carries blood to the capillaries of the adenohypophysis. These hypothalamic secretions regulate the activity of the adenohypophysis.

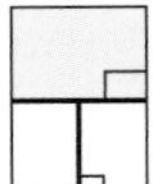

Pituitary gland, human, Mallory, x725; inset x1200.

The area of the pars distalis that is shown here reveals an almost equal distribution of acidophils and basophils. It also reveals a relatively large number of chromophobes. It should be recognized that there is a variation in the proportionate number of acidophils, basophils, and chromophobes in different areas of the pars distalis. Use of the Mallory stain results in a somewhat more distinctive coloration of the acidophils and basophils. The **acidophils** (1) appear as a deep orange-red color, whereas the **basophils** (2) appear as a deep reddish-blue color. In contrast, the **chromophobes** (3) exhibit very pale blue cytoplasm. The clusters and cords of these parenchymal cells are also better defined by the surrounding **connective tissue** (4), which stains blue. Some of the **capillaries** (5) supplying the parenchyma are markedly dilated and filled with red blood cells, which are colored yellow with the Mallory stain. The **inset** shows at much higher magnification several acidophils (1), basophils (2), and a chromophobe (3). Of particular note, the secretory vesicles are the components of the acidophils and basophils that bind the dye and appear as fine granules. In contrast, the chromophobe has a more homogeneous or nongranular cytoplasm.

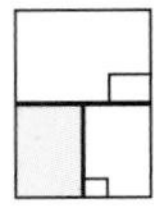

Pituitary gland, human, H&E, x325.

This micrograph shows the pars nervosa. The majority of the nuclei belong to cells called **pituicytes** (6). They are comparable to neuroglial cells of the central nervous system. The nuclei are round to oval. Their cytoplasm extends from the nuclear region of the cell in long processes. In typical H&E preparations such as this, the cytoplasm of the pituicyte cannot be distinguished from the unmyelinated nerve fibers that make up the bulk of the pars nervosa. Careful examination of these nerve fibers shows some areas where they are cut in cross-section (7), whereas in other areas they appear in longitudinal profile (8). The **Herring bodies** (9) appear as round, irregularly shaped, homogeneously stained structures of various sizes. Also worth noting are the **capillaries** (10), which are often hard to detect when coursing parallel to the axons.

Pituitary gland, human, PAS/analine blue-black, x320; inset x700.

This micrograph shows a specimen comparable to that seen in the micrograph on the bottom left and differs only in the stain used. The **Herring bodies** (11) in this micrograph appear blue-black, a result of the analine blue stain. The **inset** shows one of the Herring bodies at higher magnification. The stored neurosecretion and membrane-bound vesicles have taken up the stain. The use of the PAS stain reveals the **capillaries** (12) by staining their basement membrane.

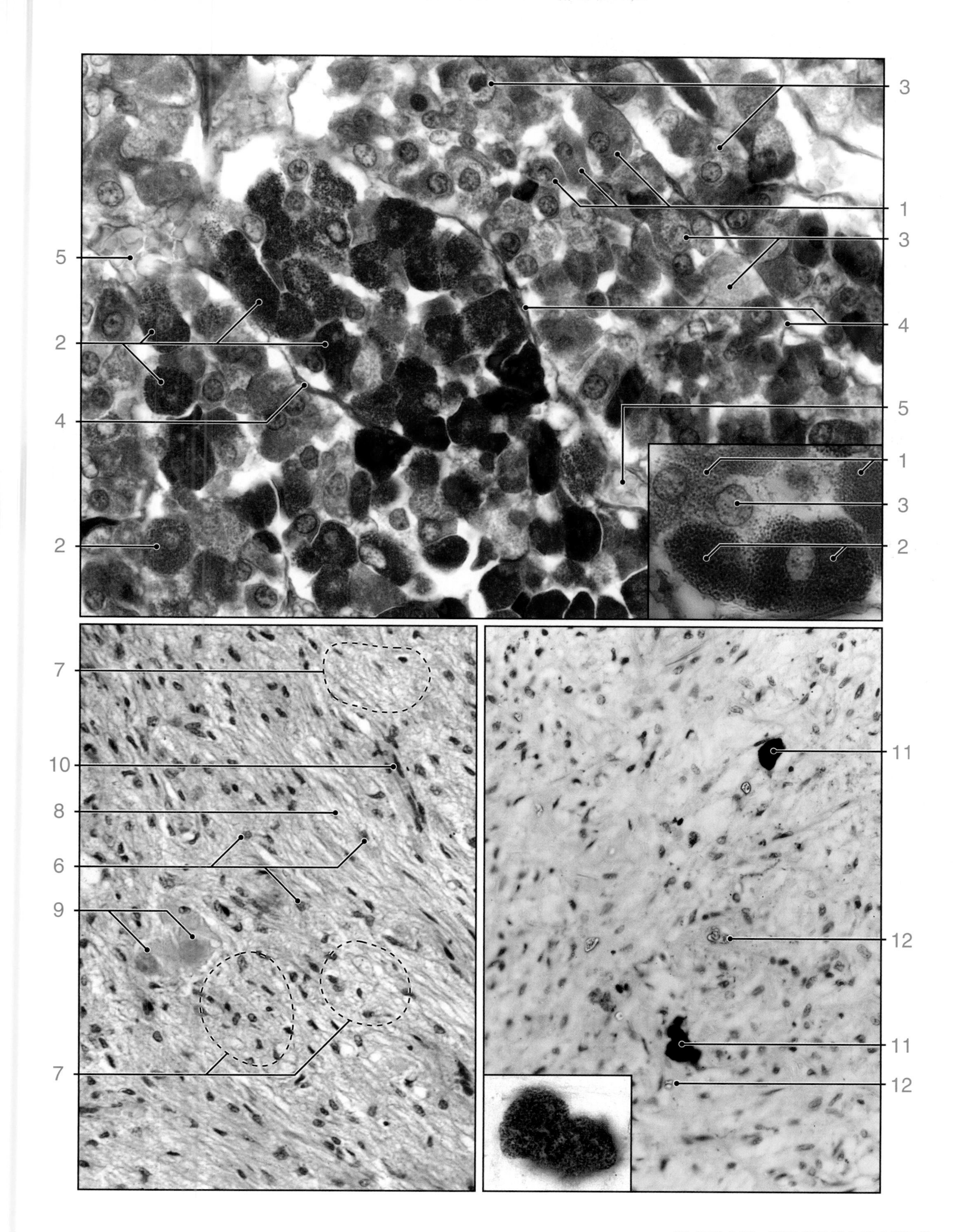

PLATE 119. PITUITARY GLAND II

PLATE 120. PINEAL GLAND

The pineal gland (pineal body, epiphysis cerebri) is located near the center of the brain above the superior colliculi. It develops from neuroectoderm, but in the adult, bears little resemblance to nerve tissue. It is a flattened, pinecone-shaped structure, hence it's name.

The two cell types found in the pineal gland are parenchymal cells and glial cells. The full extent of these cells cannot be appreciated without the application of special methods. Use of special methods would show that the glial cells and the parenchymal cells have processes and that the processes of the parenchymal cells are expanded at their periphery. The parenchymal cells are more numerous. In an H&E preparation, the nuclei of the parenchymal cells stain lightly. The nuclei of the glial cells, on the other hand, are smaller and stain more intensely.

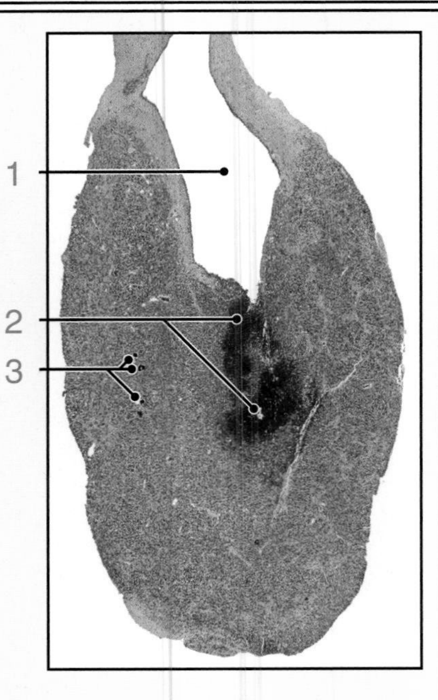

Although the physiology of the pineal gland is not well understood, the secretions of the gland evidently have an antigonadal effect. For example, hypogenitalism has been reported in patients with pineal tumors that consist chiefly of parenchymal cells, whereas sexual precocity is associated with glial cell tumors (presumably, the parenchymal cells have been destroyed). In addition, experiments with animals indicate that the pineal gland has a neuroendocrine function whereby the pineal gland serves as an intermediary that relates endocrine function (particularly gonadal function) to cycles of light and dark. The external photic stimuli are conducted via optical pathways that connect with the superior cervical ganglion, which sends postganglionic nerve fibers to the pineal gland. The extent to which findings in laboratory animals apply to humans is not yet clear.

Recent studies in humans suggest that the pineal gland plays a role in adjusting to sudden changes in day length, such as those experienced by travelers who suffer from jet lag, and a role in regulating emotional responses to reduced day length during winter in temperate and subarctic zones (seasonal affective disorder [SAD]).

ORIENTATION MICROGRAPH: The section shown here is a median cut through the pineal gland. The bottom of the micrograph is the anterior end of the gland and the **pineal recess** (1), a cleft-like space, is at the top of the micrograph. The **darker-staining areas** (2) are a result of bleeding within the gland. The characteristic concretions, the **brain sand**, or **corpora aranacea** (3), appear as very dense, round structures at this low magnification.

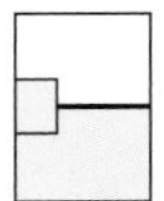

Pineal gland, human, H&E, x180.

The pineal gland is surrounded by a very thin **capsule** (1) that is formed by the pia mater. **Connective tissue trabeculae** (2) extend from the capsule into the substance of the gland, dividing it into lobules. The **lobules** (3) often appear as indistinct groups of cells of varying size surrounded by the connective tissue. Blood vessels, generally small **arteries** (4) and **veins** (5), course through the connective tissue. The arteries give rise to capillaries that surround and penetrate the lobules to supply the parenchyma of the gland. In this specimen and even at this low magnification, the **capillaries** (6) are prominent because of the red blood cells present in their lumina.

Pineal gland, human, H&E, x360; inset x700.

This micrograph shows at higher magnification the parenchyma of the pineal gland as well as a component called **brain sand**, or **corpora arenacea** (7). When viewed at even higher magnifications, the corpora arenacea are seen to have an indistinct, lamellated structure. Typically, they stain heavily with hematoxylin. The presence of these structures is an identifying feature of the pineal gland. A careful examination of the cells within the gland at the light microscopic level reveals two specific cell types. One cell type represents the parenchymal cells. These are by far the most numerous and are referred to as pinealocytes (or chief cells of the pineal gland). Pinealocytes are modified neurons. Their nuclei are spherical and relatively lightly stained because of the amount of euchromatin that they contain. The second cell type is the interstitial cell or glial cell in the gland. Their nuclei are smaller and more elongate than those of the pinealocytes. A distinction between the two cell types is best observed at higher magnification (see inset). A small **vein** (8) and several **capillaries** (9) can be recognized by virtue of the red blood cells that they exhibit. The **inset** reveals several **glial cells** (10) that can be identified by their more densely staining nuclei. The majority of the nuclei of the other cells seen here belong to pinealocytes. Also seen in the **inset** are several **fibroblasts** (11) that are present within a trabecula.

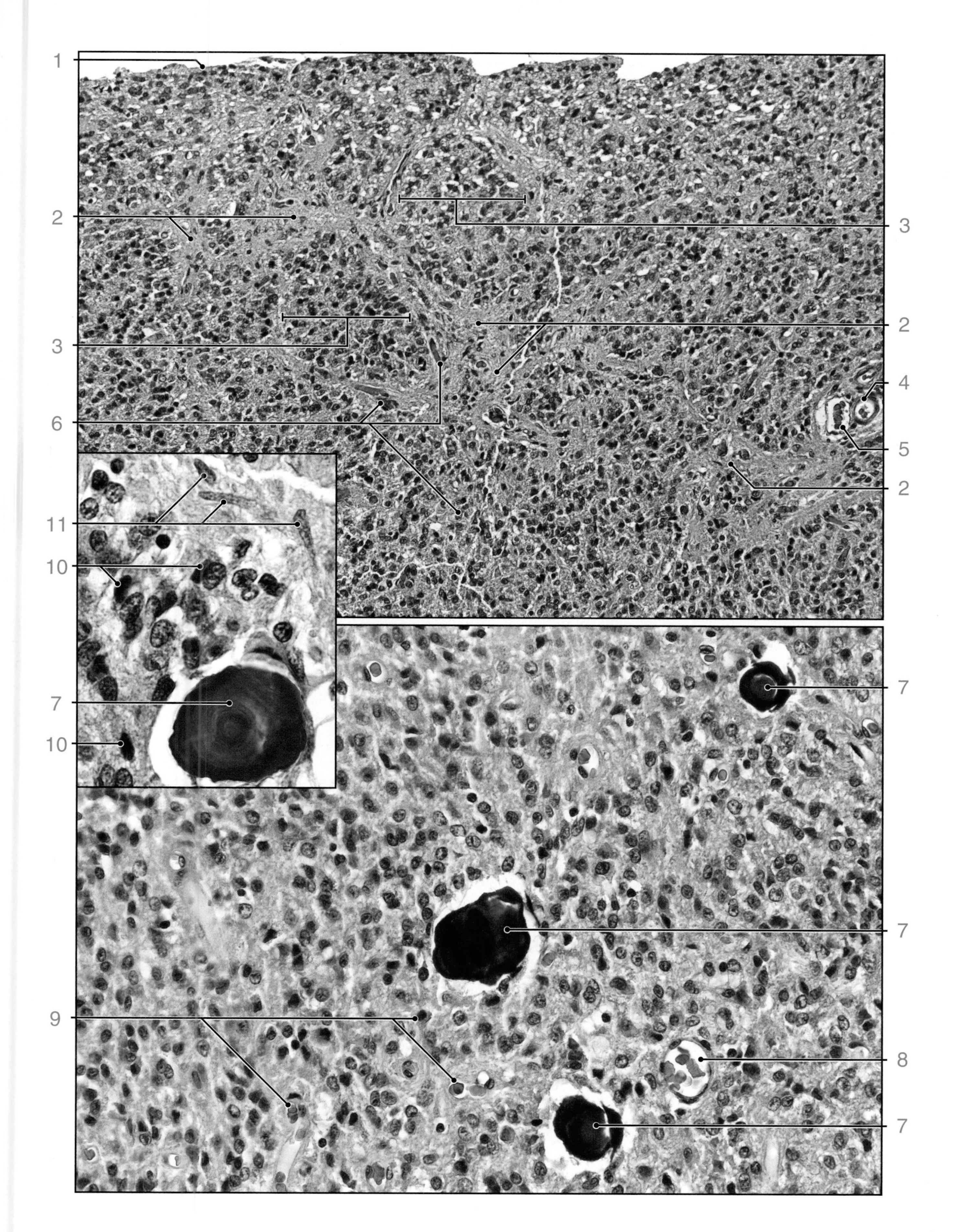

PLATE 120. PINEAL GLAND

PLATE 121. **THYROID GLAND I**

The thyroid gland is located in the lower anterior part of the neck. It consists of right and left lobes connected by a narrow band of thyroid tissue called the isthmus. The isthmus connects the lower portion of the two lobes. The gland is covered by a thin capsule of connective tissue that also penetrates the gland, forming septa that divide the parenchyma into irregular masses. The parenchyma consists of two types of secretory cells, the follicular cells that form sphere-shaped follicles and parafollicular cells. The follicles are in close proximity to one another but are separated by a sparse amount of connective tissue that contains a rich capillary network. Each follicle is filled with a viscous material referred to as colloid that contains the inactive precursor hormone thyroglobulin. Upon stimulation, the colloid produced by the follicular cells is reabsorbed by these same cells and secreted as the hormones tetraiodothyronine (T_4) and triiodothyronine (T_3). The parafollicular cells, also referred to as C-cells, are present in relatively small numbers in humans. They are located in close juxtaposition to the follicle and lie within the basement membrane of the follicle cells. The parafollicular cells are larger than the follicle cells, a reflection of their large, lighter staining nuclei and more abundant cytoplasm. They secrete the hormone thyrocalcitonin.

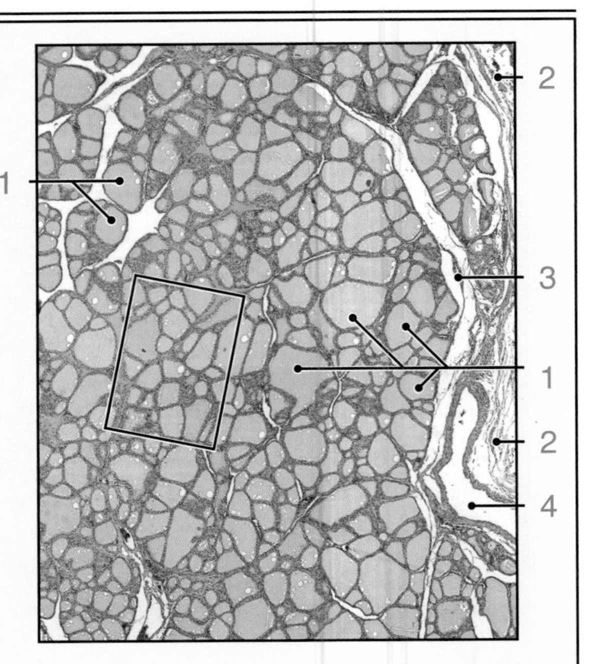

ORIENTATION MICROGRAPH: This low power micrograph shows the mass of tightly packed **follicles** (1) that make up the thyroid gland. A portion of the thin connective tissue **capsule** (2) is included in the specimen as well as **connective tissue septa** (3) that have emanated from the capsule, subdividing the gland into lobes and lobules of various sizes. Also evident is a branch of the thyroid **artery** (4) that has penetrated the capsule to supply a portion of the gland.

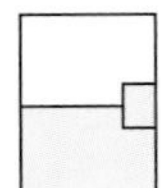

Thyroid gland, human, H&E, x180.

This micrograph is a higher magnification of the boxed area on the orientation micrograph. At this magnification it is evident that each follicle is made up of what can be generally regarded as a **simple cuboidal epithelium** (1) enclosing the **colloid** (2). Typically, the colloid has a homogeneous appearance. However, small **vacuoles** (3) are often present at the periphery of the colloid-filled space adjacent to the free surface of the follicle cells. When present, they are indicative of increased rate of colloid resorption, and they tend to be more prevalent in follicles that contain a more viscous colloid. It should be noted in observing the follicles that certain areas between follicles may show an aggregation or mass of follicle cells (*arrows*). These are a reflection of a follicle being tangentially cut to include only the wall of the follicle, thus revealing only epithelial cells.

Thyroid gland, human, H&E, x360; inset x700.

At the higher magnification shown here, the abundant **capillary network** (4) surrounding the follicles becomes recognizable, evidenced by the eosinophilia of the red blood cells in the capillaries. Although most of the epithelial cells of the follicle are cuboidal, some cells may be **columnar** (5) or even **squamous** (6). In follicles where colloid is being resorbed, the follicle becomes smaller and the epithelial cells become columnar in shape. In follicles where there is little resorption, but continued colloid production, the follicle increases in size and the follicle cells become squamous. In this specimen, also note a couple of areas revealing tangentially sectioned **follicles** (7) (clusters of epithelial cells). The **inset** shows a **C-cell** (8) that appears to lie between at least three follicles. Most likely, it belongs to the follicle on the left (*asterisk*) and as such would be included within the basement membrane of the follicle cells belonging to that follicle.

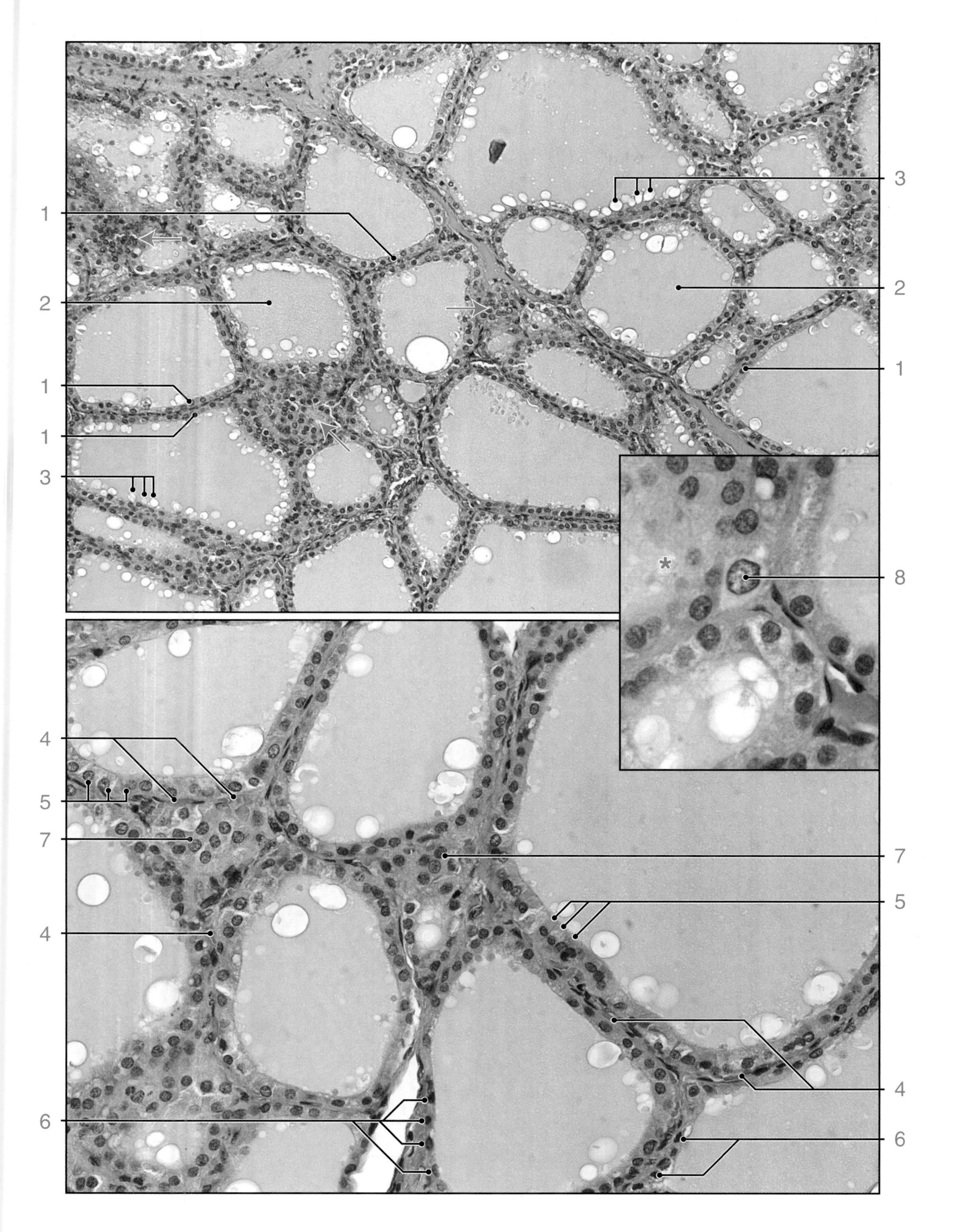
3
1
2
1
1
1
3
8
4
5
7
7
5
4
4
6
6

PLATE 122. THYROID GLAND II

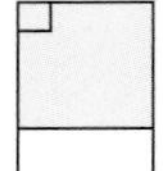

Thyroid gland, electron micrograph, x5000; inset H&E, x700.

This micrograph shows portions of two adjacent follicles. The interfollicular space includes a small **unmyelinated nerve** (1), the cytoplasmic process of a **fibroblast** (2), and a **capillary** (3). Portions of the capillary wall are extremely thin (*arrows*) and are of the fenestrated type of endothelium. The **inset** is a micrograph of a similar area from an H&E specimen to provide orientation. In the **inset**, a **capillary** (4) and several **nuclei** (5) are also present between the adjoining follicles. It is difficult to distinguish the type of cell that these nuclei represent. The follicle epithelial cells are cuboidal, and consistent with their shape, their nuclei are round. Examining the thyroid follicular cells in the electron micrograph reveals numerous profiles of **rough surfaced endoplasmic reticulum** (6). Also evident, though less conspicuous, are dilated cisternae of the **Golgi apparatus** (7). Critical examination also reveals numerous **mitochondria** (8) distributed throughout the cytoplasm. The boundaries between follicle cells are extremely difficult to discern. This is due to the interdigitation of cytoplasmic processes in the lateral aspect of the cells. The apical surface of the cell is also indistinct due to the presence of numerous small and irregular microvilli that project into the **colloid** (9). When the follicle cells are actively endocytosing colloid, large apical pseudopods extend into the colloid. The absence of such pseudopods in the follicles shown here indicates that colloid is not being actively resorbed.

Thyroid gland, dog, silver, x725.

This micrograph shows several thyroid follicles from a dog. C-cells in this species are numerous compared to human thyroid. The silver stain reacts with the numerous granules in the cell causing the cytoplasm to have a brown coloration. What appears as an aggregation of C-cells on the far right of the micrograph is a reflection of a tangential section of at least two follicles, thus giving a crowded appearance of the C-cells. A more normal section of the follicles shows the **C-cells** (10) lying within the epithelium of the follicle. The textured appearance of the C-cells is a reflection of the numerous granules within their cytoplasm. The dot-like structures along the apical surface of the follicles are terminal bars (*arrows*).

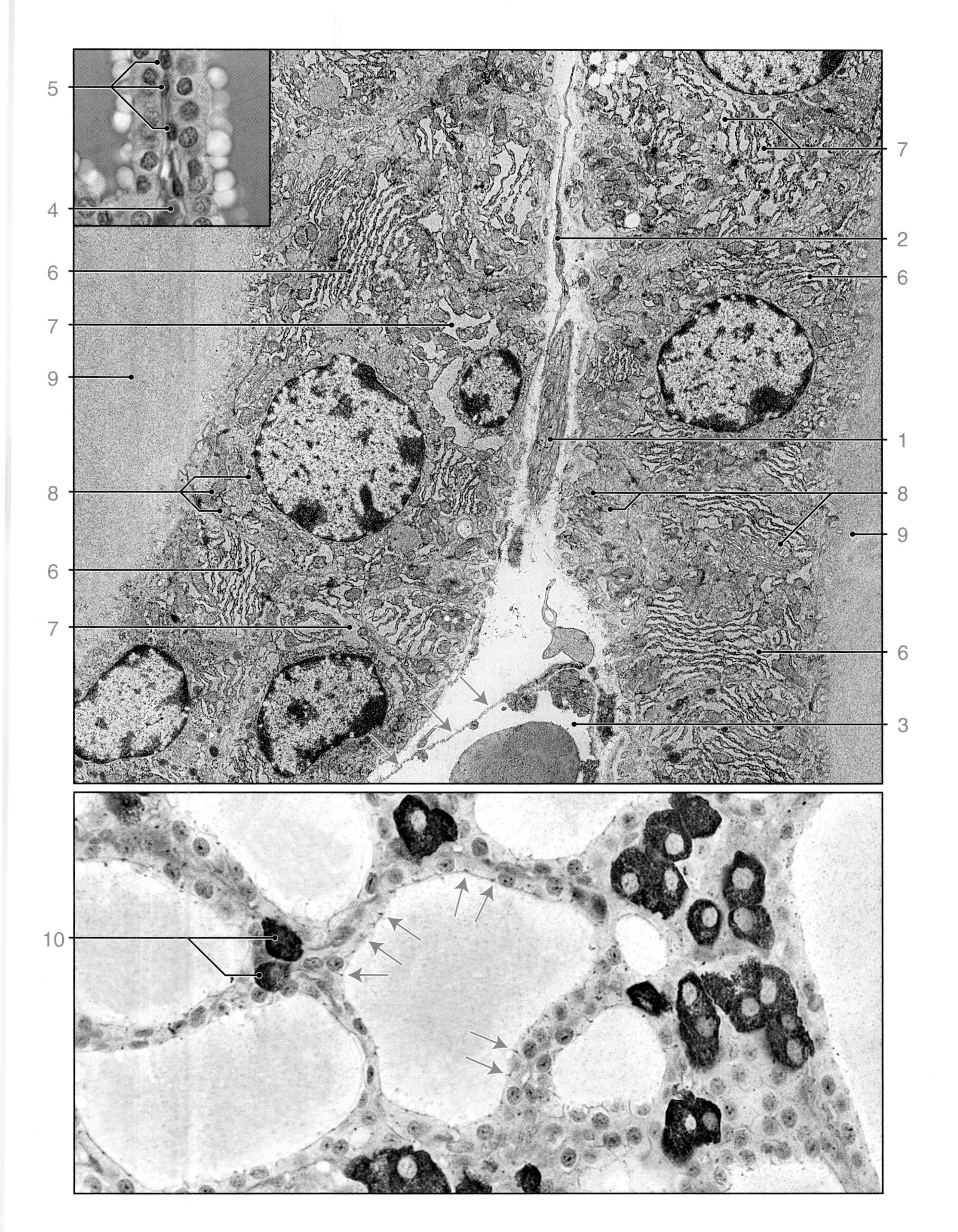

5
4
6
7
9
8
6
7
7
2
6
1
8
9
6
3
10

PLATE 123. **PARATHYROID GLAND**

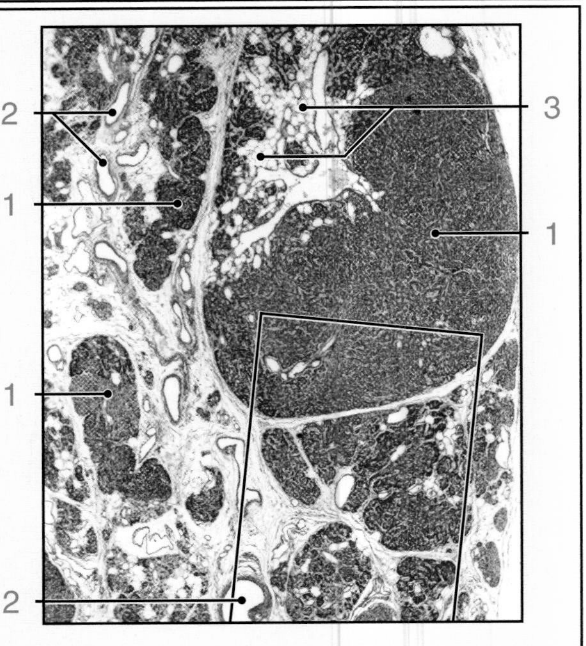

The parathyroid glands are surrounded by a connective tissue capsule and lie on or partially embedded in the thyroid gland. They are usually four in number, though occasionally a few more may be present. In most cases, they are arranged in two pairs, constituting the superior and inferior parathyroid glands. The parenchyma of the gland consists of two cell types based on histological appearance: chief (principal) cells, which are the more numerous, and oxyphil cells. The chief cells secrete the hormone parathormone (PTH), which controls blood calcium levels and is essential for life. If the parathyroid glands are removed, death ensues because muscles, including those of the respiratory system, go into tetanic contraction as the blood calcium level falls. Secretion of PTH stimulates osteocytes resulting in release of bone calcium, which enters the blood.

The oxyphil cells are found singly or in clusters. They are considerably larger than chief cells and their cytoplasm is very eosinophilic. They begin to appear around puberty. In later life, the oxyphil cells increase in number. Histochemical and ultrastructural observations reveal that the cytoplasm of the oxyphil cell is packed with large mitochondria, a feature which accounts for its eosinophilic staining. It is now generally thought that the oxyphil cell represents a modification of the chief cell. This concept is supported by the fact that only chief cells are present during early years of life as well as the fact that chief cells are found in only a few mammalian species.

ORIENTATION MICROGRAPH: This low magnification micrograph shows a portion of one of the parathyroid glands. The parenchyma of the gland is subdivided into **lobules** (1) of varying size by connective tissue septa. Within the connective tissue septa are numerous **blood vessels** (2) and an area containing **adipose tissue** (3).

Parathyroid gland, human, H&E, x90.
This micrograph is a higher magnification of the boxed area on the orientation micrograph. At this magnification, it is difficult to distinguish between chief cells and oxyphil cells. However, one of the lobules, based on the light eosinophilic staining, appears to be made up entirely of **oxyphil cells** (1). The only other prominent feature is the presence of numerous **vessels** (2) within the lobules as well as coursing in the connective tissue space. The vessels seen here are **venules** and **small veins** (3), an **artery** (4), and several **lymphatic vessels** (5).

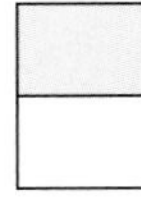

Parathyroid gland, human, H&E, x360.
At this higher magnification of the boxed area on the upper micrograph, chief cells and oxyphil cells can be readily identified. The **chief cells** (6) are arranged in cords and clumps. Their nuclei are in close proximity to one another because of the small amount of surrounding cytoplasm. In contrast, the **oxyphil cells** (7) are readily distinguished from the chief cells due to the wider dispersion of their nuclei and their eosinophilic cytoplasm. Also readily identified are **adipocytes** (8) and a number of **blood vessels** (9) within the connective tissue septa.

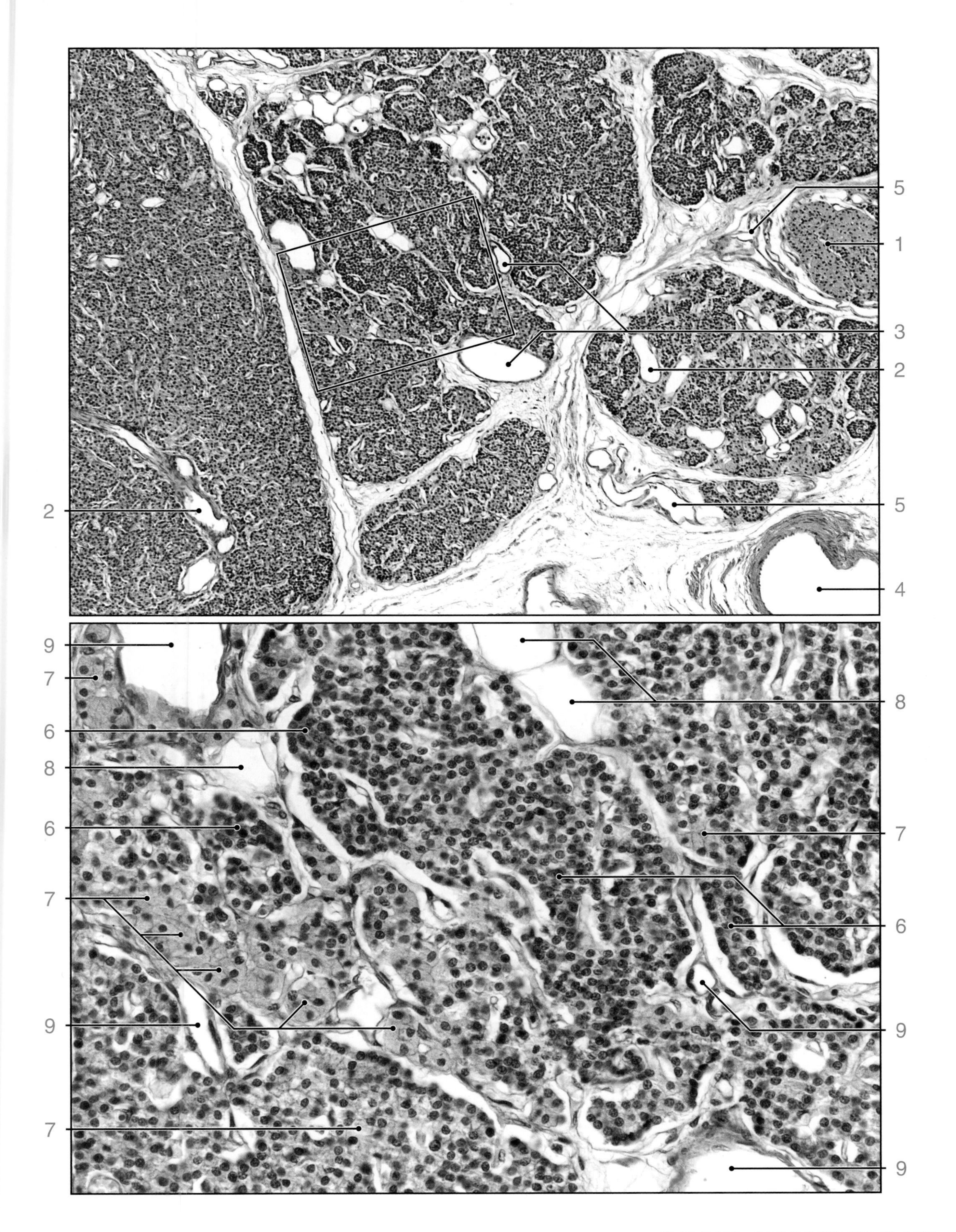
5
1
3
2
2
5
4
9
7
6
8
6
7
9
7
8
7
6
9
9

PLATE 124. ADRENAL GLAND

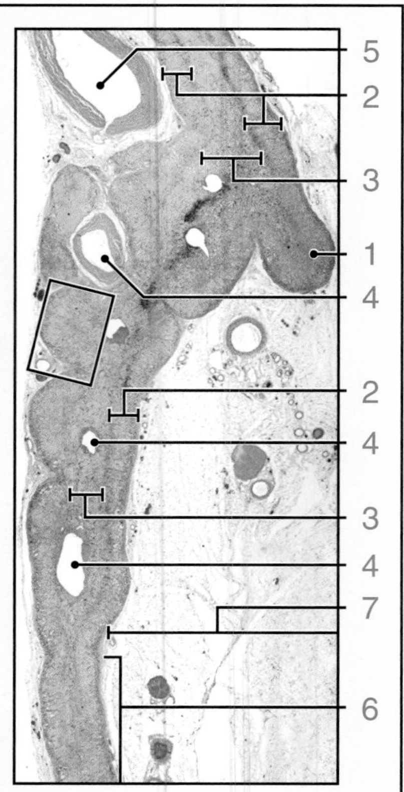

The paired adrenal, or suprarenal, glands in humans are roughly elongated pyramidal bodies. They are embedded in the perirenal fat at the superior pole of the kidney. The cortex, or outer portion of the gland, comprises the greater part of the gland and is organized in three defined layers based upon the organization of the cells. The outer layer, the zona glomerulosa, consists of cuboidal to short columnar cells arranged in ovoid groups. The cells are relatively small. Their spherical nuclei are closely packed and stain densely. The small size of these cells is due mainly to their relatively scant cytoplasm. They secrete mineralocorticoids that function in the regulation of sodium–potassium homeostasis and water balance. The underlying zona fasciculata consists of large, polyhedral cells arranged in long, straight cords, one or two cells thick, that are separated by sinusoidal capillaries. They stain lightly due in part to their numerous lipid droplets. The principal secretion of these cells is glucocorticoids that regulate glucose and fatty acid metabolism. The deepest zone, the zona reticularis, consists of small cells arranged in anastomosing cords. Weak androgens and glucocorticoids are their principal secretions. The central portion of the gland, the medulla, consists of pale staining cells referred to as chromaffin cells or medullary cells. The cells are of two types. One type contains large, dense-core vesicles seen at the electron microscopic level that secrete norepinephrine. The second cell type contains vesicles also seen at the electron microscopic level that are small and less dense. These cells secrete epinephrine. The vein that drains the capillaries of the adrenal cortex and medulla join to form the large central vein from which blood leaves the adrenal gland. These veins are unusual in that they possess a media of smooth muscle cells arranged in groups that are organized parallel to the vein rather than circumferentially.

ORIENTATION MICROGRAPH: This section is a slice through an adrenal gland showing its superior surface and **apex** (1) on the right and its basal aspect facing the kidney on the left. The **cortex** (2) of the gland can be distinguished from the **medulla** (3). Within the medulla, several profiles of the **central vein** (4) are evident. This vessel drains into the **medullary vein** (5). Note that at the gland's periphery, the medullary tissue is absent, leaving both sides of the cortex in apposition (6). The remainder of the micrograph consists of **perirenal adipose tissue** (7).

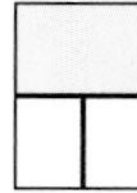

Adrenal gland, human, H&E, x90.
The boxed area on the orientation micrograph is shown here at higher magnification. The gland is covered by a moderately thick **capsule** (1). Beneath this is the **zona glomerulosa** (2). It is of somewhat irregular thickness. Because of the smaller size cells in this part of the cortex, the nuclei are in relatively close proximity, giving this area a more basophilic appearance. The **zona fasciculata** (3) consists of linear cords of cells separated by **capillaries** (4), many of which are filled with red blood cells and thus are visible. The **zona reticularis** (5) consists of irregular, anastomosing cords of cells, thus distinguishing it from the zona fasciulata. A small area of the **medulla** (6) is visible at the bottom of the micrograph.

Adrenal gland, human, H&E, x160.
This higher magnification shows the **capsule** (7) and **zona glomerulosa** (8). Note how the cells of the zona glomerulosa are arranged mostly in ovoid groups and are small compared to the cells of the **zona fasciculata** (9). Also, note that the cells of the zona fasciculata, in addition to being relatively large, have a clear cytoplasm, a reflection of the numerous lipid droplets they possess. In the deeper portion of the **zona fasciculata** (10), the cells contain less lipid and are smaller.

Adrenal gland, human, H&E, x160.
This micrograph reveals part of the **zona reticularis** (11) and below it the **medulla** (12), which includes a section through the **central vein** (13). The medullary cells are organized in ovoid clusters and short interconnecting cords. The central vein seen here has a thick layer of longitudinally disposed **smooth muscle cells** (14) on one side. Much of the remainder of the wall of the vein exhibits little or no smooth muscle.

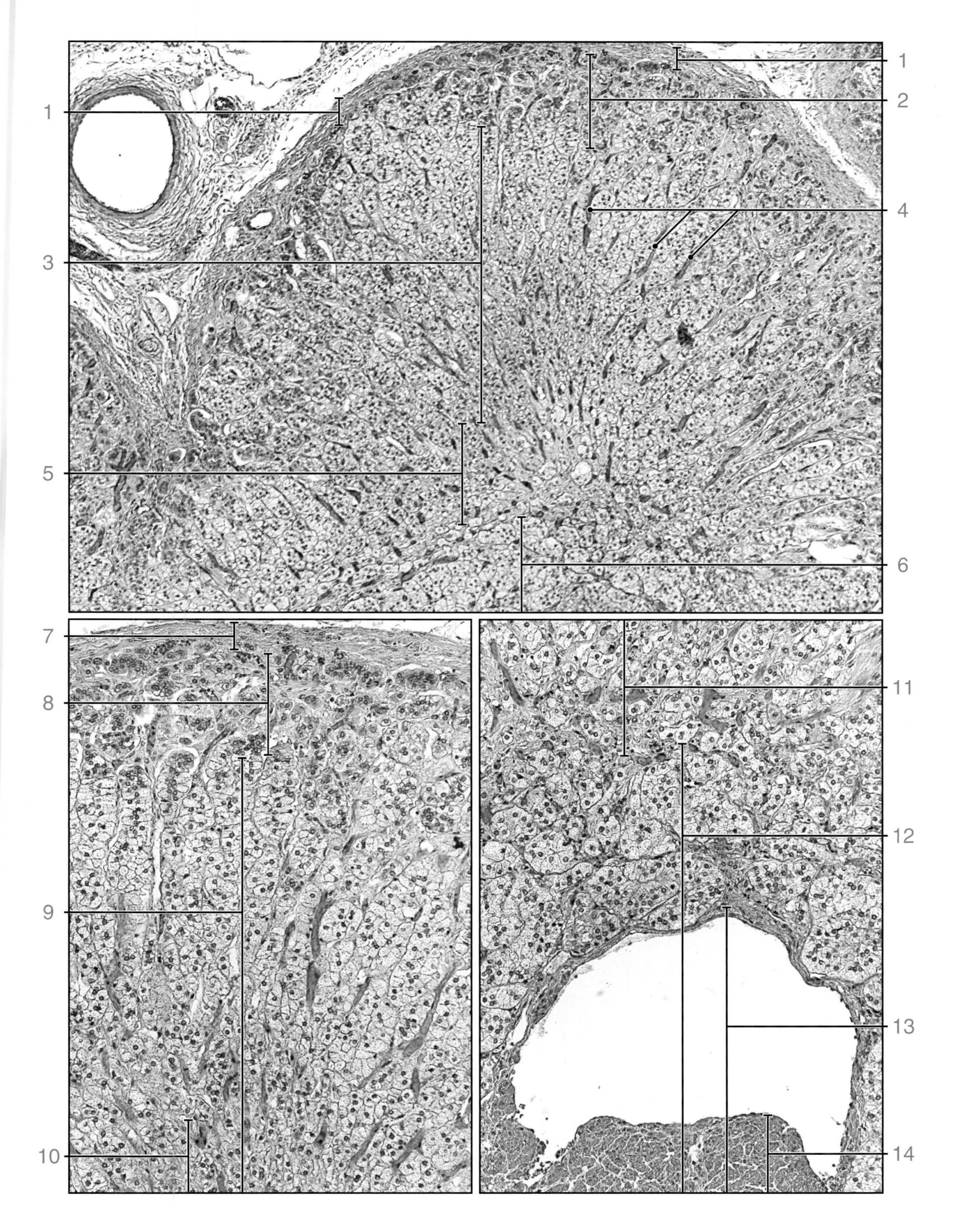

PLATE 124. ADRENAL GLAND

PLATE 125. FETAL ADRENAL

An unusual feature of the fetal adrenal gland is its organization and large size relative to other developing organs. It arises from mesodermal cells at around 4 to 5 weeks, forming a large, eosinophilic cell mass that becomes the functional fetal adrenal. A second wave of cells proliferates later from the mesenchyme and surrounds the primary cell mass. At around 6 to 7 weeks of fetal life, cells from the neural crest migrate into the developing fetal adrenal gland and aggregate to form the adrenal medulla. Around the fourth fetal month, the gland reaches its maximum mass in terms of body weight and is only slightly smaller than the kidney. At birth, the adrenal glands are equal in size and weight to those of an adult. They produce 100 to 200 mg of steroid compounds daily, approximately twice that of adult adrenals. The appearance of the fetal adrenal gland, once developed, is similar to that of the adult. The outer portion of the gland, the part that arose from the secondary mesodermal cell migration, is referred to as the permanent cortex. It has the general appearance of the adult zona glomerulosa. The cells are arranged in arched groups, extending into short cords. They become continuous with the cords of cells that developed from the initial mesodermal cell migration. The portion of the cortex that formed these cords of cells is referred to as the fetal cortex (zone). At this late stage of development, the chromaffin cells that have invaded the fetal zone are present in small, scattered cell groups but are difficult to recognize in H&E preparations. The fetal cortex at birth undergoes a rapid involution through the loss of its cells that reduces the gland to a quarter of its birth size within the first postnatal month. The permanent cortex grows and matures to form the characteristic zonation of the adult cortex. With the disappearance of the fetal zone, the chromaffin cells aggregate to form the medulla.

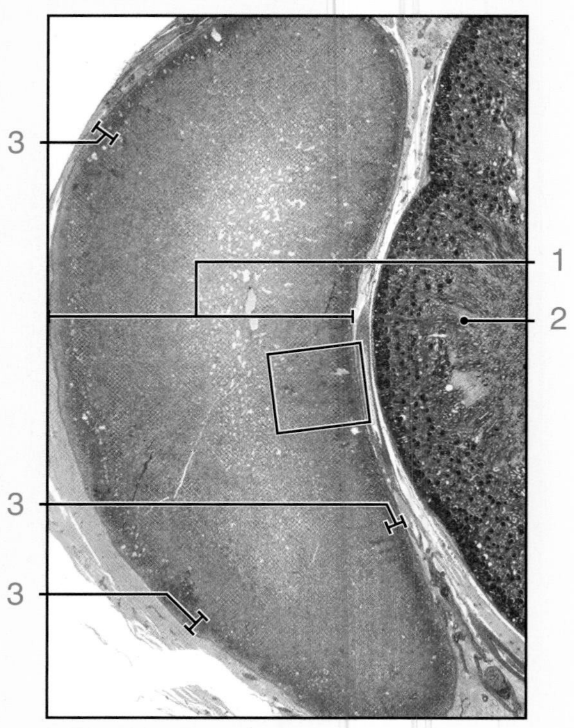

ORIENTATION MICROGRAPH: This section is from a **fetal adrenal** (1) near late term. The darker-staining structure is the **kidney** (2). The adrenal gland displays an outer basophilic-staining region, which represents the **permanent cortex** (3). The remainder of the gland consists of the fetal zone.

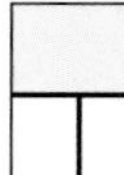

Fetal adrenal gland, human, H&E, x90.
The boxed area on the orientation micrograph is shown here at higher magnification. It shows the **capsule** (1), the underlying **permanent cortex** (2), and the **fetal zone** (3) that occupies the major portion of the micrograph. The basophilia of the permanent cortex is due largely to the closely spaced nuclei and relatively scant, poorly staining cytoplasm of these cells. In contrast, the fetal zone cells are larger, their nuclei are more dispersed, and their cytoplasm, which is more abundant, stains with eosin. The clear spaces, which are seen in the deeper portion of the fetal zone, are dilated **blood sinuses** (4).

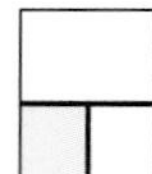

Fetal adrenal gland, human, H&E, x255.
The boxed area on the top micrograph is shown here at higher magnification. It reveals the **capsule** (5), the **permanent cortex** (6), and the very upper portion of the **fetal zone** (7). Note the upper part of the permanent cortex consists of very small cells; their nuclei are in close apposition. The cells close to the capsule are arranged in ovoid groups, whereas those deeper in the permanent cortex are arranged in cord-like fashion. Occasional **capillaries** (8) can be recognized where red blood cells fill the vessels. Compare the cells of the fetal zone with those of the permanent cortex, particularly those in the upper region of the permanent cortex.

Adrenal gland, human, H&E, x255.
This micrograph shows the cells of the fetal zone near the center of the gland. Note their relatively large size. The light staining areas in these cells are due to the presence of numerous **lipid droplets** (9), whereas the dark, eosinophilic areas are due to the presence of an extensive system of **smooth endoplasmic reticulum** (10). Between groups and cords of these fetal zone cells are **blood sinuses** (11).

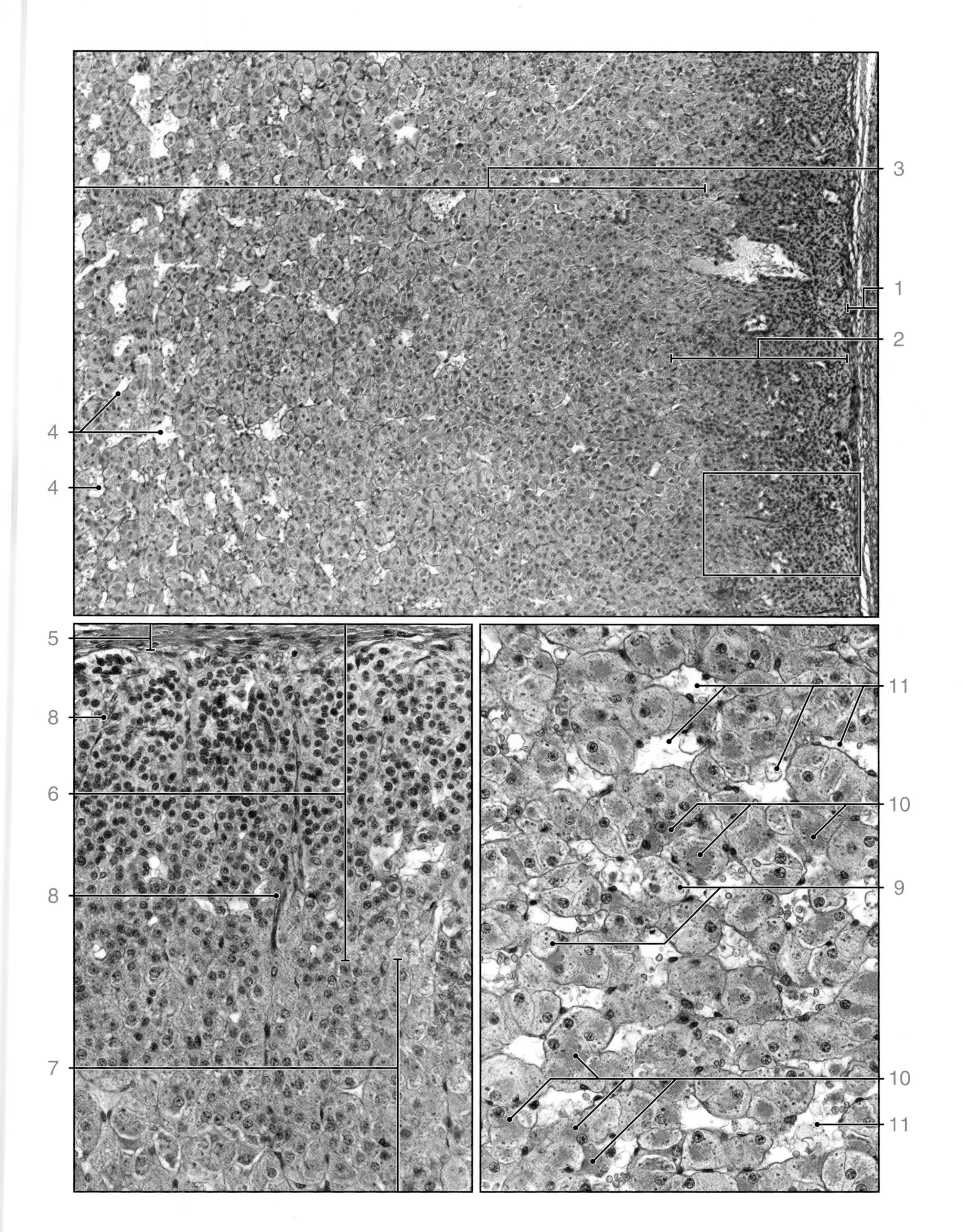
3
1
2
4
4
5
8
6
8
7
11
10
9
10
11

CHAPTER 18

Male Reproductive System

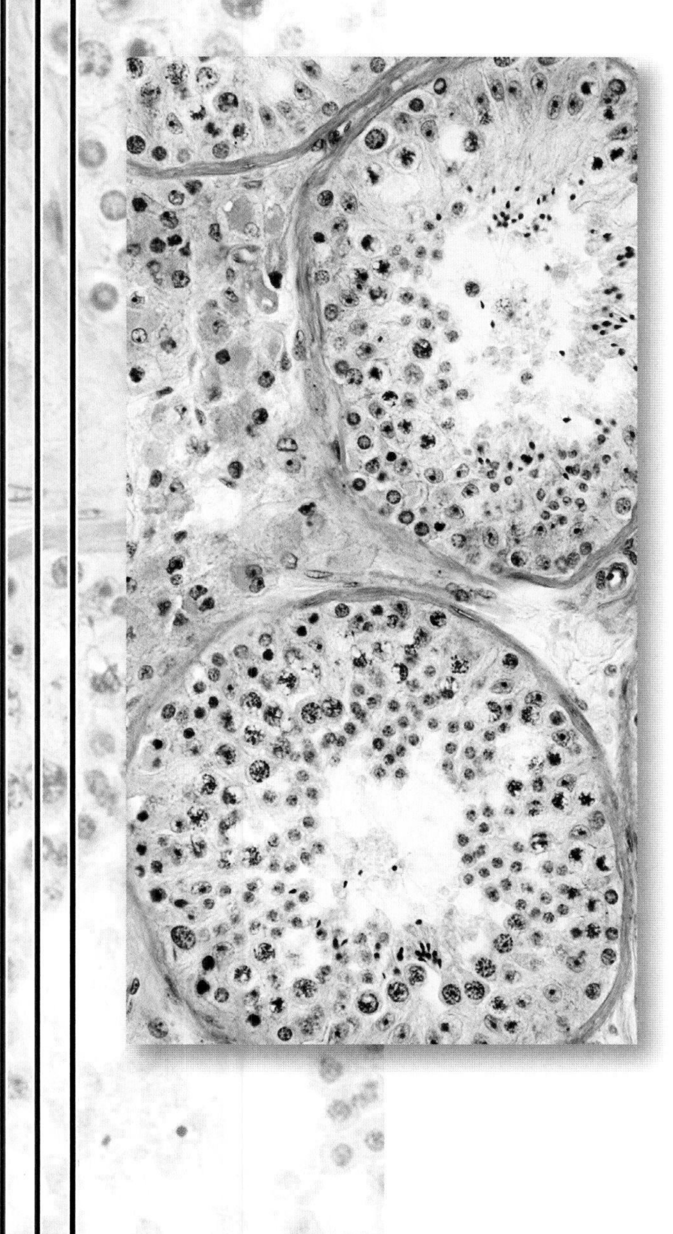

PLATE 126. **TESTIS I**

The male reproductive system consists of a number of tubular structures that include the seminiferous tubules and rete testis, which are contained within each of the testes, and an excurrent duct system outside of the testis, which includes the ducts within the epididymis and the ductus (vas) deferens. Also part of the system are the accessory glands: the prostate, the seminal vesicles, and the bulbourethral (Cowpers) glands.

The testes produce the male gametes (the spermatozoa), which are created by the spermatogonia that line the seminiferous tubules. By means of a series of mitotic divisions and one meiotic division, a large number of spermatids are produced from each spermatogonium. The spermatids mature into spermatozoa and are then released into the tubule lumen and carried by fluid and peristaltic activity to the ampulla of the ductus deferens where they are stored.

Within the testes, the interstitial cells of Leydig, which are aggregated in groups of varying size between the seminiferous tubules, produce the hormone testosterone.

ORIENTATION MICROGRAPH: This micrograph shows part of a hemisection of a **human testis** (1), including part of the overlying **epididymis** (2). The testis possesses an unusually thick connective tissue capsule, the **tunica albuginea** (3), which is conspicuous even at this very low magnification. Compare its thickness with that of the capsule of the **epididymis** (4). The bulk of the testis consists of the seminiferous tubules, which are just barely visible at this magnification. The wider diameter duct of the epididymis takes an extremely tortuous course and, consequently, appears as numerous profiles, all of which represent a single duct. The boxed area is seen at higher magnification in the top micrograph on the adjacent page.

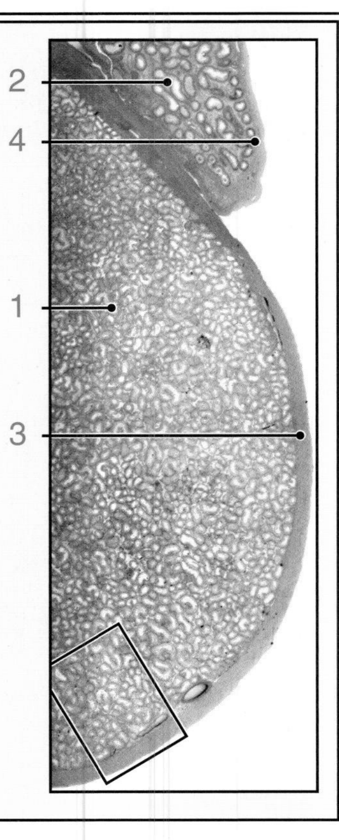

Testis, human, H&E, x50; inset x285.

This micrograph of the boxed area in the orientation micrograph shows the **tunica albuginea** (1) and the **seminiferous tubules** (2). A less dense layer of connective tissue, the **tunica vasculosa** (3), lies below the tunica albuginea and contains most of the larger **blood vessels** (4). This layer extends into the interior of the testis, filling the space between the tubules. It contains the smaller blood vessels. In histological section, the seminiferous tubules exhibit a variety of profiles; some appear as cross-sectioned, circular profiles, whereas others are C- or S-shaped. This pattern is a reflection of the tortuous course of the tubules. The **inset** shows the tunica albuginea at higher magnification. Note the paucity of fibroblasts within the densely packed collagen. The surface of the testis is covered by a **simple cuboidal epithelium** (5), the visceral layer of the tunica vaginalis that surrounds most of the testis.

Testis, human, H&E, x365.

This micrograph shows portions of several seminiferous tubules, two of which are cut in cross-section, at higher magnification. Also included is a prominent intertubular area, containing many **Leydig cells** (6) and **blood vessels** (7) of various size. The seminiferous epithelium of each tubule is surrounded by a tunica propria consisting of two to three layers of **myoid cells** (8). At this magnification, it is possible to recognize various developmental steps of the spermatogenic cells. The **nearly mature spermatids** (9) have densely stained, small, elongate nuclei. They are located at the lumen. **Early spermatids** (10) have round nuclei and though they also are seen in proximity to the lumen, they occur in stratified layers. The **spermatocytes** (11) are more removed from the lumen and are present in lesser number. They have larger round nuclei with a distinctive threadlike nuclear chromatin pattern. The **spermatogonia** (12) are fewest in number, with small round nuclei, and are located at the periphery of the tubule against the basement membrane. The nonspermatogenic cells, the **Sertoli cells** (13), also rest on the basement membrane. They have elongate nuclei, almost pyramidal in shape, and display a dense, round nucleolus.

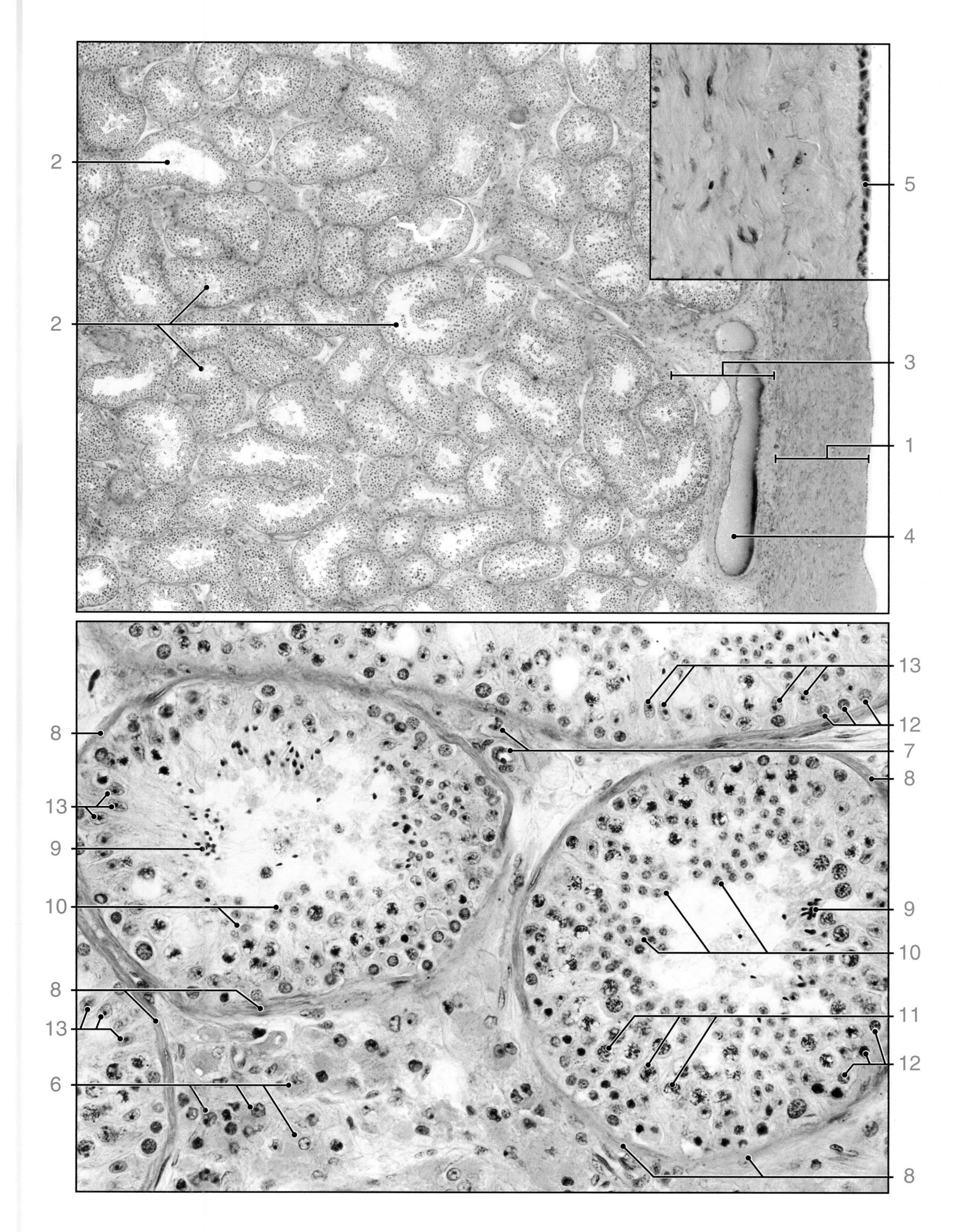

2
2
5
3
1
4
13
12
8
7
8
13
9
9
10
10
8
11
13
12
6
8

PLATE 127. **TESTIS II**

Seminiferous epithelium consists of two specific cell types: Sertoli cells and spermatogenic cells. The latter undergo a regular pattern of replication and differentiate into sperm. The Sertoli cells make up the true epithelium of the tubule. They are tall, columnar cells that rest on the basal lamina and extend to the tubule lumen. The spermatogenic cells are interspersed between the Sertoli cells and exhibit various stages of differentiation. They are surrounded, to varying degrees, by complex processes of the Sertoli cells throughout their process of differentiation. The least differentiated spermatogenic cells are the spermatogonia. Like the Sertoli cells, they rest on the basal lamina. Spermatogonia undergo division to produce either new spermatogonia that maintain the spermatogonial population as stem cells, or they undergo a series of divisions in which they produce a population of new spermatogonia that ultimately give rise to primary spermatocytes. The primary spermatocytes then divide meiotically to form secondary spermatocytes. The secondary spermatocytes divide again to produce spermatids. Finally, the spermatids undergo a maturation process that results in spermatozoa. During the process of spermatogenesis, the spermatogenic cells, after passing through the blood–testis barrier as primary spermatocytes, become physically attached to the Sertoli cells by specialized junctions until they are ready for release into the tubule lumen as spermatozoa.

Testis, human, H&E, x725.
Shown here are profiles of two adjoining seminiferous tubules that include the lumen of each. At the bottom of the micrograph is the outermost portion of a third tubule profile. At this higher magnification, the nuclei of the **peritubular myoid cells** (1) are readily recognized. Peripheral to these cells, **fibroblast nuclei** (2) can sometimes be recognized. They provide for the collagenous component of the interstitium, whereas the myoid cells are responsible for peristaltic movement of the seminiferous tubules. Also evident in this micrograph are several **Leydig cells** (3), which represent the endocrine component in the testis. The **Sertoli cells** (4), which rest on the basement membrane, are best recognized by their nuclei. The nucleus often appears pyramidal in shape and contains a small, dense nucleolus with prominent round-bodied karyosomes. Also located immediately adjacent to the basement membrane of the tubules and the myoid cell layer are the **spermatogonia** (5) of the seminiferous epithelium. Their nuclei are spherical in shape. The last division of the spermatogonia results in the production of **primary spermatocytes** (6), which can be recognized by their relatively large nuclei and the threadlike appearance of their chromatin material in preparation for meiosis. The cells derived from the first meiotic division are the secondary spermatocytes. These cells are present for an exceedingly short time compared to primary spermatocytes and thus are rarely seen. Division of a secondary spermatocyte results in the production of two **spermatids** (7). These cells, located close to the lumen, have small, round nuclei. Maturation of these early spermatids involves, among other processes, condensation of the nucleus, forming the **late spermatid** (8). They are identified by their elongate shape and very dense looking nuclei.

Testis, human, H&E, x890.
The micrographs shown here represent three different stages of the six recognizable cell-associations that occur in the cycle of the human seminiferous epithelium. The micrograph on the left represents stage III. The cells with the small, round nuclei are **early spermatids** (9). They are seen in association with **pachytene spermatocytes** (10). The micrograph in the middle represents stage V. Note that the spermatids near the lumen are considerably advanced in development. The nuclei have become elongate and have a shape approaching that of a mature spermatid. The large, round nuclei indicate spermatocytes that are still at the pachytene stage. Finally, the micrograph on the right represents stage I. The **spermatids** (11) closest to the lumen exhibit highly condensed nuclei. They are almost ready for release into the lumen. The cytoplasm in proximity to these spermatids consists of remnants of the spermatid cytoplasm, which will become residual bodies on release of the spermatids at stage II. Also present at this stage is a new generation of **very early spermatids** (12). Two of the remaining nuclei belong to **pachytene spermatocytes** (13).

Testis, human, H&E, x510.
This micrograph is from the same testis as the specimens shown above. The seminiferous tubule is in an area that reflects testicular dysfunction. Most of the cells in this tubule are **Sertoli cells** (14). The spermatogenic cells are few in number, and those that can be readily identified appear to be late spermatids.

Testis, human, H&E, x350.
This micrograph, taken from the same general area as the previous illustration, shows further degradation of some of the seminiferous tubules. Here, there is a virtual absence of Sertoli cells. For the most part, the cellular component of the tubule has been replaced by **fibrous material** (15). The remaining seminiferous tubule in the micrograph shows a population of **spermatogonia** (16) but an absence of succeeding cell types. The **interstitium** (17) contains numerous cells of Leydig.

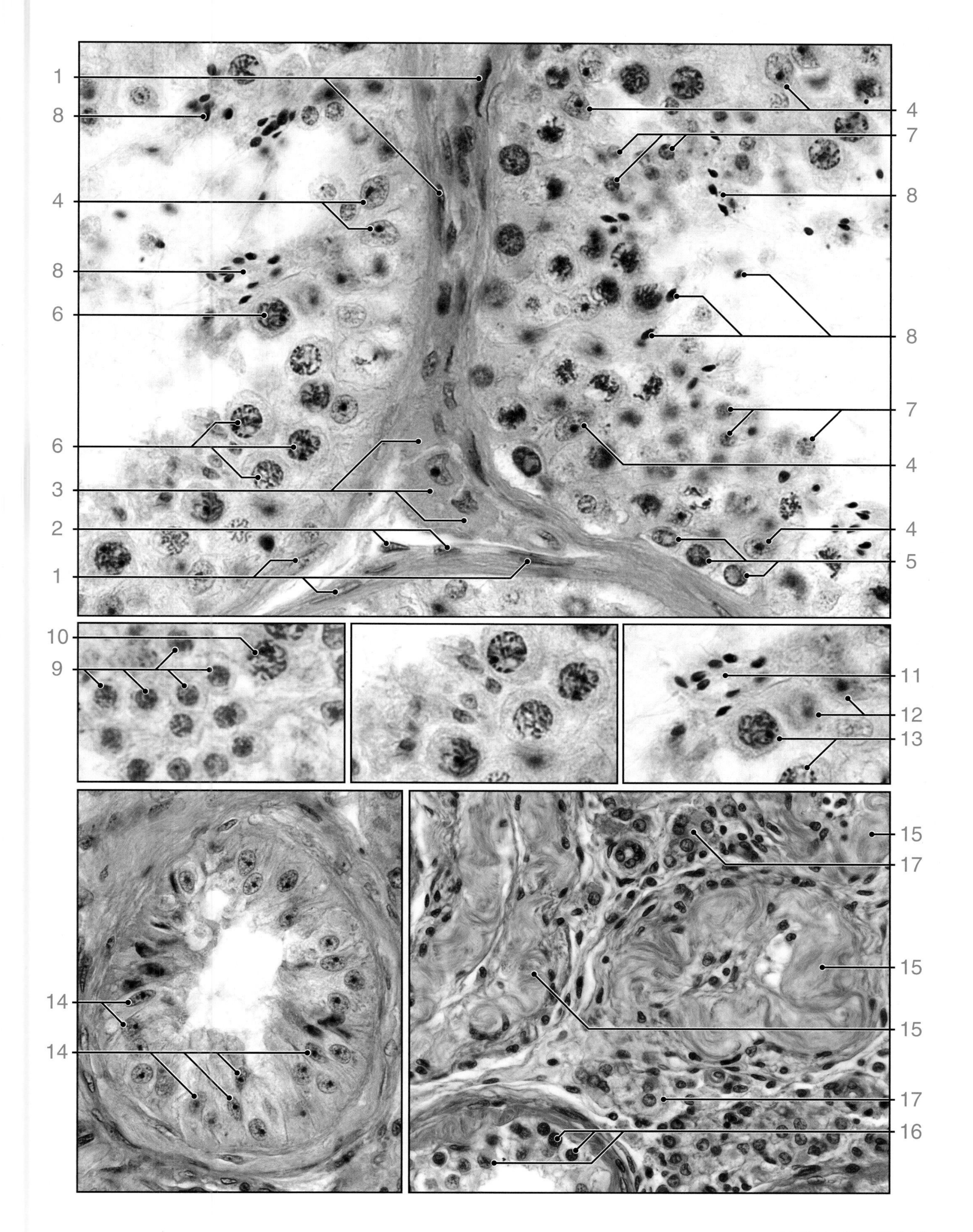
1
8
4
8
6
6
3
2
1
4
7
8
8
7
4
4
5
10
9
11
12
13
14
14
15
17
15
15
17
16

PLATE 128. **TESTES III, PREPUBERTAL TESTIS AND RETE TESTIS**

The prepubertal testis is characterized by cords of cells consisting of sustentacular (Presumptive Sertoli) cells and gonocytes. The gonocytes are the precursors of spermatagonia and are derived from primordial germ cells that invaded the developing gonad. At puberty, these cords become canalized to form tubules and the gonocytes undergo multiple divisions that ultimately give rise to spermatagonia.

In the mature individual, the seminiferous tubules terminate as straight tubules (tubuli recti) lined only by Sertoli cells. The tubuli recti lead to the rete testis, a series of anastomosing channels in the mediastinum testis. The rete testis is the termination of the intratesticular tubule system.

ORIENTATION MICROGRAPH: This micrograph shows a midsagittal section through the **testis** (1) of a newborn. The anterior portion of the testis is on the right. At the top of the micrograph are the **efferent ductules** (2) and the initial portion of the **ductus epididymis** (3). These two segments of the excurrent duct system make up the head of the epididymis. Below and to the left is part of the **body of the epididymis** (4). The **mediastinum testis** (5) is located in the concavity of the testis.

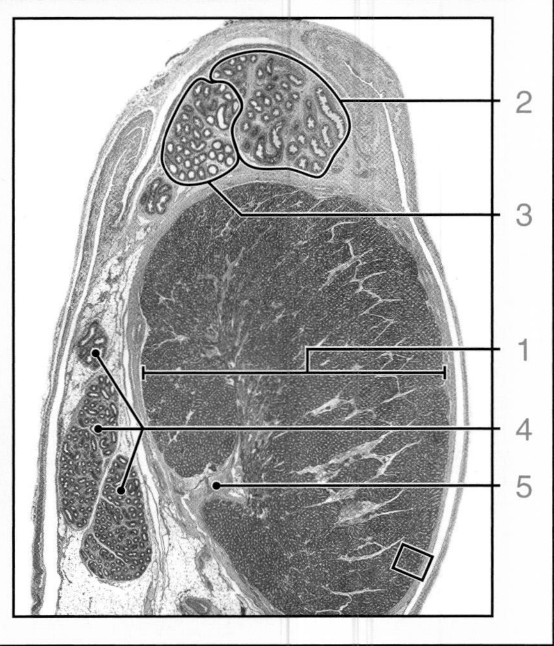

Testis, newborn human, H&E, x125.
This micrograph from the boxed area on the orientation micrograph shows the outer part of the testis and includes the **tunica albuginea** (1) and the seminiferous cords and surrounding **interstitial tissue** (2). The diameters of the cords are smaller than of the tubules of the adult testis. Compare the diameters of the seminiferous cords with those of the seminiferous tubules in Plate 126 (in determining relative size, note the lower magnification of the tubules in Plate 126). Another distinction between prepubertal and adult tunica albuginea is that the former is more cellular and contains fewer collagen fibers.

Testis, human, prepubertal, H&E, x250.
At this higher magnification, the seminiferous cords can be seen to consist of two cell types. The principal cell type, the **presumptive Sertoli cell** (3), is the predominant cell type present. The nuclei of these cells reside at the base of the cells. The second cell type, the **gonocyte** (4), is identified by its dense nucleus and surrounding clear cytoplasm. These cells undergo periodic division; some of the cells then move towards the center of the cord and ultimately degenerate, whereas others remain at the periphery of the tubule to maintain the gonocyte population. At puberty, the gonocytes are stimulated to divide at a more rapid rate and become spermatogonia. At the same time, the presumptive Sertoli cells undergo further differentiation and begin to secrete fluid, thus creating a fluid-containing lumen. Surrounding each cord is a thin layer of cells that look like **fibroblasts** (5). Some of these cells will ultimately differentiate into the contractile myoid cells that surround the seminiferous epithelium; others retain fibroblast activity in the interstitium. In addition to the small blood vessels in the interstitium, **Leydig cells** (6) are also present. They are the largest cells in the prepubertal testis.

Testis, monkey, H&E, x65.
The portion of the testis shown here includes part of the mediastinum testis. At the top of the micrograph are several **seminiferous tubules** (7) and below is part of the **rete testis** (8) within the mediastinum testis. One of the **seminiferous tubules** (9) has been sectioned at its terminal region, where it becomes a straight tubule (tubulus rectus). It contains only Sertoli-like cells. At the point where the Sertoli-like cells are no longer present, the tubule empties into the rete testis. In addition to the dense connective tissue, blood vessels, and nerves, areas of **adipose tissue** (10) are also present in the mediastinum testis.

Testis, monkey, H&E, x400.
This micrograph is a higher magnification of the boxed area on the micrograph immediately above. The **seminiferous tubule** (11) is seen in continuity with a **straight tubule** (12). It is relatively short and terminates at the point where it enters a channel of the **rete testis** (13). The rete testis is identified by its simple cuboidal epithelium. Note that because of the angle at which the straight tubule was sectioned, it appears that the epithelium characterizing the rete testis begins on the upper side of the tubule before it is seen on the lower side of the tubule.

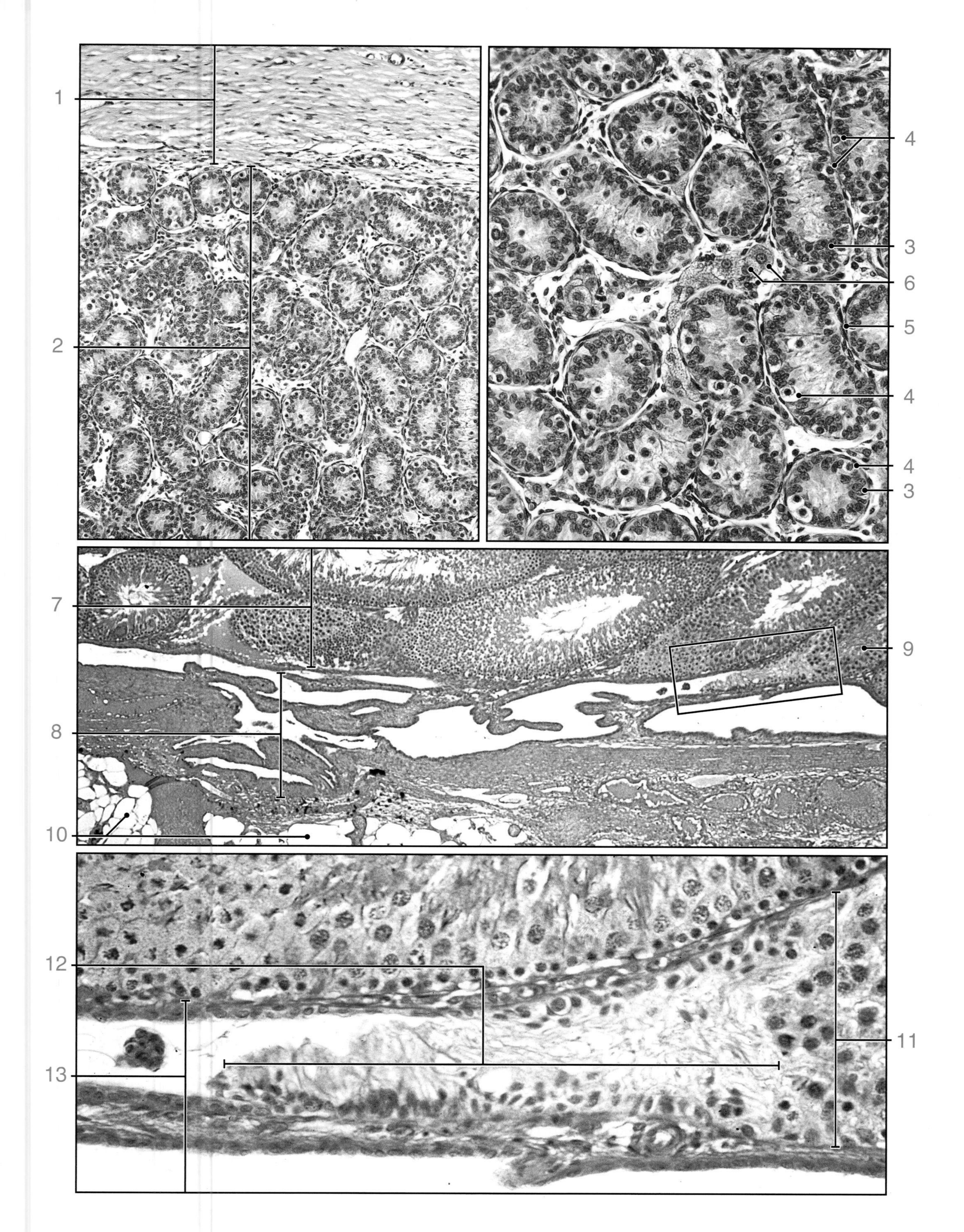

PLATE 128. TESTES III, PREPUBERTAL TESTIS AND RETE TESTIS

PLATE 133. SEMINAL VESICLE

The seminal vesicles are paired structures that develop as evaginations of each ductus deferens. They form a tightly coiled tube and rest on the posterior wall of the urinary bladder, parallel to the ampulla of the ductus deferens. A short excretory duct extending from each seminal vesicle joins the ampulla of the ductus deferens to form the ejaculatory duct. A cross section of the seminal vesicle typically shows many lumina; however, they are all profiles of a single, continuous, tortuous, tubular lumen. The lumen is lined with a pseudostratified columnar epithelium, closely resembling that of the prostate gland. Their secretion is a whitish-yellow, viscous material containing fructose, other simple sugars, amino acids, ascorbic acid, and prostaglandins. The latter, although secreted by the prostate gland, is synthesized in larger amounts in the seminal vesicles. Fructose is the major nutrient source for sperm in the semen. The mucosa rests on a thick layer of smooth muscle, the muscularis, which is directly continuous with that of the ductus deferens. The smooth muscle consists of an indistinct inner circular layer and an outer longitudinal layer. Contraction of the smooth muscle during ejaculation forces the secretions of the seminal vesicles into the ejaculatory ducts. The smooth muscle is surrounded by connective tissue, forming the adventitia of the gland.

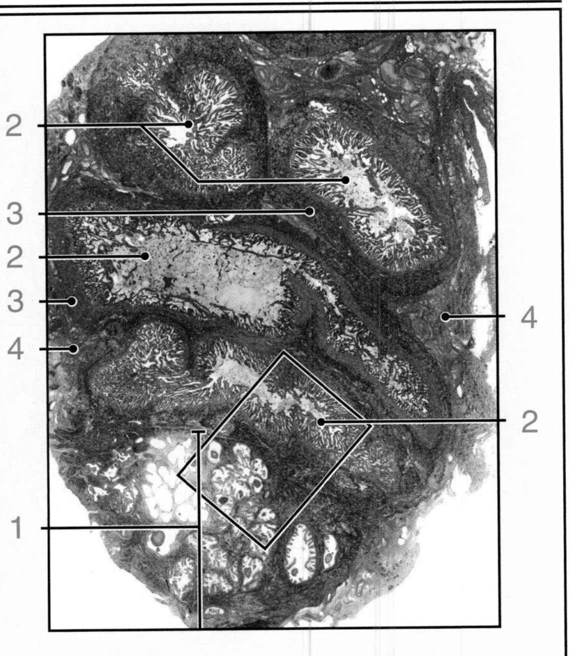

ORIENTATION MICROGRAPH: This is a cross section through a seminal vesicle near its junction with the ejaculatory duct. A small part of the **prostate gland** (1) is included in the section. The remainder of the micrograph reveals four profiles of the coiled **seminal vesicle** (2). The smooth muscle component, the **muscularis** (3), appears red, whereas the connective tissue, the **adventitia** (4), binding the coils of the structure together, appears blue.

Seminal vesicle and prostate, human, Mallory-Azan, x45; inset x175.

The boxed area on the orientation micrograph is shown here at higher magnification. Note the **prostatic concretions** (1), which readily identify the prostate tissue that is adherent to the seminal vesicle. The right side of the micrograph reveals the seminal vesicle. The **mucosa** (2) consists of an extensively folded or ridged structure. Each of these ridges has a core of connective tissue extending to the base of each ridge where it is continuous with the underlying connective tissue. Surrounding the mucosa is smooth muscle, the **muscularis** (3). The **inset** shows, at higher magnification, the basal aspect of the ridges with their **epithelial surface** (4), the underlying **connective tissue** (5), and a portion of the **muscularis** (6).

Seminal vesicle, human, Mallory-Azan, x365; inset x675.

The mucosa of the seminal vesicle is shown here at higher magnification. The **epithelium** (7) of the mucosal folds rests on a very thin layer of **connective tissue** (8). **Capillaries** (9) within this connective tissue core are abundant and can be readily observed. The irregular folds of the mucosa give the appearance of **enclosed spaces** (10) similar to those seen in the prostate gland. In actuality, these spaces are continuous with, and part of, the lumen of the seminal vesicle. The **inset** shows the epithelium at much higher magnification. The epithelium is pseudostratified and consists of **basal cells** (11) and tall **columnar cells** (12). The nuclei of the basal cells are close to the basement membrane and usually appear somewhat flattened or oval in shape. The nuclei of the columnar cells are elongate and assume the shape of the columnar cell. In examining the basal cells, note that they are prevalent in some areas whereas they seem to be sparse in others. The **inset** micrograph also prominently shows the **terminal bars** (13) of the columnar cells, which appear as dot-like structures.

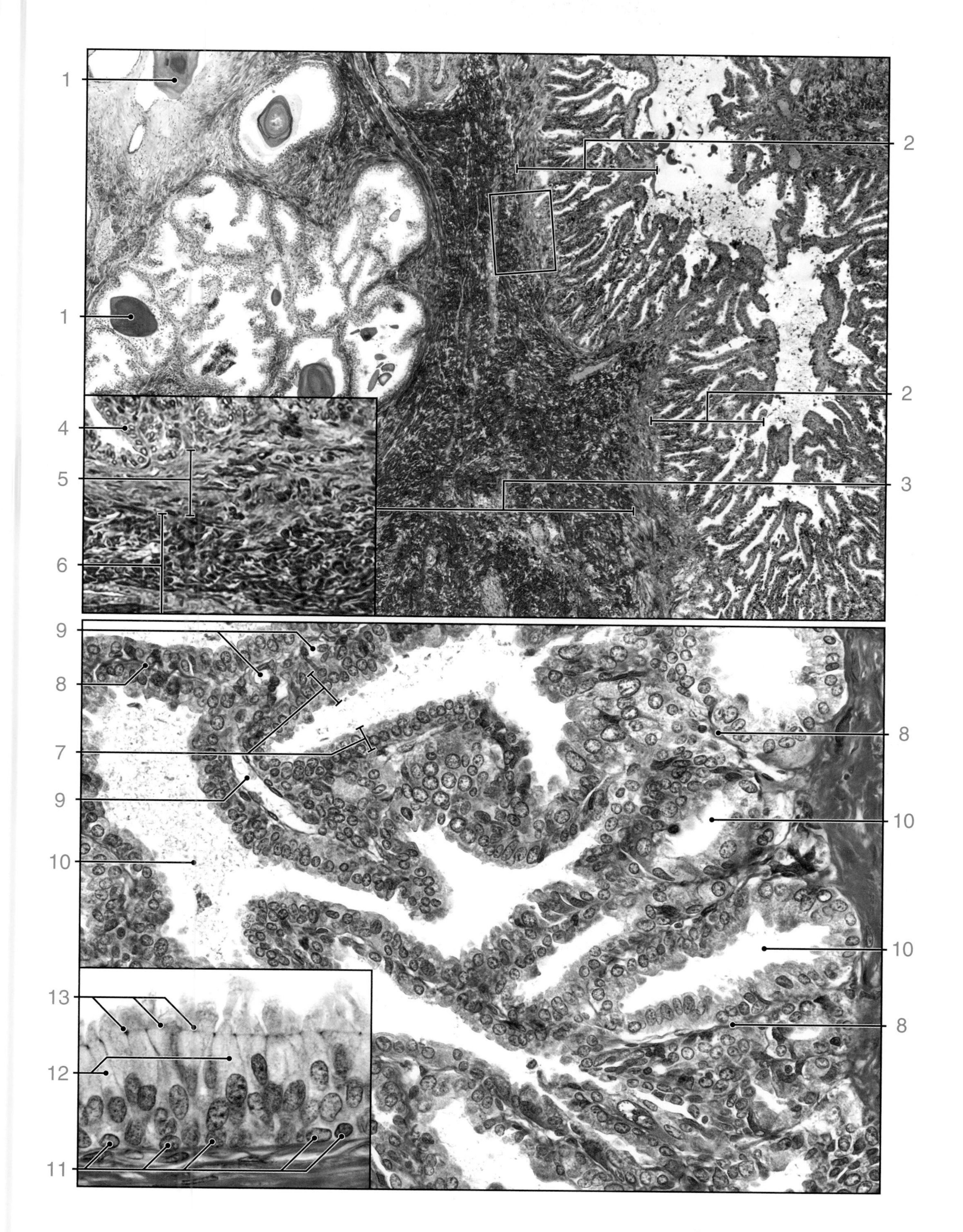
1
1
2
2
3
4
5
6
9
8
7
9
10
8
10
10
8
13
12
11

CHAPTER 19

Female Reproductive System

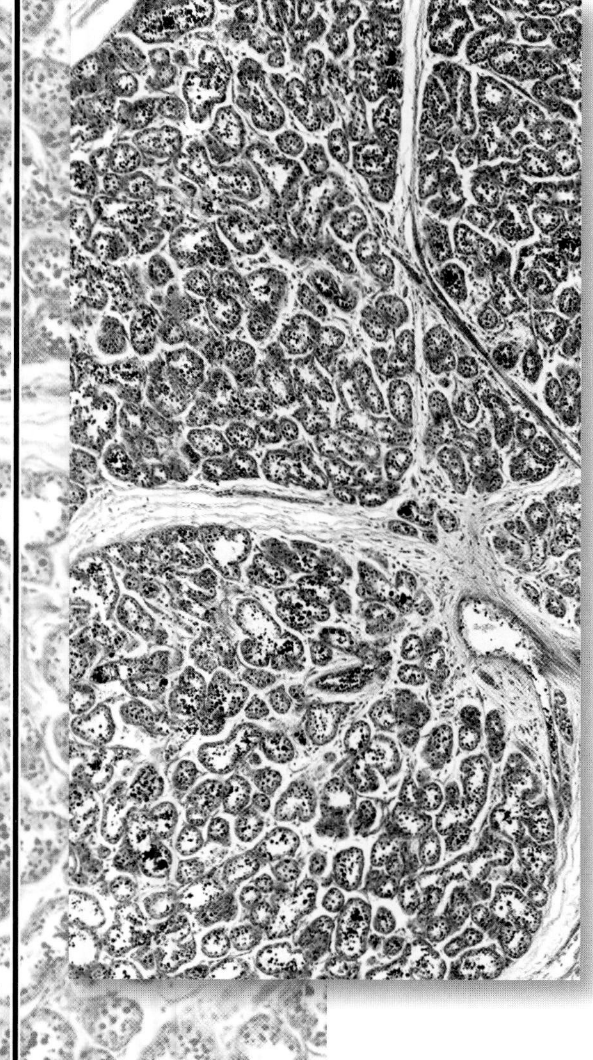

PLATE 134. OVARY I

The ovaries are small, paired, ovoid organs whose functions are the production of gametes (oocytes) and the synthesis and secretion of steroid hormones (estrogens and progestogens). Each ovary is attached to the broad ligament by the peritoneal fold, the mesovarium, that extends from the broad ligament into the hilum of the ovary. The hilum is the structure through which neurovascular components pass into the ovarian stoma.

The outer part of the ovary, the cortex, contains numerous primordial follicles. Primordial follicles are oocytes arrested in the prophase of the first meiotic division. A primordial follicle is surrounded by a single layer of squamous follicular cells bounded by a basal lamina on their outer surface. Towards the end of puberty, under the influence of pituitary gonadotropins, the ovaries begin to undergo cyclical changes known as the ovarian cycle. At the beginning of the ovarian cycle, influenced by the pituitary's follicle-stimulating hormone (FSH), some of the primordial follicles begin to undergo changes that lead to the development of a mature (Graafian) follicle and an oocyte capable of ovulating. Development of the follicle involves enlargement of the oocyte as well as proliferation of the surrounding follicular cells, which become cuboidal; the follicle is then identified as a primary follicle.

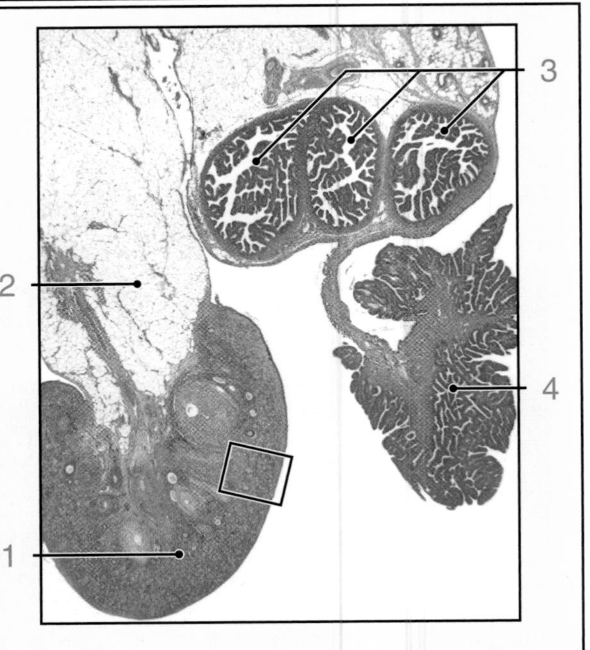

ORIENTATION MICROGRAPH: A section through an **ovary** (1), its **mesovarium** (2), the **uterine tube** (3), and the **infundibulum** (4) of the uterine tube is shown here. The infundibulum is the funnel-shaped, distal part of the uterine tube that opens into the peritoneal cavity. Its opening is surrounded by fimbriae, fingerlike extensions, which reach toward the ovarian surface at the time of ovulation.

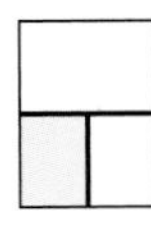

Ovary, cortex, monkey, H&E x120.

This low power micrograph shows part of the ovarian cortex from the boxed area on the orientation micrograph. The **germinal epithelium** (1) consists of a layer of simple cuboidal epithelium at the ovarian surface. It is continuous with the simple squamous epithelium (mesothelium) of the mesovarium (contrary to its name, it does not give rise to germ cells). The germinal epithelium covers a dense irregular connective tissue layer known as the **tunica albuginea** (2). The **cortex of the ovary** (3) exhibits follicles in different stages of development. The **primordial follicles** (4) are found at varying distances below the tunica albuginea, but are more numerous in the peripheral part of the cortex. **Early primary follicles** (5) are seen in greater numbers in the deeper parts of the cortex. In the deepest part of the cortex are the advanced follicles including a **late primary follicle** (6), in which the oocyte is surrounded by multilayered granulosa cells. Also present is a later stage of development, a **secondary follicle** (7). It has been tangentially sectioned; thus, the oocyte, or antrum, is not revealed.

Ovary, cortex, monkey, H&E x240.

A higher magnification of the cortex of the above specimen is shown here. It clearly shows the nature of the **germinal epithelium** (8). It is usually simple cuboidal, though some of the cells seen here, on the left side, are slightly taller than they are wide. Also, in some instances, this epithelium may have a squamous character in localized areas due to surface stretching, as in the case of cystic formations or development of the corpus luteum of pregnancy. The **tunica albuginea** (9) consists of thick collagen fibers with numerous small cells. The tunica albuginea and the germinal epithelium together are equivalent to the serosa or peritoneum of the abdominal cavity. This outer part of the cortex shows **primordial** (10) and **early primary follicles** (11). Note the squamous cells surrounding the oocyte in the primordial follicles. The primary follicles, in contrast, exhibit more numerous and more closely spaced round, follicle cell nuclei. The oocyte is also slightly larger. Some of the primary oocytes appear to lack a nucleus because as the size of the oocyte increases, the probability that its nucleus will be included in the section decreases.

Ovary, early primary follicles, monkey, H&E x280.

This micrograph reveals several early primary follicles from deeper in the cortex. Note the significantly increased size of the **oocytes** (12). The surrounding **follicle cells** (13) are also more numerous. The very small follicle with its eccentric nucleus, right, appears to have undergone very limited growth and will probably degenerate. It is common to observe small clusters of seemingly isolated **follicle cell nuclei** (14) in the ovarian stroma. Such profiles represent follicle cells seen *enface* as a result of a tangential section through the periphery of a follicle. The numerous cells that possess elongated nuclei, appearing much like those of fibroblasts, are stromal cells. They give rise to the theca interna and theca externa of more advanced follicles (described later).

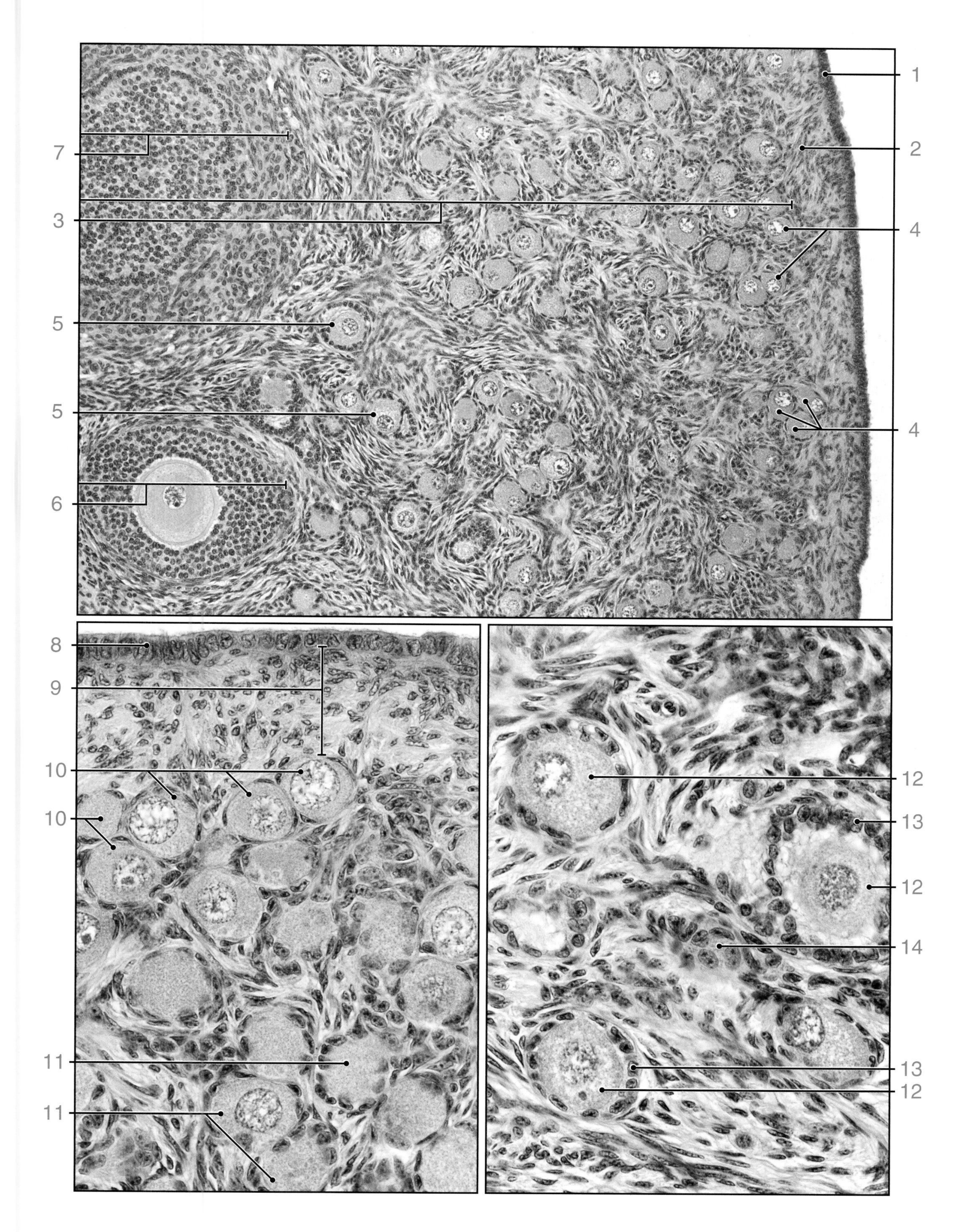
1
2
3
4
4
5
5
6
7
8
9
10
10
11
11
12
13
12
14
13
12

PLATE 136. **OVARY III**

Growth and maturation of follicles are always accompanied by atresia. Atresia is a process mediated by apoptosis of granulosa cells and subsequent resorption of follicles. It is a regular event in the ovary that begins in embryonic life. In any section through the ovary, follicles of various stages can be seen undergoing atresia. In all cases, the follicular cells show pycnosis of the nucleus and dissolution of the cytoplasm; the follicle is invaded by macrophages and other connective tissue cells. The oocyte degenerates, leaving behind the prominent zona pellucida. This may be folded inwardly or collapsed, but it usually retains its thickness and staining characteristics. Thus, disturbance in the normal architecture of the follicle and a partially disrupted zona pellucida, when included in the plane of section, serve as reliable diagnostic features of an atretic follicle.

Ovary, mature (Graafian) follicle, monkey, H&E x77; inset x420.

This micrograph shows a mature (Graafian) follicle that occupies almost the entire left side of the image. The **stratum granulosum** (1), which is formed by several layers of granulosa cells, encloses the large **antrum** (2). The antrum is filled with an eosinophilic precipitate, a remnant of liquor folliculi. The wall of the Graafian follicle possesses a mound made of granulosa cells called the **cumulus oophorus** (3). The cumulus oophorus projects into the antrum and contains a **mature oocyte** (4) surrounded by the zona pellucida. The cells of the cumulus oophorus, immediately surrounding the oocyte, remain with it after ovulation and are referred to as the corona radiata. The first layer of stromal cells surrounding to the stratum granulosum represents a highly vascular, steroid-producing layer referred to as the **theca interna** (5). The second layer of connective tissue cells, outside the theca interna, represents the **theca externa** (6). The **inset** shows layers of the follicular wall from the boxed area at higher magnification. The surface of the ovary is visible on the right. Multiple **primordial follicles** (7) are located beneath the germinal epithelium. In addition, in the deep cortex two **primary follicles** (8) are visible in their early stages of development.

Ovary, atretic primary follicle, monkey, H&E x77.

This micrograph shows a single atretic primary follicle at higher magnification. As atresia progresses, the oocyte becomes smaller and degenerates. Often, only parts of zona pellucida remain in the cavity once occupied by the oocyte. The follicle in this micrograph can be identified by virtue of the retained **zona pellucida** (9). In more advanced atresia, the **granulosa cell layer** (10) tends to degenerate more rapidly than the cells of the **theca interna** (11). Characteristics of atretic follicles include the absence of mitotic cell profiles and the invasion of the granulosa cell layer by strands of connective tissue.

Ovary, atretic mature (Graafian) follicle, monkey, H&E x77; inset x200.

This micrograph shows two atretic follicles. The larger one is the atretic **mature (Graafian) follicle** (12), and the smaller one, adjacent to the large one, represents the **atretic primary follicle** (13). Noticeable changes in the stratum granulosum of the Graafian follicle include the area *(arrows)* where granulosa cells are sloughing off into the antrum of the follicle. In addition, degeneration of the oocyte and cumulus oophorus is evident in this micrograph. Usually atresia of the oocyte within larger and more mature follicles is delayed and appears secondary to changes in the follicular wall. Note the more advanced changes in the oocyte of the primary follicle. The **inset** shows a higher magnification of the cells immediately surrounding the **oocyte** (14) from the boxed area on this micrograph. They have already lost their attachment to the rest of the cumulus cells. The zona pellucida and the border of the oocyte are difficult to discern.

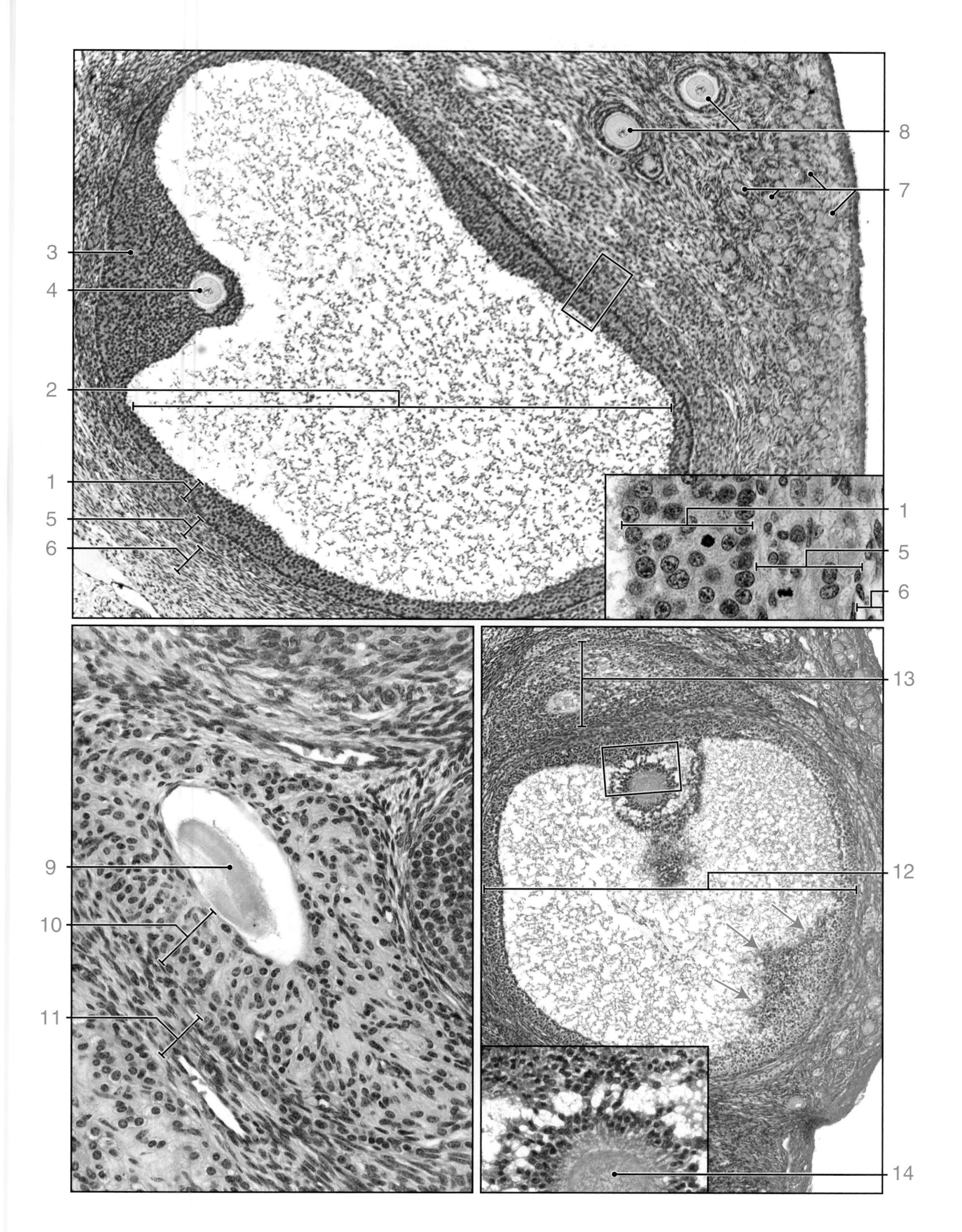
8
7
3
4
2
1
5
6
1
5
6
13
9
12
10
11
14

PLATE 138. CORPUS LUTEUM AND CORPUS ALBICANS

The corpus luteum of menstruation degenerates and undergoes a slow regression and involution after menstruation to form the corpus albicans. The corpus luteum becomes invaded by connective tissue stroma and gradually transforms into a scar. Large lutein cells in the degenerating corpus luteum decrease in size and undergo apoptosis. An immature corpus albicans may contain macrophages loaded with hemosiderin pigment. A mature corpus albicans is a well-defined structure with convoluted borders, formed by dense connective tissue, and occasional fibroblasts and intercellular hyaline material that accumulate among the degenerating cells of the former corpus luteum. The corpus albicans sinks deeper into the ovarian cortex as it is slowly resorbed over a period of several months. Persistent corpora albicantia are typically found in the medulla of the ovary and are usually remnants of former corpora lutea of pregnancy.

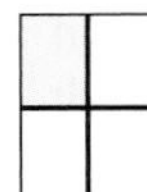

Ovary, corpus luteum of menstruation, human, H&E x255.

This micrograph shows a wall of the corpus luteum at higher magnification. The majority of cells on this section represent **granulosa lutein cells** (1). Each has a large spherical nucleus and a large amount of cytoplasm. The cytoplasm contains yellow pigment (usually not evident in routine H&E sections), hence the name, corpus luteum. The **theca lutein cells** (2) also have spherical nuclei, but the cells are considerably smaller than the granulosa lutein cells. The nuclei of adjacent theca lutein cells appear closer to each other than nuclei of adjacent granulosa lutein cells. The **connective tissue** (3) and small blood vessels that invaded the mass of corpus luteum can be identified as the flattened and elongated components between the lutein cells.

Ovary, corpus luteum of menstruation, human, H&E x255.

This micrograph shows the corpus luteum with a portion of the fibrous central core. On the upper part of the image are the **theca lutein cells** (4), below them is **loose connective tissue** (5), occupying the former antrum of a mature follicle that became a corpus luteum. In the process of corpus luteum formation, connective tissue from the ovarian stroma invades the former antrum. Accumulation of fibroblasts near the theca lutein cell border forms an **inner fibrous layer** (6) that lines the central cavity of the corpus luteum.

Ovary, atretic corpus luteum, human, H&E x45.

This micrograph shows two atretic corpora lutea of menstruation. If pregnancy occurs, the corpus luteum of menstruation undergoes further growth and development and becomes the corpus luteum of pregnancy; if pregnancy does not occur, the corpus luteum becomes atretic and is finally replaced by the corpus albicans. **Large blood vessels** (7), originating in the **connective tissue stroma** (8), surround the degenerating corpus luteum and penetrate its layers to reach the central cavity. The connective tissue that accompanies the vessels usually forms a **dense fibrous network** (9) between the granulosa lutein cells, which at the same time undergo apoptosis. The cells decrease in size, develop pycnotic nuclei, and accumulate lipids in their cytoplasm. Connective tissue also invades the central cavity to form a **fibrous central core** (10). This inner fibrous layer will contribute to the development of a hyaline membrane. Theca lutein cells are not visible at this stage.

Ovary, mature corpus albicans, human, H&E x35.

This micrograph shows a **mature corpus albicans** (11) at higher magnification. This is a connective tissue scar, a remnant of a former corpus luteum. It is a well-circumscribed, dense connective tissue structure, surrounded by the **ovarian stroma** (12). It contains remnants of hyalinized membranes *(arrows)* and sparsely distributed fibroblasts.

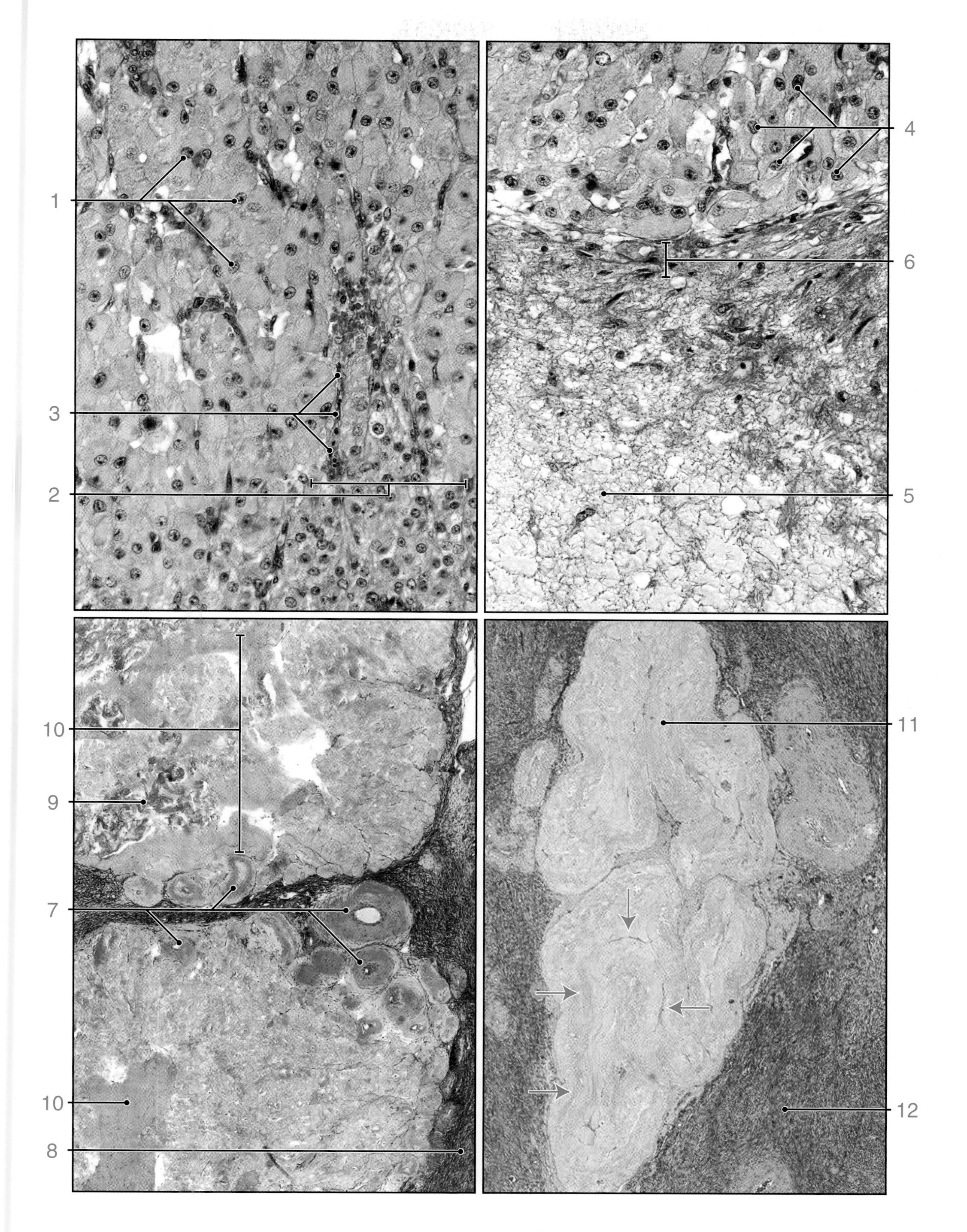

PLATE 138. CORPUS LUTEUM AND CORPUS ALBICANS

PLATE 139. UTERINE TUBE

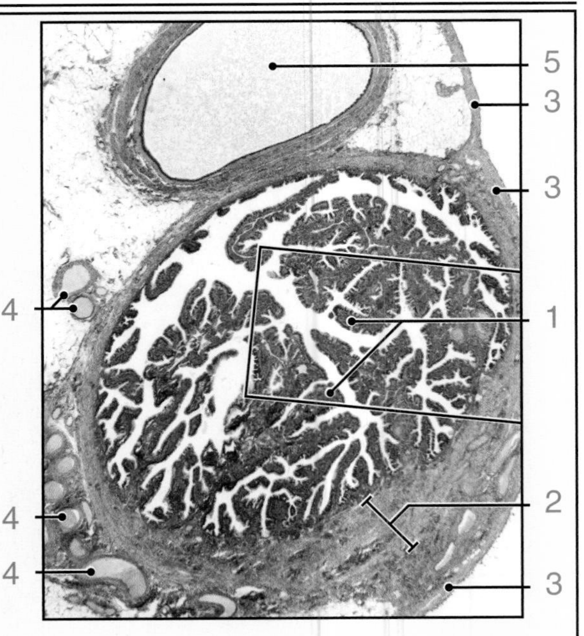

The uterine tubes (oviducts, Fallopian tubes) are paired, tubular structures that extend from each side of the superior end of the body of the uterus to the ovaries, where they present an open, flared end (abdominal ostium) for entry of the ovum at ovulation. The uterine tube is lined by a simple columnar epithelium, which, under hormonal influence, undergoes cyclic changes. The epithelial cells increase in height during the middle of the cycle and decrease in height during the premenstrual period. Also, the ratio of ciliated to nonciliated cells increases during the follicular phase of the ovarian cycle.

The uterine tube varies in size and degree of mucosal folding along its length. Mucosal folds are evident in the distal portion, near the abdominal opening, where the tube flares outward and is called the infundibulum. It has fringed, folded projections called fimbriae that facilitate the collection of the ovulated oocyte from the ovary. The infundibulum leads proximally to the ampulla, which constitutes about two thirds of the length of the uterine tube. It has the most numerous and complex mucosal folds and is the site of fertilization. Mucosal folds are least numerous at the proximal end of the uterine tube, near the uterus, where the tube is narrow and referred to as the isthmus. The uterine, or intramural, portion measures about 1 cm in length and passes through the uterine wall to empty into the uterine cavity.

ORIENTATION MICROGRAPH: This is a low magnification view of a cross section of the ampulla of a uterine tube. The **mucosa** (1) of the uterine tube exhibits relatively thin longitudinal folds that project into the lumen. The **muscular layer** (2) is composed of inner circular and outer longitudinal layers of smooth muscles. The **serosa** (3), which is part of the peritoneal broad ligament of the uterus, covers the uterine tube from outside. The tubal branches of the ovarian and **uterine arteries** and **veins** (4) travel along the uterine tube. The **epoophoron** (5), a remnant of the mesonephric (Wolffian) duct, is adjacent to the uterine tube and is usually found near the attachment of the broad ligament to the uterine tube.

Uterine tube, human, H&E x45.

This micrograph shows a cross section through the ampulla of the uterine tube. Many tall, slender **mucosal folds** (1) project into the **lumen** (2). Due to the variety of mucosal folds' profiles, the lumen of the ampulla is highly irregular. The mucosal folds are lined by **simple columnar epithelium** (3). The underlying **lamina propria** (4) forms a connective tissue core that contains numerous **blood vessels** (5). The **muscularis** (6) consists of smooth muscle that forms a relatively thick layer of circular fibers and a thinner outer layer of longitudinal fibers. The layers are not clearly delineated, and no sharp boundary separates them. The outermost layer of the uterine tube is formed by the peritoneum of the broad ligament, which possesses all the characteristics of a **serosa** (7).

Uterine tube, human, H&E x175.

This is a higher magnification of the wall of the uterine tube within the box on the top micrograph. The outermost layer of the wall (on the right) is **serosa** (8), which possesses simple squamous mesothelial cells and a thin layer of underlying connective tissue. The muscularis is organized into an outer, thinner, **longitudinal layer** (9) and an inner, relatively thick, **circular layer** (10); their boundaries are often difficult to discern. Numerous **blood vessels** (11) are visible in this layer. Mucosal folds (on the left) rest on the muscularis, from which the connective tissue extends into the **lamina propria** (12). The character of the connective tissue is essentially the same from the muscularis to the **simple columnar epithelium** (13), and for this reason, no submucosa is identified.

Uterine tube, human, H&E x255.

This micrograph shows the tip of a mucosal fold. The **lamina propria** (14) in the core of the mucosal fold contains a highly cellular connective tissue. The connective tissue contains cells whose nuclei are typically arranged in a random manner. They vary in shape, being elongated, oval, or round. Their cytoplasm cannot be distinguished from the extracellular matrix. The epithelium lining the **mucosa** (15) consists of ciliated cells that are identified by the presence of well-formed cilia. Nonciliated cells (*arrows*), also called peg cells, are identified by the absence of cilia and sometimes appear to be squeezed between the ciliated cells.

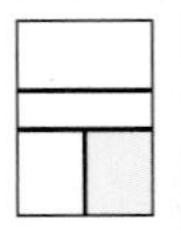

Uterine tube, human, H&E x515.

This is a higher magnification of the **epithelium** (16), lining the uterine tube, and the underlying **lamina propria** (17). Most of the epithelial cells in this section are ciliated and are readily identified by the presence of well-formed cilia. These cells also possess a darker stained line near their apical surface, an indication of basal bodies. A single nonciliated cell (*asterisk*) is identified by the absence of cilia.

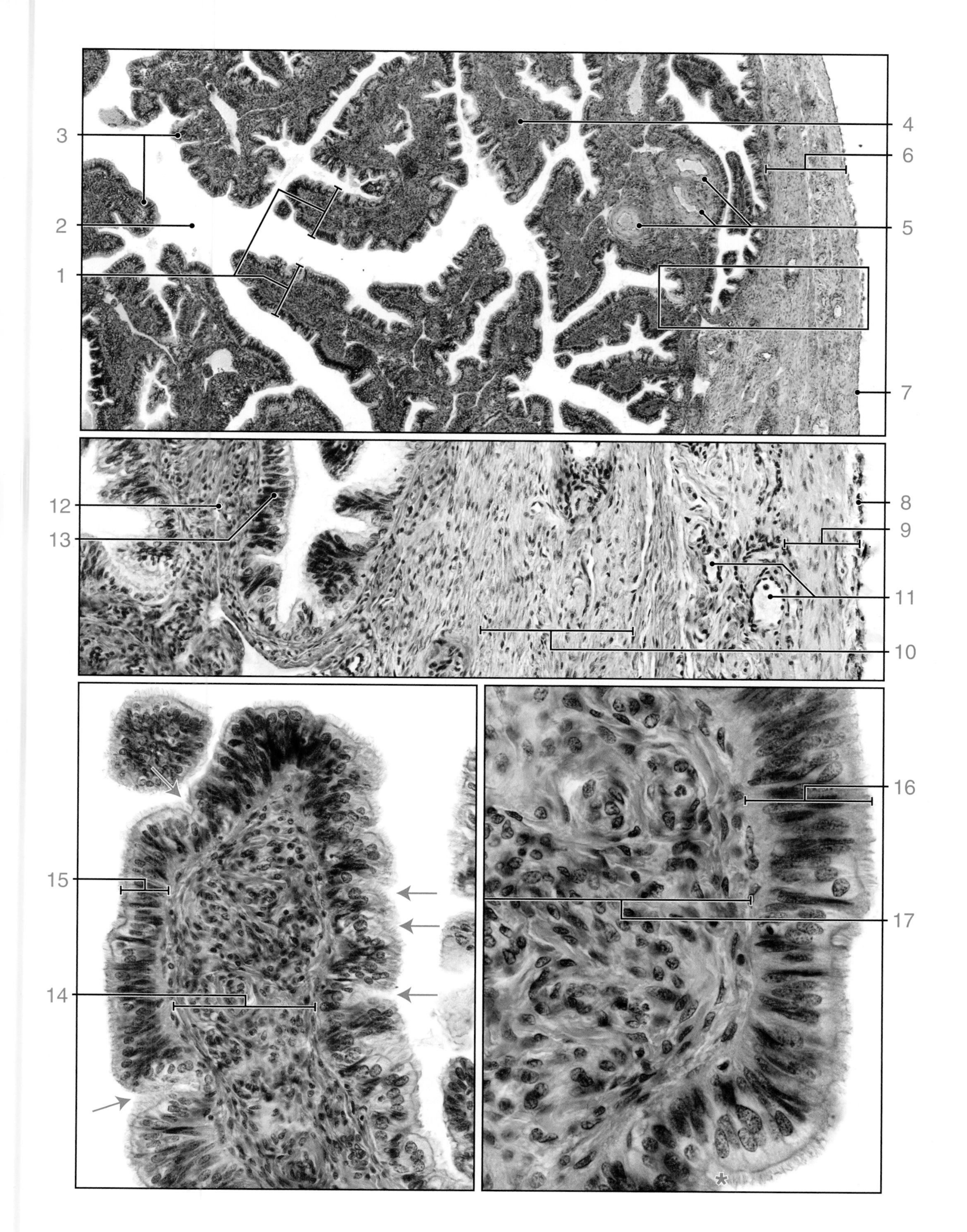

1
2
3
4
5
6
7
8
9
10
11
12
13
14
15
16
17
*

PLATE 140. UTERUS I

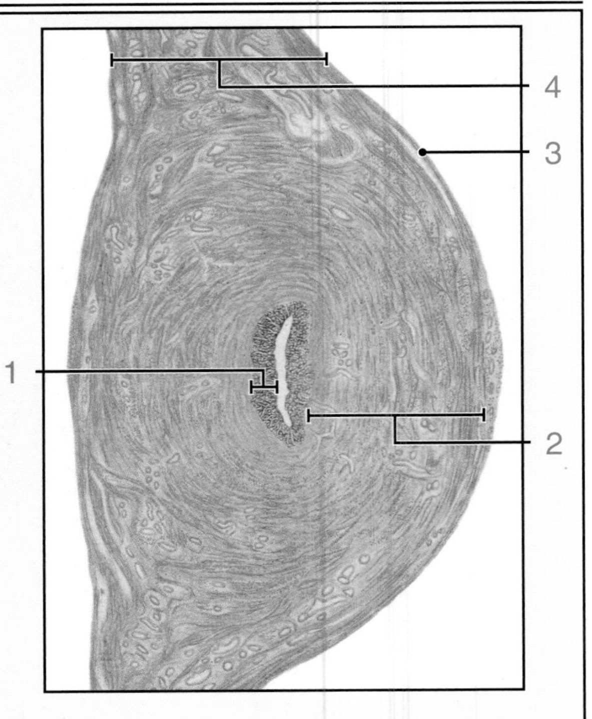

The uterus is a hollow, pear-shaped, muscular organ that consists of an extended body and a smaller cervix. The uterine cavity is narrow and triangular in shape; it continues superiorly to the lumina of ovarian tubes and communicates inferiorly, by an internal os, with the lumen of the cervical canal. The uterine wall is composed of the mucosa, which lines the uterine cavity and is referred to as the endometrium; a muscularis in the middle that is referred to as a myometrium; and, externally, a serosa, or the perimetrium. The myometrium consists of an outer longitudinal, inner circular, and an interposed, thick, middle layer composed of randomly interdigitating fibers. Smooth muscle bundles of the myometrium are separated by connective tissue. This middle layer also contains large blood vessels that give rise to the vessels that supply the endometrium. The endometrium undergoes hormonally controlled, cyclical changes during the ovarian cycle in preparation for implantation of the embryo. If implantation does not occur, the more superficial part of the endometrium, referred to as the stratum functionale, degenerates and is sloughed off, constituting the menstrual flow. Under the influence of estrogen, smooth muscle and connective tissue go through less pronounced proliferative changes. The degenerative changes of fibroblasts and smooth muscle cells occur in the secretory phase of the menstrual cycle.

ORIENTATION MICROGRAPH: This is a low magnification view of a cross section of a human uterus near the internal os. The uterine wall is composed of a mucosal layer, or **endometrium** (1), that lines a cavity of the uterus and cervical canal; a muscular layer, or **myometrium** (2), that occupies almost the entire thickness of the uterine wall; and an outermost, serosal layer, or **perimetrium** (3), that is formed by peritoneum. The peritoneum extends laterally and forms the **broad ligament of the uterus** (4). It attaches the uterus to the sides of the pelvis and contains nerves as well as blood and lymphatic vessels passing to the uterus.

Uterus, myometrium, human, H&E x15; inset x255. This micrograph shows a higher magnification of a muscular arrangement within the myometrium of the uterine wall. The **endometrium** (1) is present at the top of this image. The underlying myometrium consists of three distinct layers: a submucosal, inner, **circular layer** (2); an interposed, thick, **middle layer** (3) composed of randomly interdigitating muscle bundles and containing numerous large **blood vessels** (4); and an outer, **longitudinal layer** (5) of smooth muscle covered by a serous layer. The **inset** shows a higher magnification of the **perimetrium** (6), which consists of mesothelium and a thin layer of underlying loose connective tissue. Note the presence of the distinct, longitudinal smooth muscle bundles just beneath the serosa. These are referred to as cervicoangular fascicles and represent vestigial remnants of the mesonephric (Wolffian) duct.

Uterus, myometrium, human, H&E x125. This is a higher magnification of the middle layer of the myometrium from the boxed area on the top left micrograph. As indicated earlier, the middle layer of the myometrium is composed of randomly interdigitating **smooth muscle bundles** (7), which represent their spiral arrangement. Muscle bundles in this layer are separated by loose connective tissue that contains **blood vessels** (8).

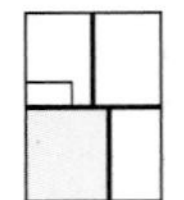

Uterus, early proliferative phase of the endometrium, human, H&E x25. This micrograph shows a higher magnification of an **endometrium** (9) and underlying **myometrium** (10). The **stratum functionale** (11), which was sloughed off during menstruation, is being rebuilt. The epithelial resurfacing originates from proliferation of the glands that remain in the **stratum basale** (12). The glandular epithelium proliferates and grows over the surface. Proliferation of **endometrial glands** (13), connective tissue stroma, and vessels occur as a response to increasing levels of estrogen. In early proliferation, glands and vessels are straight (noncoiled). This micrograph shows the endometrium as it appears in the early proliferative stage when resurfacing is complete. At this stage the endometrium is relatively thin, and more than half of it consists of the stratum basale.

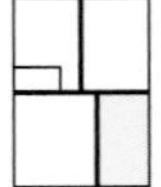

Uterus, early proliferative phase of the endometrium, human, H&E x125. This is a higher magnification of an **endometrial gland** (14) within the stratum functionale from the boxed area on the bottom left micrograph. Under the influence of estrogen, the endometrial glands, connective tissue stroma, and blood vessels proliferate (proliferative stage), so that the total thickness of the endometrium increases. Note that the **simple columnar epithelium** (15) that covers the endometrial surface is similar to the epithelium lining the endometrial glands.

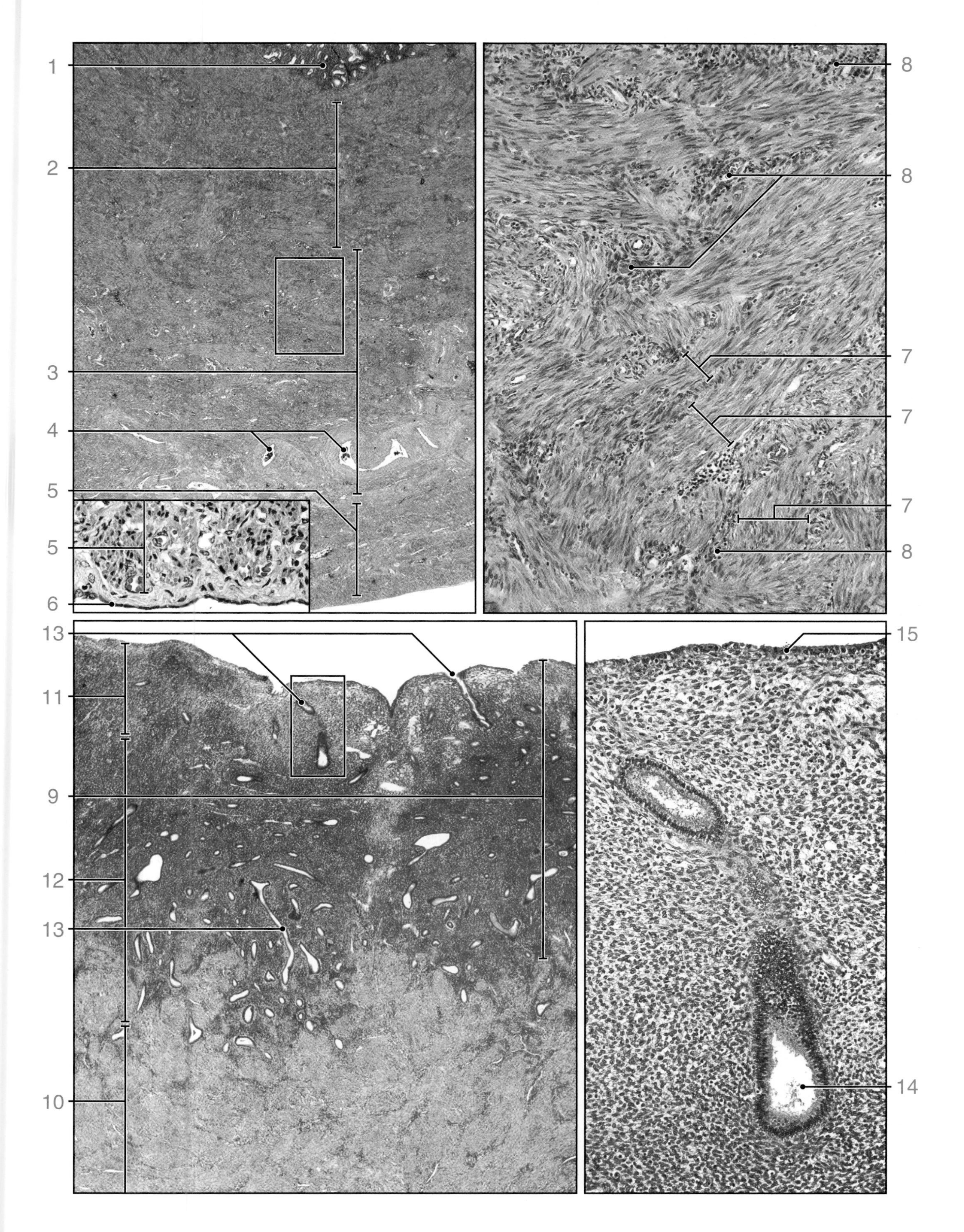

1
2
3
4
5
5
6
8
8
7
7
7
8
13
11
9
12
13
10
15
14

PLATE 141. UTERUS II

Cyclic changes during the menstrual cycle are reflected in the dramatic transformation that occurs in the functional layer of the endometrium. There are three recognizable phases of endometrial change: proliferative, secretory, and menstrual, altogether lasting an average of 28 days. From the biological perspective, it is logical to describe the phases of the cycle starting with the proliferative, then secretory, and ending with the menstrual phase. This sequence follows growth, differentiation, and subsequent destruction of the stratum functionale of the endometrium. However, clinically, the first day of the menstrual cycle is conventionally identified as the first day of the menstrual flow.

The proliferative phase, which lasts from 9 to 20 days, is influenced by estrogen secreted from the growing ovarian follicles; the secretory phase (having a constant length of 14 days) is influenced by progesterone secreted from the corpus luteum; and the menstrual phase, which usually lasts less than 5 days, reflects a decline in hormone secretion and the degeneration of the corpus luteum. In the proliferative phase, glands are being reconstructed by the proliferation of cells from the stratum basale. The glands have a narrow lumen and are relatively straight, with a slightly wavy appearance. In the secretory phase, the glands take on a more pronounced corkscrew shape and secrete mucus that accumulates in sacculations along their length. In the menstrual phase, preceded by contractions of the spiral arteries, the stratum functionale becomes ischemic, causing disruption of the surface epithelium and rupture of the blood vessels. Blood, uterine fluid, and sloughing stromal and epithelial cells from the stratum functionale constitute the menstrual discharge.

Uterus, late secretory phase of the endometrium, human, H&E x25; inset x175.

This micrograph shows a high magnification of an endometrium in the late secretory stage. At the top, a thick layer of the **stratum functionale** (1) is clearly visible. Just below is a thin layer of the **stratum basale** (2), and, at the bottom of the micrograph, a small area of the **myometrium** (3) is visible. The **uterine glands** (4) have been cut in a longitudinal plane. These glands are classified as simple glands (they do not branch). At this phase, uterine glands exhibit a prominently coiled shape with numerous, shallow sacculations that give them a saw-toothed appearance. The luminal secretion is often present. In contrast to the characteristic coiled course of the glands in the stratum functionale, glands of the stratum basale more closely resemble those in the proliferative stage (**inset**).

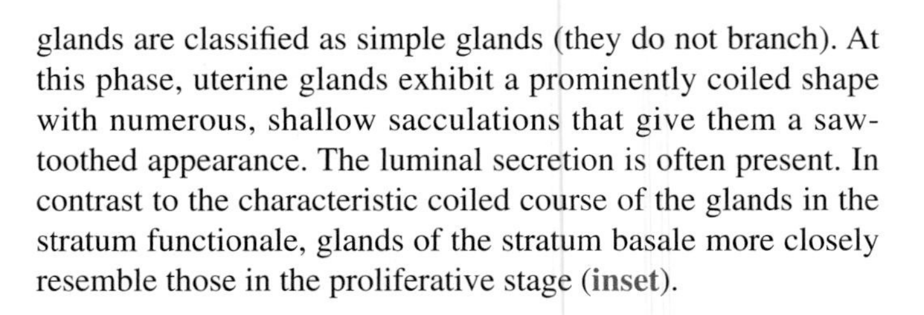

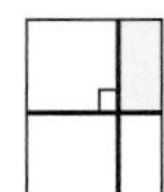

Uterus, late secretory phase of the endometrium, human, H&E x175.

This is a higher magnification of the endometrial gland from the boxed area on the top left micrograph. **Endometrial glands** (5) in this late secretory stage become enlarged and coiled, and their lumen, which is filled with secretory products, appears sacculated. The **endometrial stroma** (6) becomes edematous and is infiltrated by stromal granulocytes. Stromal cells near the vessels become rounded; they acquire more cytoplasm and their borders become fully visible. These cells form islands of **predecidual cells** (7). The spiral arteries lengthen and become more coiled; they extend to the surface of the endometrium.

Uterus, menstrual phase of the endometrium, human, H&E x25.

This micrograph shows a high magnification of an endometrium and underlying **myometrium** (8) during the menstrual phase. The **stratum functionale** (9) undergoes disruption and dissolution. These dramatic changes are characterized by disruption of blood vessels, interstitial hemorrhages (extravasations of blood into the connective tissue stroma), fragmentation of the endometrium, infiltration by neutrophils, and necrosis. In contrast, the **stratum basale** (10) maintains a relatively constant appearance throughout the entire menstrual cycle.

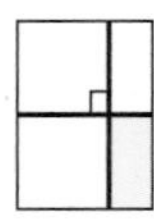

Uterus, menstrual phase of the endometrium, human, H&E x175.

This is a higher magnification of the stratum functionale from the boxed area on the bottom left micrograph. Note the interstitial **hemorrhages** (11), the disruption of the **uterine glands** (12), and the fragmentation of the endometrium with accompanying necrosis at the luminal surface.

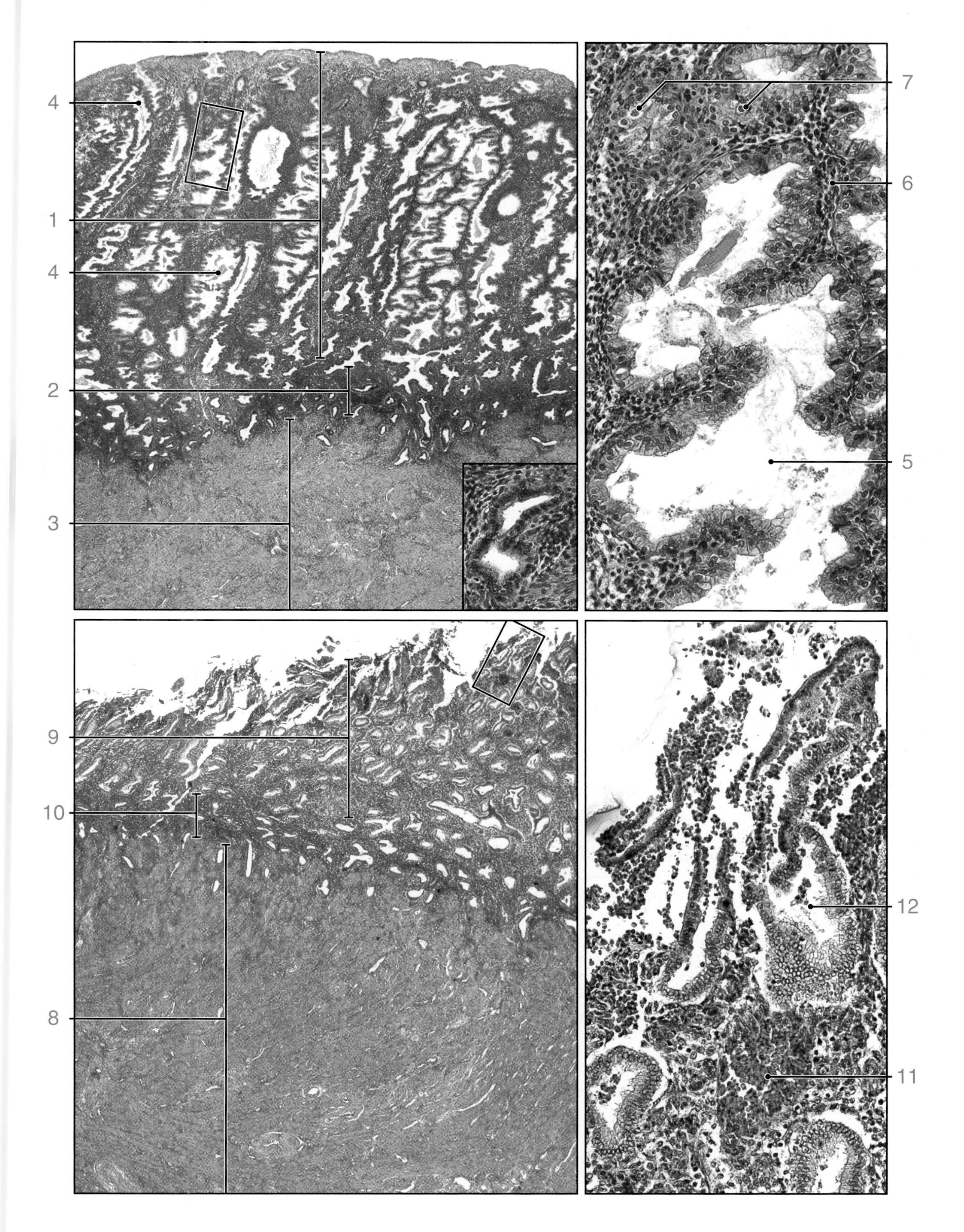
4
1
4
2
3
7
6
5
9
10
8
12
11

PLATE 142. CERVIX

The cervix is the narrow, or constricted, inferior portion of the uterus, part of which projects into the vagina. The cervical canal traverses the cervix and provides a channel that connects the vagina and the uterine cavity. The structure of the cervix resembles the rest of the uterus in that it consists of a mucosa (endometrium) and a myometrium. There are, however, some important differences in the mucosa.

The endometrium of the cervix does not undergo the cyclical growth and loss of tissue that is characteristic of the body and fundus of the uterus. Instead, the amount and character of the mucous secretions of the cervical's simple columnar epithelium vary throughout the uterine cycle under the influence of ovarian hormones. At midcycle, there is a 10-fold increase in the amount of mucus produced; this mucus is less viscous and provides a favorable environment for sperm migration. At other times in the cycle, the mucus restricts the passage of sperm into the uterus.

The myometrium is the thickest layer of the cervix. It consists of interweaving bundles of smooth muscle cells in an extensive, continuous network of fibrous connective tissue.

Cervix, vaginal part, human, H&E x15.

The vaginal part of the **cervix** (1), projecting into the vagina (ectocervix), is represented by the upper two thirds of this micrograph. This part of the cervix is covered by nonkeratinized, stratified squamous epithelium that extends from the walls of the vagina. The lower third of the micrograph reveals the supravaginal part of the cervix, or **endocervix** (2). It includes the **cervical canal** (3), represented by the narrowed connection between the internal os (which opens to the uterine cavity) and the **external os** (4) (which opens into the vagina). The cervical canal is lined by simple columnar epithelium. At the **transformation zone** (5), near the external os, there is an abrupt change in the epithelial lining. The plane of this section passes through the long axis of the cervical canal. Note that only one side of the longitudinal section of the cervix is shown in these micrographs. In the actual specimen, if seen in cross-section, would present a mirror image on the other side of the cervical canal.

Cervix, supravaginal part, human, H&E x25.

This micrograph shows the supravaginal part of the cervix cut through the long axis of the **cervical canal** (6). The mucosa of the cervical canal differs from the covering of the vaginal part of the cervix. The cervical canal contains **cervical glands** (7); these glands are absent in the vaginal part of the cervix. These glands differ from those of the uterus in that they branch extensively and secrete a mucous substance into the cervical canal that lubricates the vagina.

Cervix, vaginal part, human, H&E x240.

This micrograph shows the **nonkeratinized stratified squamous epithelium** (8) of the vaginal part of the cervix from the upper boxed area on the top left micrograph. The epithelium–connective tissue junction presents a relatively even contour in contrast to the irregular profile seen in the vagina. In other respects, this epithelium has the same general features as vaginal epithelium; it is also devoid of glands. Another similarity is that the epithelial surface of the vaginal part of the cervix undergoes cyclical changes similar to those of the vagina in response to ovarian hormones.

Cervix, external os, transformation zone, human, H&E x240.

This micrograph is a higher magnification of the lower boxed area from the top left micrograph. It shows an abrupt change from **nonkeratinized stratified squamous epithelium** (9) to **simple columnar epithelium** (10). This change occurs within the **transformation zone** (11) at the vaginal opening of the cervical canal (external os). The mucosa of the cervical canal is covered with columnar epithelium. Note the large number of lymphocytes present in the transformation zone region.

Cervix, cervical glands, human, H&E x240.

This micrograph shows portions of the **cervical gland** (12) at high magnification. Note the tall epithelial cells and the lightly staining supranuclear cytoplasm, a reflection of the mucin dissolved from the cell during tissue preparation. The crowding and the change in shape of the nuclei (*asterisk*) seen at the lower part of one of the glands in this micrograph are due to the tangential cut through the wall of the gland. It is not uncommon for cervical glands to develop into cysts as a result of an obstruction in the duct; such cysts are referred to as Nabothian cysts.

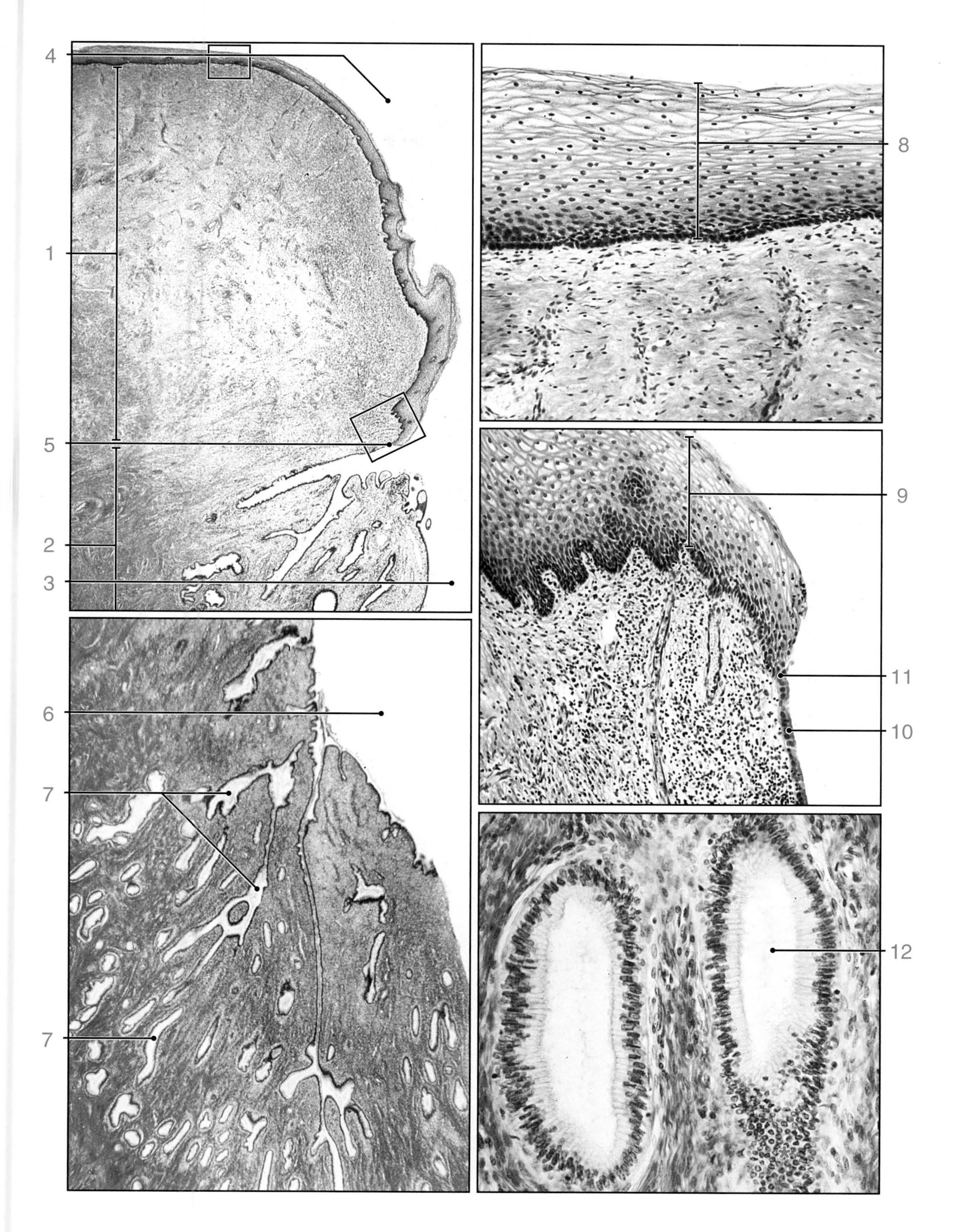
4
1
5
2
3
6
7
7
8
9
11
10
12

PLATE 143. PLACENTA I

The placenta is a large, discoid-shaped fetomaternal organ that develops during pregnancy from fetal and maternal tissues. It averages 20 cm in diameter and 3 cm in thickness at term. The umbilical cord normally inserts near the center of the placenta and extends to the umbilicus of the fetus. The placenta functions not only as a transport organ for nutrients and waste products exchanged between the fetal and maternal circulations but also as an important endocrine organ. It has two portions: a fetal portion that derives from the chorionic sac and a maternal portion that develops from the decidua basalis (a part of the transformed endometrium deep to the developing embryo). The fetal portion of the placenta, facing the amniotic cavity, is called the chorion frondosum and the maternal portion, embedded in the uterine wall, is called the basal plate. After birth, the placenta separates from the wall of the uterus and is discharged along with the contiguous amniotic membranes.

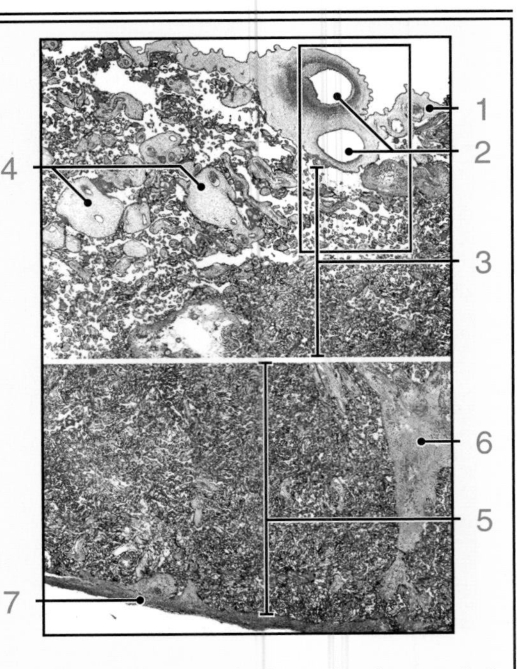

ORIENTATION MICROGRAPH: The upper micrograph shows a cross section of the fetal portion of the placenta. At this magnification, the amniotic surface of the **chorionic plate** (1) is clearly visible. Large tributaries of the umbilical vein with accompanied branches of the **umbilical arteries** (2) travel on the surface of the chorionic plate. The area deep to the chorionic plate contains various-sized profiles of **chorionic villi** (3) that originate from the chorionic plate. A few sections of **placental septa** (4) are noticeable. The lower micrograph shows a cross section of the maternal side of the placenta. Almost the entire thickness of this section is occupied by **chorionic villi** (5). A small section of the **placental septum** (6) that originates from the stratum basale is clearly visible. The **stratum basale** (7) is a connective tissue layer that forms the surface that communicates with the uterus.

Placenta, fetal portion, human, H&E x70; inset x220.

This micrograph is a higher magnification of the boxed area on the upper orientation micrograph. It shows a cross section through the fetal portion of the placenta, extending from the amniotic surface (on the right) into the intervillous space containing chorionic villi (on the left). The amniotic surface of this portion is covered by **amnion** (1), which consists of a layer of simple cuboidal epithelium, and the underlying connective tissue of the **chorionic plate** (2) (**inset**). The chorionic plate is a thick layer of connective tissue that contains chorionic vessels. **Chorionic veins** (3) are tributaries of the umbilical vein and are accompanied by **chorionic arteries** (4), branches of the umbilical arteries. The umbilical arteries carry poorly oxygenated blood away from the developing fetus. At the site of umbilical cord attachment, they divide into chorionic arteries, which have much thicker muscular walls than their accompanying veins. Chorionic arteries and veins are found in the chorionic plate; they branch freely before entering chorionic villi. The **intervillous space** (5), visible deep to the chorionic plate, is occupied by **chorionic villi** (6) of different sizes. These emerge from the chorionic plate as stem chorionic villi and project into the intervillous space, which contains maternal blood. Some of them extend from the chorionic plate to the basal plate; these villi are called anchoring villi. Other stem villi simply arborize within the intervillous space without anchoring to the maternal side and are called free villi. The intervillous space is traversed by several wedge-shaped projections of stratum basale that extend toward the chorionic plate. These **placental septa** (7) divide the intervillous space into irregular convex regions called cotyledons. Placental septa are infiltrated by **decidual cells** (8); they do not contain branches of the chorionic vessels, and on this basis, they can frequently be distinguished from stem or free villi.

Placenta, maternal portion, human, H&E x120; inset x240.

This micrograph shows a cross section through the maternal portion (basal plate) of the placenta, extending from the stratum basale (on the right) into the intervillous space containing chorionic villi (on the left). The **stratum basale** (9) originates from the decidua basalis, which represents transformed endometrium in response to increasing levels of progesterone during pregnancy. This transformation is commonly referred to as the decidual reaction, or decidualisation of the endometrum. Along with the usual connective tissue elements, the stratum basale contains large, often polygonal-shaped cells called **decidual cells** (10). They resemble epithelial cells and are referred to as having an epithelioid appearance. The **inset** shows a higher magnification of some decidual cells. Their cytoplasm contains areas of glycogen accumulation and lipid droplets. These cells derive from endometrial stromal cells and are usually found in clusters; because of their features, they are easily identified. Placental septa, originating from the basal plate, often contain decidual cells, a helpful feature in identification of these structures.

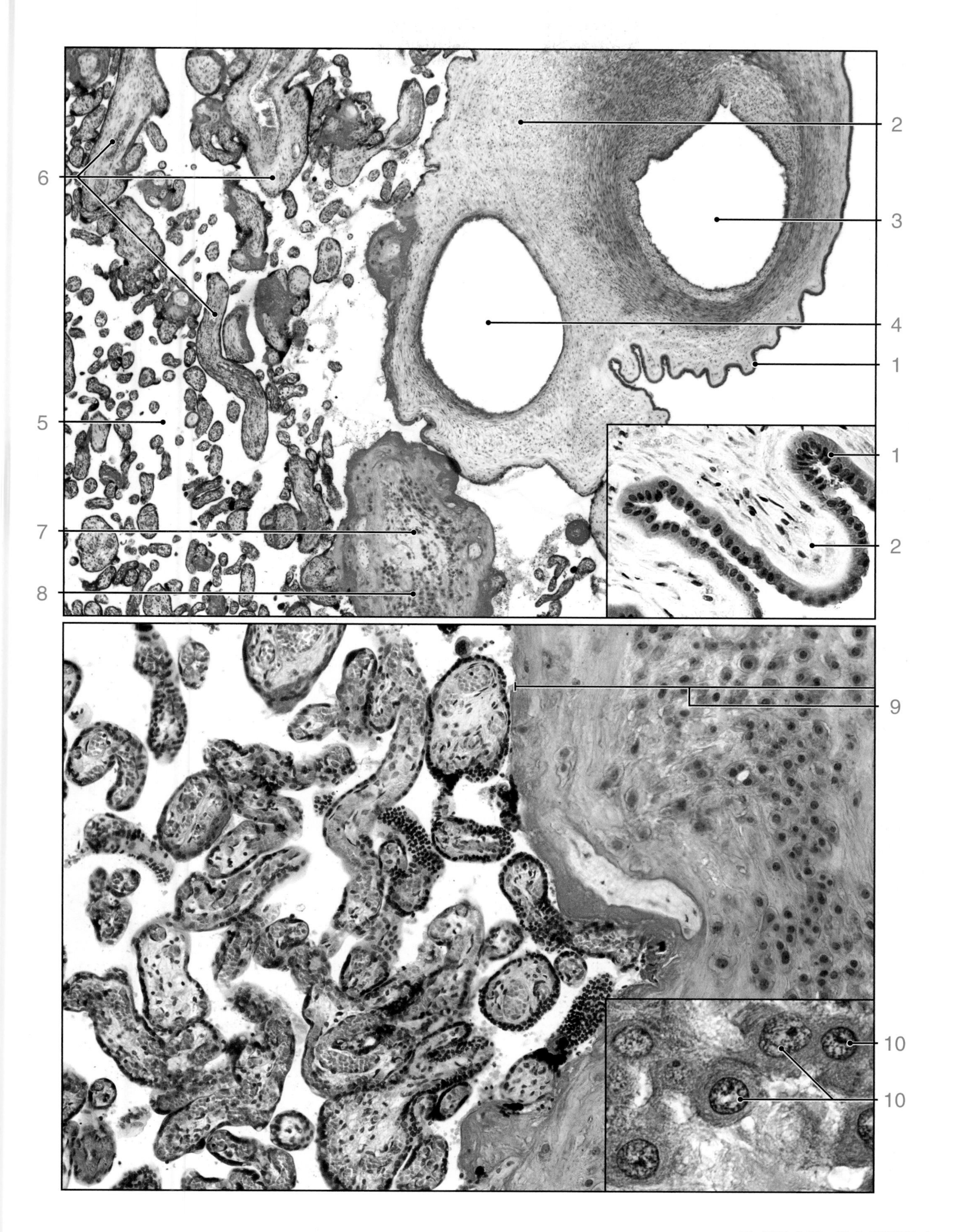

PLATE 143. PLACENTA I

PLATE 144. **PLACENTA II**

During the first two weeks of development, the embryo is nourished through simple diffusion. Due to its rapid growth, a utero-placental circulation system develops to allow for exchange between maternal and fetal circulation without direct contact with each other. This separation of fetal and maternal blood remains throughout the pregnancy and is referred to as the placental barrier. As the embryo develops, the invasive activity of the syncytiotrophoblast erodes the maternal capillaries and anastomoses them with the trophoblast lacunae, forming the maternal blood sinusoids. These communicate with each other and form a single blood compartment, lined by syncytiotrophoblasts, called the intervillous space. At the end of the second week of development, cytotrophoblast cells form primary chorionic villi. They project into the maternal blood space. In the third week of development, invasion of the extra-embryonic mesenchyme into the primary chorionic villi creates secondary chorionic villi. At the end of the third week, core mesenchyme differentiates into connective tissue and blood vessels that connect with the embryonic circulation. These tertiary chorionic villi constitute functional units for gas, nutrient, and waste product exchanges between maternal and fetal circulation. Thus, each tertiary villus consists of a connective tissue core surrounded by two distinct layers of trophoblast-derived cells. The outermost layer consists of the syncytiotrophoblast; immediately under it is a layer of cytotrophoblast cells. Starting at the fourth month, these layers become very thin to facilitate the exchange of products across the placental barrier. The thinning of the wall of the villus is due to the loss of the inner, cytotrophoblastic layer. The tertiary chorionic villi grow first in length, and they also increase in density and width. The villi divide further and become very thin. At this stage, the syncytiotrophoblast forms numerous trophoblastic buds that resemble the primary chorionic villi; however, the cytotrophoblast and the connective tissue grow very rapidly into these structures, transforming them into tertiary villi. Villi arising out of the tertiary villi either remain free and project into the intervillous space, or they anchor themselves to the basal plate. At term, the placental barrier consists of the syncytiotrophoblasts; a sparse, discontinuous, inner cytotrophoblast layer; the basal lamina of the trophoblast; the connective tissue of the villus; the basal lamina of the endothelium; and the endothelium of the fetal placental capillary in the tertiary villus. This barrier bears a strong resemblance to the air–blood barrier of the lung, with which it has an important parallel function.

Placenta, chorionic villi, human, H&E x280.

This micrograph shows a cross section through the intervillous space of the placenta at term. It includes **chorionic villi** (1) of different sizes and the surrounding **intervillous space** (2). The connective tissue of the villi contains branches and tributaries of the **umbilical vein** (3) and arteries. The intervillous space usually contains maternal blood. The maternal blood was drained from this specimen before preparation; therefore, only a few maternal blood cells are seen. The outermost layer of each chorionic villus derives from the fusion of cytotrophoblast cells. This layer, known as the **syncytiotrophoblast** (4), has no intercellular boundaries, and its nuclei are rather evenly distributed, giving this layer an appearance similar to that of cuboidal epithelium. In some areas, nuclei are gathered in clusters *(arrowheads)*; in other regions the syncytiotrophoblast layer appears relatively free of nuclei *(arrows)*. These stretches of the syncytiotrophoblast may be so attenuated in places that the villous surface appears devoid of a covering. The syncytiotrophoblast contains microvilli that project into the intervillous space. In well-preserved specimens they may appear as a striated border *(see inset below)*.

The cytotrophoblast consists of an irregular layer of mononucleated cells that lies beneath the syncytiotrophoblast. In immature placentas, the cytotrophoblasts form an almost complete layer of cells. In this full-term placenta, only occasional **cytotrophoblast cells** (5) can be discerned. Most of the cells within the core of the villus are typical connective tissue fibroblasts and endothelial cells. The nuclei of these cells stain well with hematoxylin, but the cytoplasm blends with the extracellular matrix. Other cells have a visible amount of cytoplasm that surrounds the nucleus. These are considered to be **placental macrophages** (6), also called Hofbauer cells.

Placenta, chorionic villi, human, H&E x320; inset x640.

This micrograph shows the secondary chorionic villi in the third week of embryonic development. These villi are composed of a **mesenchymal core** (7) surrounded by two distinct layers of the trophoblast. Secondary villi have a much larger number of **cytotrophoblast cells** (8) than the mature tertiary villi and form an almost complete layer of cells immediately deep to the **syncytiotrophoblast** (9). The syncytiotrophoblast covers not only the surface of the chorionic villi but also extends into the chorionic plate. Maternal red blood cells are present in the intervillous space. See also **inset**.

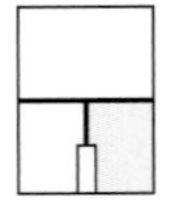

Placenta, chorionic villi, human, H&E x320.

This higher magnification micrograph shows a cross section through immature chorionic tertiary villi in a mid-term placenta, surrounded by the **intervillous space** (10). At this stage, chorionic villi are growing by proliferation of their core mesenchyme, **syncytiotrophoblast** (11), and fetal endothelial cells. The syncytiotrophoblast surrounding the chorionic villus *(center of the image)* forms a **trophoblastic bud** (12), which will be invaded subsequently by cells of the **cytotrophoblast** (13), connective tissue, and rapidly developing new blood vessels. In addition to fibroblasts, several **placental macrophages** (**Hofbauer cells**) (14) can be identified by the amount of cytoplasm surrounding their nuclei.

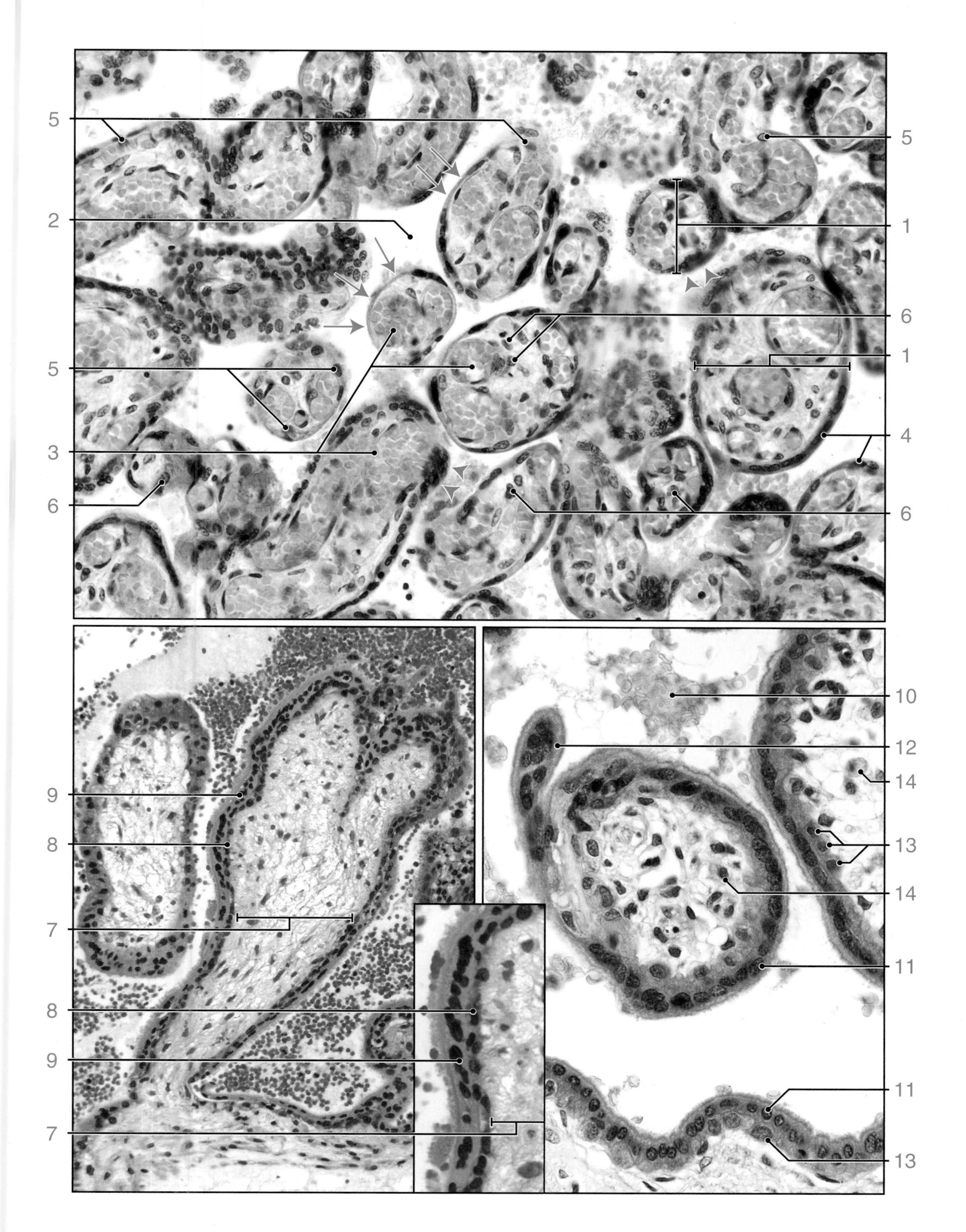

5
5
2
1
6
5
1
3
4
6
6
10
12
9
14
8
13
14
7
11
8
9
11
7
13

PLATE 145. **UMBILICAL CORD**

The umbilical cord in humans connects the developing embryo (fetus, in later stages) to the placenta. It develops from embryonic tissue and originates from the connecting stalk that contains remnants of the yolk sac and allantois. The umbilical cord extends from the umbilicus of the fetus to the amniotic surface of the placenta. At delivery it resembles a translucent, shiny, white cord approximately 60 cm in length and 2 cm in diameter. It consists mostly of embryonic, mucous connective tissue, referred to as Wharton's jelly, which characterizes the cord at later stages in fetal development. The cord is covered by a simple cuboidal epithelium, except near the fetus where it becomes a stratified squamous epithelium. The extracellular matrix of Wharton's jelly consists of a ground substance that contains various hydrated mucopolysaccharides. Its collagen fibers, arranged in a three-dimensional meshwork, are extremely fine. Fibroblasts are the only cell type within Wharton's jelly, and they are widely dispersed. In the early stages of fetal development, there are two arteries and two veins within the cord. The arteries carry deoxygenated, nutrient-depleted blood from the fetus to the placenta. The veins carry oxygenated, nutrient-rich blood to the fetus. As fetal development continues, one of the umbilical veins regresses and becomes almost completely obliterated. Thus, at term, the umbilical cord contains two arteries and only one vein. There are no nerves, smaller blood vessels, or lymphatic vessels within the umbilical cord.

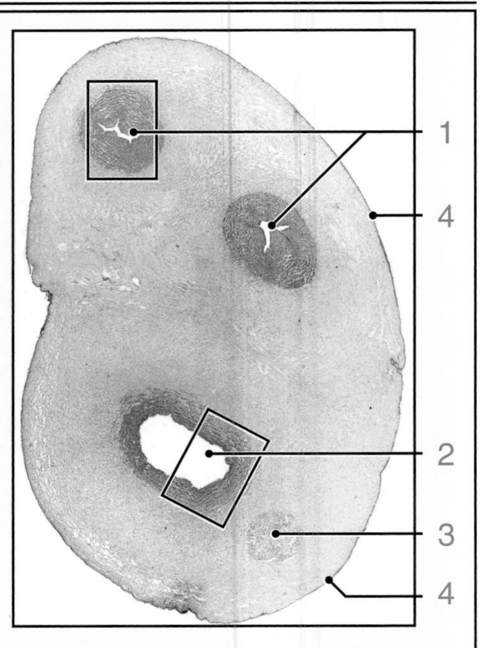

ORIENTATION MICROGRAPH: This cross section of an umbilical cord at term reveals two **umbilical arteries** (1) and a single **umbilical vein** (2). Since the walls of the arteries contain several layers of smooth muscle, their lumen is markedly reduced after fixation. The umbilical vein usually remains open. The nearly obliterated **umbilical vein** (3) is represented by the collagen fibers that are more densely aggregated than those in the surrounding mucous connective tissue. At this magnification, the **epithelium** (4) of the umbilical cord appears as a thin, deeper-stained external border.

Umbilical cord, artery, human, H&E, x65.

This micrograph is a higher magnification of the umbilical artery in the upper boxed area on the orientation micrograph. Its relatively thick tunica media is composed of smooth muscle bundles arranged in two distinct layers: an **inner longitudinal layer** (1) and an **outer circular layer** (2). Both layers are heavily infiltrated by extracellular matrix, which separates the muscle bundles from each other. The umbilical artery does not possess an internal elastic membrane, therefore it is difficult to visualize the boundary between the tunica intima and the tunica media. An external elastic membrane is also absent. The wall of the artery contains a delicate network of elastic fibers. Unlike other arteries, there is no tunica adventitia. The bulk of the cord consists of **mucous connective tissue** (3).

Umbilical cord, vein, human, H&E, x65.

This micrograph is a higher magnification of the umbilical vein from the lower boxed area on the orientation micrograph. The umbilical vein generally has a larger diameter lumen and a thinner **muscular wall** (4) compared to the umbilical artery. It is composed of a single layer of circularly arranged smooth muscle fibers intermingled with muscle fibers running in oblique and longitudinal directions. Like the umbilical arteries, the umbilical vein lacks a tunica adventitia.

Umbilical cord, artery, human, H&E, x255.

This micrograph shows a high magnification of a portion of the artery wall. The nuclei at the surface belong to **endothelial cells** (5). The nuclei below the endothelium belong to **smooth muscle cells** (6) that have been cross-sectioned. Note there is no internal elastic membrane between the endothelium and the muscle cells below. These muscle cells belong to the inner longitudinal layer of the muscle in the media. There is a narrow, intermediate region where **obliquely cut muscle cells** (7) are visible. In the lower part of the micrograph are **longitudinally sectioned muscle fibers** (8). These belong to the circular layer of smooth muscle in the media. **Mucous connective tissue matrix** (9) occupies the space between the muscle cells.

Umbilical cord, surface, human, H&E, x255; inset x510.

This is a higher magnification of the umbilical cord surface and underlying connective tissue. A single layer of amnionic epithelium covers the surface of the umbilical cord. However, in the region shown here, which is close to the fetal cord insertion, the simple cuboidal epithelium becomes **stratified** (10) and resembles the fetal epidermis. The **inset** shows the boxed area at higher magnification. Note the stratification of the epithelium. The underlying **mucous connective tissue** (11) contains **spindle-shaped fibroblasts** (12) that have been cut in varying planes. They are separated by an intercellular matrix that contains wispy collagen fibers embedded in the gelatin-like ground substance.

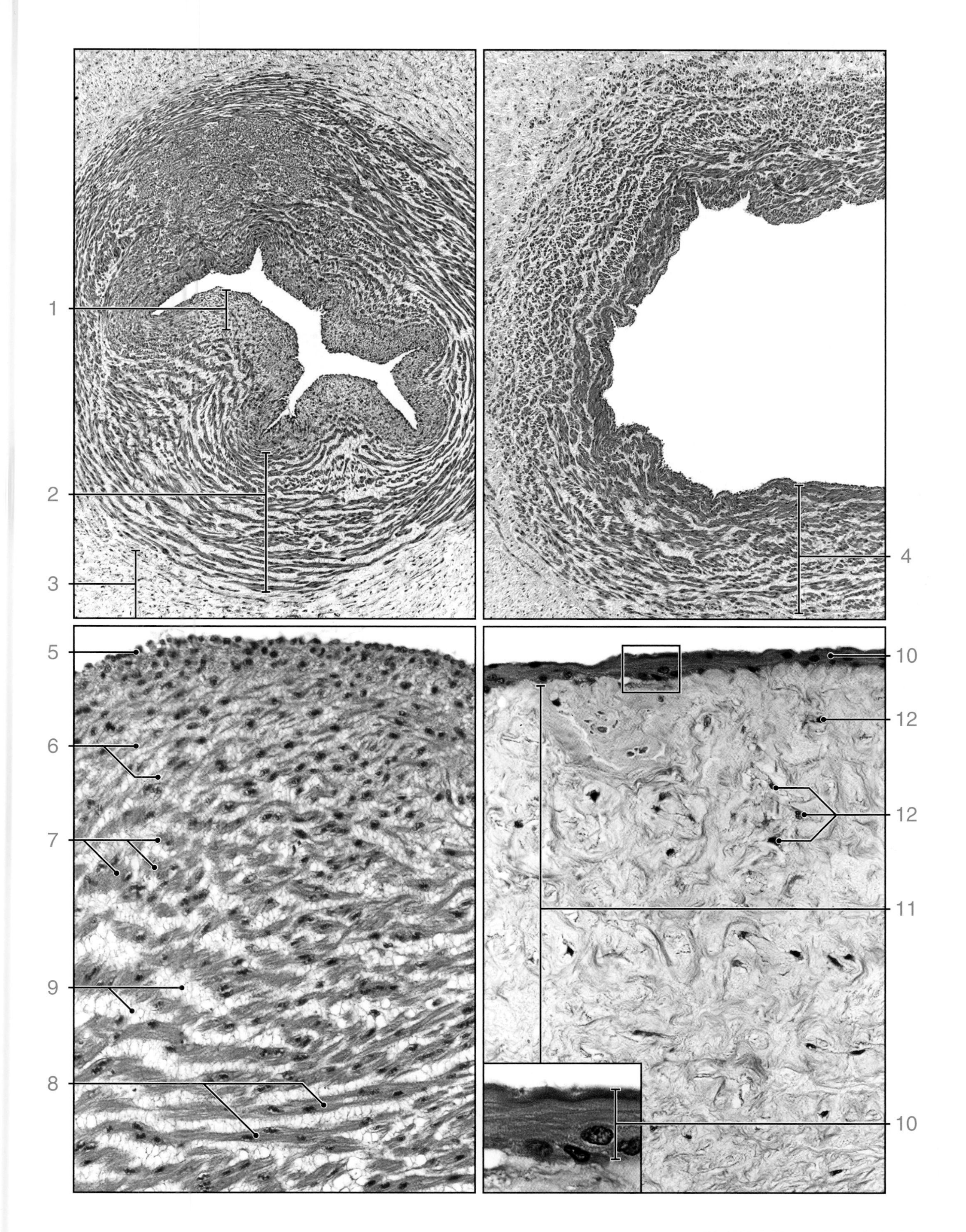

PLATE 145. UMBILICAL CORD

PLATE 146. **VAGINA**

The vagina is a distensible, fibromuscular tube that extends from the vestibule of the vagina in the perineum through the pelvic floor and into the pelvic cavity. The vagina is attached to the circular margin of the uterine cervix. The wall of the vagina consists of three layers: the inner mucosal layer, the intermediate muscular layer, and the outer adventitial layer. The mucosal layer consists of nonkeratinized stratified squamous epithelium and the underlying lamina propria, which projects into the epithelial layer forming connective tissue papillae. The epithelium of the vagina undergoes cyclical changes that correspond to the ovarian cycle. The lighter staining of the vaginal surface reflects the large amount of glycogen stored within the epithelial cells. (During routine H&E slide preparation, glycogen washes out of the tissue.) The amount of glycogen (highest during ovulation) in the epithelial cells changes due to the influence of estrogen, whereas the rate of desquamation is influenced by progesterone. Glycogen released from the desquamated cells is used by lactobacillus vaginalis, producing lactic acid. Lactic acid acidifies the vaginal transudate fluid that moistens the vaginal surface and inhibits colonization by yeasts and potentially pathogenic bacteria.

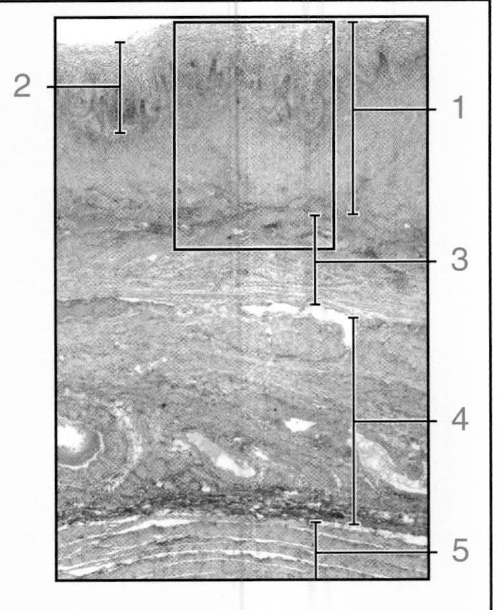

ORIENTATION MICROGRAPH: This micrograph shows a section of the posterior vaginal wall. The vaginal surface (top) is lined by the **inner mucosal layer** (1). The outermost layer contains a pale-stained **nonkeratinized stratified squamous epithelium** (2). Deep to the mucosal layer are the **muscular layer** (3) and the **adventitial layer** (4). In this section, the adventitial layer contains numerous blood and lymphatic vessels. The adventitial layer blends with the adventitia of the rectum. The **muscularis externa** (5) of the rectum is visible at the bottom of the image.

Vagina, human, H&E x90.
The mucosa of the vagina consists of a **nonkeratinized stratified squamous epithelium** (1) and an underlying, **dense irregular connective tissue** (2) that often appears more cellular than in other organs. The connective tissue is rich in elastic fibers and contains numerous **blood vessels** (3) and lymphatic vessels. The boundary between the nonkeratinized stratified squamous epithelium and the dense irregular connective tissue is identified by the closely packed, small cells of the **stratum basale** (4). Connective tissue papillae project into the underside of the epithelium, giving the epithelial–connective tissue junction an uneven appearance. The papillae may be cut obliquely or in cross-section and thus may appear as connective tissue islands *(arrows)* within the deeper portion of the epithelium. The epithelium is characteristically thick, and although keratohyalin granules may be found in the superficial cells, keratinization does not occur in human vaginal epithelium. Nuclei can be observed throughout the entire thickness of the epithelium. The **muscular layer** (5) of the vaginal wall consists of smooth muscle arranged in two ill-defined layers. The outer layer is longitudinally arranged and the inner layer is circularly arranged. Smooth muscle fibers are usually organized as interlacing bundles surrounded by connective tissue.

Vagina, human, H&E x70.
This higher-magnification micrograph of the muscular layer of the vaginal wall emphasizes the irregularity of the arrangement of the muscle bundles. At the right edge of the micrograph is a narrow bundle of **longitudinally sectioned smooth muscle** (6). Adjacent is a **cross-sectioned bundle of smooth muscle** (7). This bundle abuts a longitudinally sectioned **lymphatic vessel** (8) with a recognizable valve. To the left of the lymphatic vessel is another **longitudinally sectioned bundle of smooth muscle** (9). A small **blood vessel** (10) is present next to the lymphatic vessel.

Vagina, human, H&E x110.
This is a higher magnification of the **vaginal epithelium** (11) from the boxed area on the top micrograph. The obliquely cut and cross-sectioned portions of connective tissue papillae appear as connective tissue islands in the epithelium (*arrows*). Note that the epithelial cells, even at the surface, still retain their nuclei, and there is no evidence of keratinization.

Vagina, human, H&E x225.
This is a higher magnification micrograph of the basal portion of the **epithelium** (12) between connective tissue papillae. Note the regular arrangement of the basal epithelial cells. Following divisions, these cells differentiate and migrate toward the surface of the epithelium. They begin to accumulate glycogen and become less regularly arranged as they move toward the surface. The highly cellular connective tissue immediately beneath the **stratum basale** (13) of the epithelium is typically infiltrated by **lymphocytes** (14), whose numbers vary with the stage of the ovarian cycle. Lymphocytes invade the epithelium around the time of menstruation.

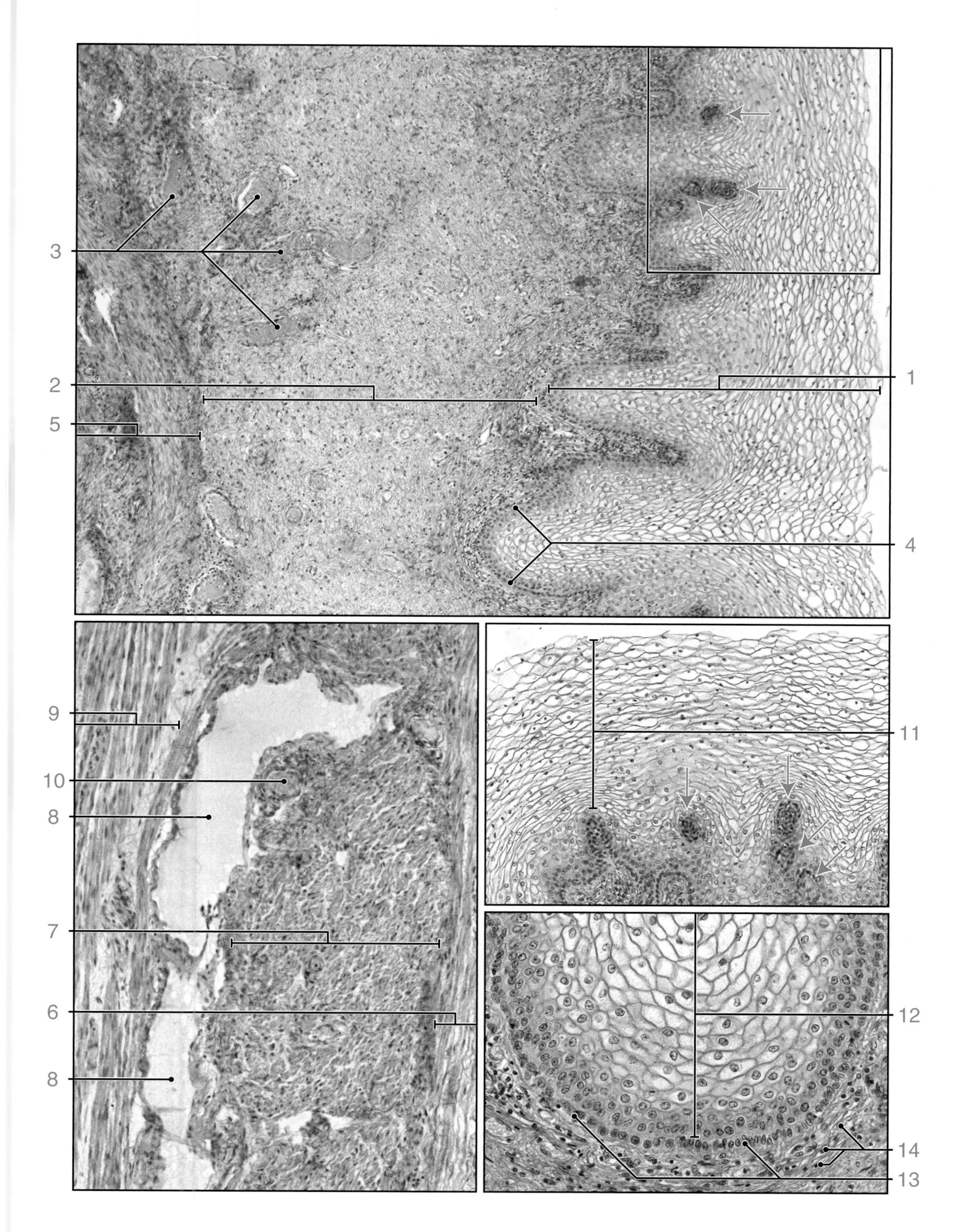
3
2
5
1
4
9
10
8
7
6
8
11
12
14
13

PLATE 147. VULVA

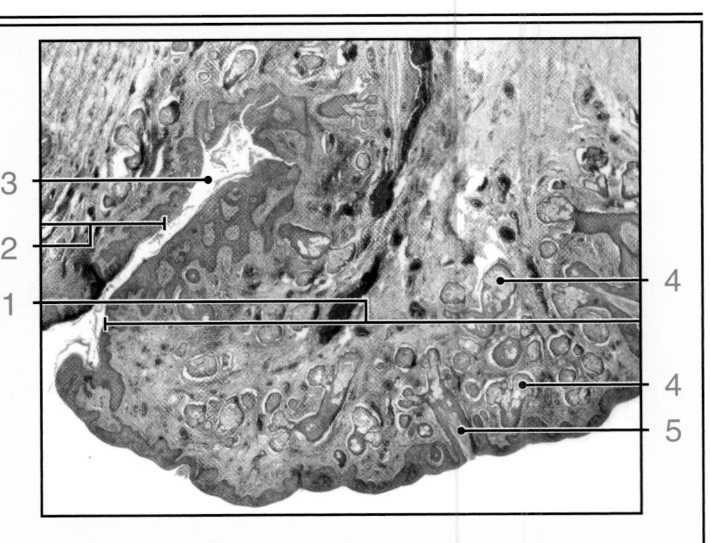

The vulva represents the urogenital part of the female perineum containing the female external genitalia. It includes the mons pubis, labia majora, labia minora, clitoris, and vesibule of the vagina. Lateral boundries of the vulva are formed by the labia majora. The labia majora consist of two large folds of adipose and connective tissue. Medial to the labia majora and separated by the interlabial sulcus are the labia minora. The labia majora contain a thin layer of smooth muscle that resembles the dartos muscle of the scrotum. The labia majora are covered by stratified squamous epithelium derived from two germinal layers. The inner surfaces of the labia majora are covered by endodermally derived, nonkeratinized stratified squamous epithelium, which abuts the ectodermally derived, keratinized stratified squamous epithelium covering the outer surface of labia majora. The edges of the posterior junctions of the labia majora form a distinct line, however nonkeratinized epithelium may be present (such as in this specimen) on the outer labial surfaces. The outer surface of the labia majora contains numerous hair follicles and sebaceous, eccrine, and apocrine sweat glands; the inner surface is smooth, devoid of hair, and contains sebaceous and sweat glands.

ORIENTATION MICROGRAPH: This is a low magnification of a frontal section of a **labium majus** (1). The inner surface of the labium majus is separated from the **labium minus** (2) by the **interlabial sulcus** (3). The inner surface of the labium majus is hairless and contains **sebaceous glands** (4) whose ducts open directly onto the epithelial surface. The outer surface of the labium majus (bottom of the image) contains numerous **hair follicles** (5) with associated sebaceous glands. The underlying dense irregular connective tissue is highly vascular.

Labium majus, human, H&E x180.
This micrograph shows a higher magnification of the **epidermis** (1) covering the outer surface of the labia majora. This section represents the transitional zone between the keratinized stratified squamous epithelium of the skin covering the outer surface of the labia majora and the nonkeratinized epithelium found on the inner surface of the labia. The darker appearance of the skin of the labia majora is a result of the larger number of melanocytes and pigmented keratinocytes in the **stratum basale** (2) of the epidermis. Two **hair follicles** (3) containing **hair shafts** (4) are evident in this section. Their histological structure is similar to other hair follicles in the body; they contain a hair shaft surrounded by the inner and outer root sheaths. The hair follicles are associated with numerous **sebaceous glands** (5) and form the pilosebaceous unit. Sebum, a product of the holocrine secretion of the sebaceous gland, is discharged into the infundibulum of a hair follicle via the short **pilosebaceous canal** (6). The area beneath the epithelium is composed of dense irregular connective tissue and contains blood vessels and bundles of smooth muscles.

Labium majus, pilosebaceous unit, human, H&E x125.
This is a higher magnification of the pilosebaceous unit from the boxed area on the top micrograph. Secretions of the sebaceous gland include lipid-rich secretory product and the cell debris that is discharged into the short **pilosebaceus canal** (7). It empties this secretion, called sebum, into the infundibulum of the hair folicle. At this magnification, the hair follicle reveals the **external root sheath** (8) surrounding the **hair shaft** (9). Several cross-sectional profiles of the sebaceous glands are found within the dermis of the skin covering the labia majora.

Labium majus, human, H&E x300; inset x240.
This micrograph shows a higher magnification of nonkeratinized stratified squamous epithelium. The epithelium is pigmented due to the presence of **melanocytes** (10) and **pigmented ketatinocytes** (11) in the stratum basale of the epidermis. Keratinocytes in this epithelium exhibit **perinuclear halos** (12), a characteristic feature of the epithelium of the vulva. Note that the nuclei of keratinocytes are present in the top layer of the epidermis. This is a transition area between nonkeratinized and keratinized epithelium. The presence of keratohyalin granules in the top layer of epithelium (**inset**) is a first step in the keratinization process.

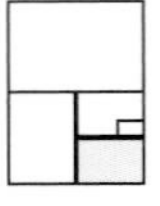

Labium majus, human, H&E x320.
This is a higher magnification of a sebaceous gland, which appears as a cluster of cells with a washed-out or finely reticulated cytoplasm. These cells contain numerous lipid droplets that are lost in routine H&E preparations. The **basal cells** (13) at the periphery of the gland are best preserved. As the cells mature they accumulate increasingly greater amounts of lipids and move progressively closer to the pilosebaceous canal. Sebaceous secretions also include entire cells; therefore, cells need to be replaced constantly in a functional gland.

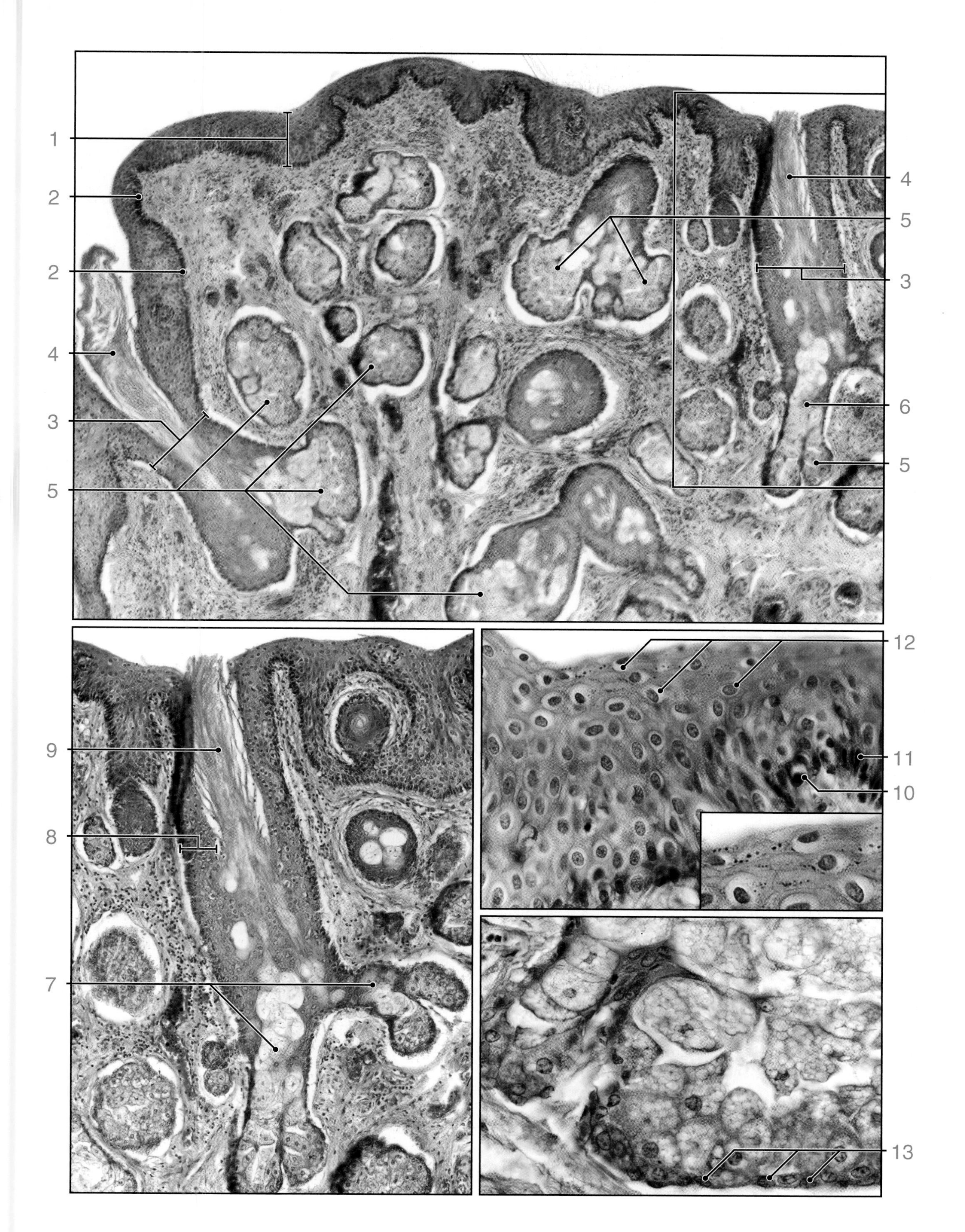

PLATE 147. VULVA

PLATE 148. MAMMARY GLAND, RESTING

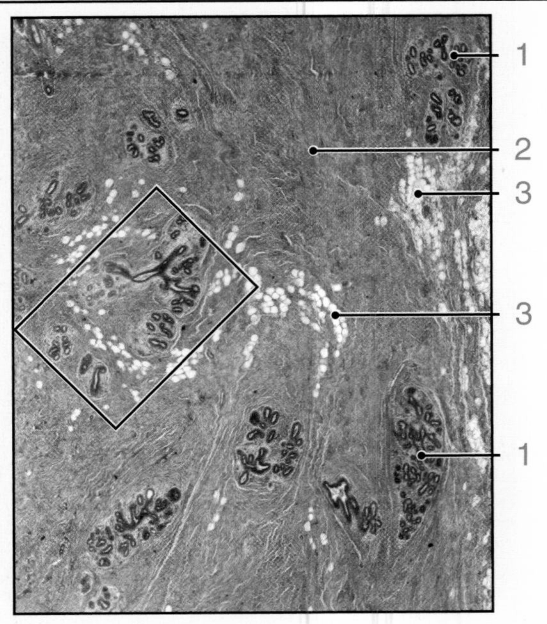

The mammary glands are organs that undergo change with age, menstrual cycle, and reproductive status. They are classified as branched, tubuloalveolar glands that develop from epithelial cells in the epidermis of the thoracic region. They lie in the subcutaneous tissue of the pectoral region on each side of the anterior thoracic wall. The structure of the mammary gland is essentially the same in both sexes until puberty. At puberty, under the influence of estrogen, the mammary glands begin to develop in females. They increase in size, mainly due to the deposition of adipose tissue. The duct systems of the glands extend and branch in the connective tissue stroma at puberty but do not reach a fully functional state until after pregnancy. In pubertal males, testosterone inhibits mammary gland growth.

In females, each mammary gland is organized into 15 to 20 irregular lobes. Each lobe further divides into a number of lobules containing secretory lobular units. Each lobular unit consists of mammary gland parenchyma (secretory alveoli in the lactating mammary gland and terminal ductules in the resting mammary gland) and connective tissue stroma. Alveoli and terminal ductules drain into intralobular ducts. Intralobular ducts drain into interlobular ducts that empty into a lactiferous sinus below the nipple. The interlobular stroma of the mammary gland consists of dense irregular connective tissue infiltrated by varying amounts of adipose tissue. In contrast, the intralobular stroma consists of loose connective tissue that surrounds the lobular epithelial components of the gland.

ORIENTATION MICROGRAPH: This micrograph shows a section of a resting, postpubertal female mammary gland. At this low magnification, a number of **lobules** (1) can be seen. Each lobule consists of terminal ductules. They are surrounded by **loose connective tissue** (2) contained within a dense connective tissue stroma. **Adipose cells** (3) are visible within the connective tissue.

Mammary gland, resting, human, H&E x90; inset x350.

This micrograph shows a higher magnification of the boxed area on the orientation micrograph. The parenchyma of the resting mammary gland is formed by branching of the lactiferous ducts leading to the **terminal duct lobular unit** (1). Each unit represents a cluster of branched **terminal ductules** (2) that form a lobule. During pregnancy and after birth, the cells of the terminal ductules differentiate into functional secretory alveolar cells that produce milk. Each lobule has an **intralobular collecting duct** (3) that carries the secretions into the **interlobular duct** (4). Structures in the terminal duct lobular unit are surrounded by an intralobular stroma, which is a specialized, hormonally sensitive loose connective tissue surrounding the terminal ductules. The **inset** shows the terminal ductules (2) surrounded by the loose connective tissue of the **intralobular stroma** (5) at higher magnification. The epithelium of the terminal ductule in the resting mammary gland is cuboidal; in addition, myoepithelial cells at the base of the epithelium are present (*arrows*). As elsewhere, the myoepithelial cells are on the epithelial side of the basement membrane. **Dense irregular connective tissue** (6) forms interlobular septa and contains sparsely distributed **adipose cells** (7).

Mammary gland, resting, terminal duct lobular unit, human, H&E x180.

This higher magnification of the area in the right quadrant of the top micrograph shows two terminal duct lobular units. The **intralobular stroma** (8) surrounding the **terminal ductules** (9) consists of a specialized, loose connective tissue that contains far more cells per unit area than other regions of the body and a greater variety of cell types. Note the cluster of **lymphocytes** (10) and, in the **inset**, plasma cells (*arrowheads*) and fibroblasts with elongated nuclei. The ductules from each lobule drain to the **intralobular ducts** (11), which then empty into the **interlobular duct** (12).

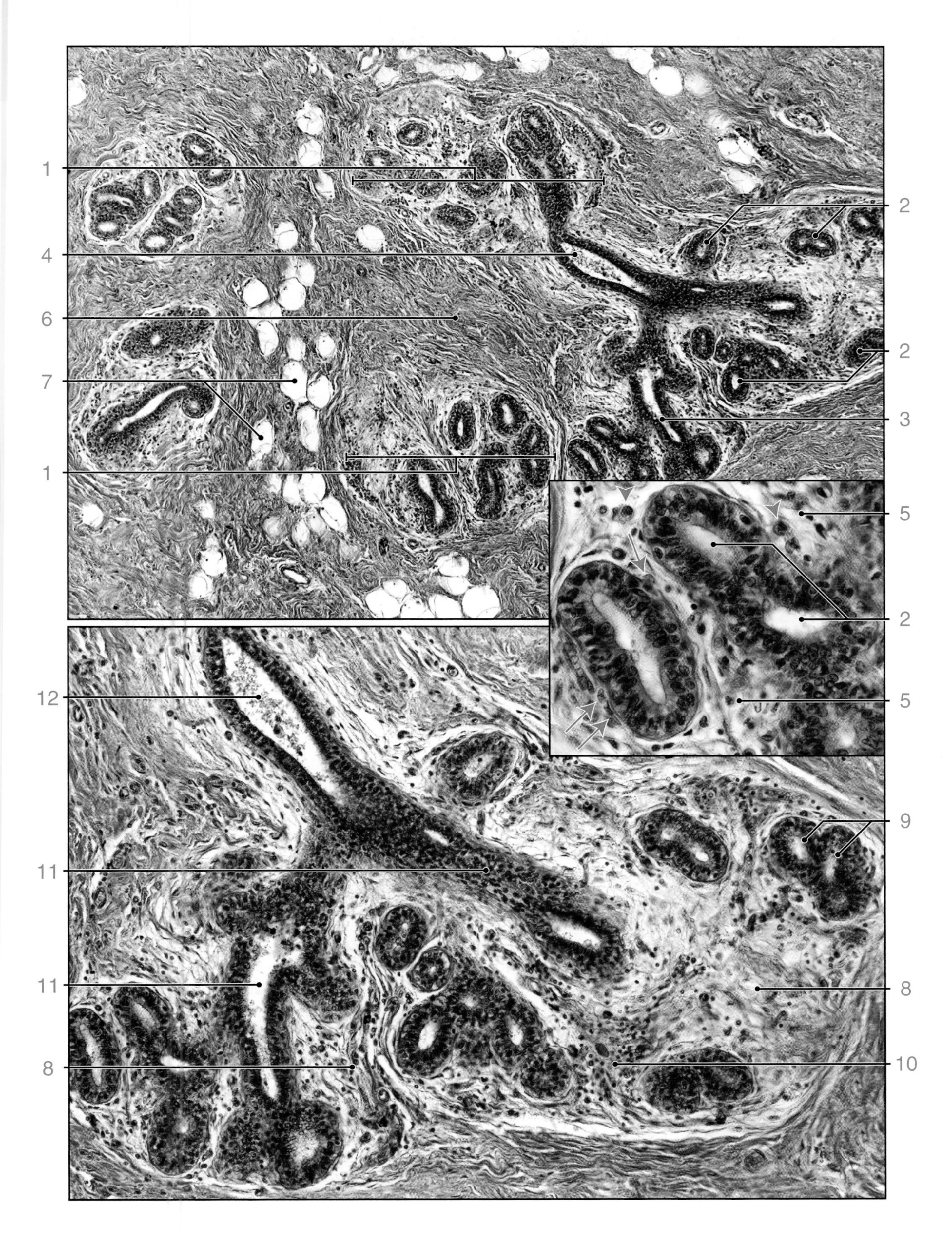

PLATE 148. MAMMARY GLAND, RESTING

PLATE 149. MAMMARY GLAND, PROLIFERATIVE

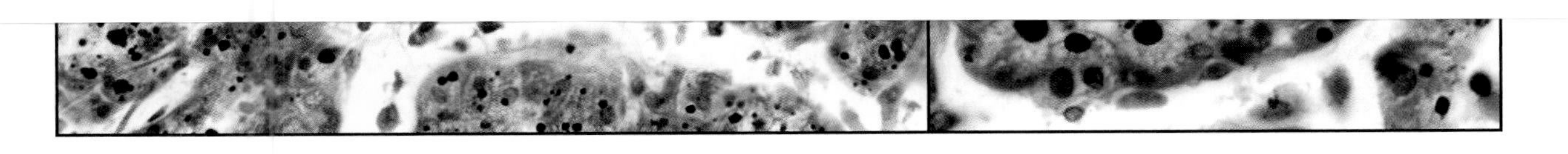

PLATE 150. MAMMARY GLAND, LACTATING

PLATE 149. MAMMARY GLAND, PROLIFERATIVE

Mammary glands experience their most rapid phase of proliferation during pregnancy in preparation for lactation. This occurs in response to

PLATE 150. MAMMARY GLAND, LACTATING

Both merocrine and apocrine secretion mechanisms are involved in the production of milk during the lactation period. The protein component of the milk is synthesized, concentrated, and secreted by exocytosis. The lipid component begins as droplets in the cytoplasm in the alveolar glandular cells. These droplets coalesce into larger droplets in the apical cytoplasm of the alveolar glandular cells and cause the apical plasma membrane to bulge into the alveolar lumen. The large droplets are surrounded by a thin layer of cytoplasm and are enveloped in plasma membrane as they are released. During this period, the milk is expelled

PLATE 151. NIPPLE

The nipple, also known as the mammary papilla, represents a small conical or cylindrical projection of the skin in the middle of the areola, the pigmented area surrounding the nipple. The tip of the nipple contains the openings of 15 to 20 lactiferous ducts arranged in cylindrical fashion.

The epidermis of an adult nipple and areola is highly pigmented and wrinkled due the to contraction of smooth muscle fibers that are arranged radially and circumferentially in the dense connective tissue and longitudinally along the lactiferous ducts. Numerous sensory nerve endings are present in the nipple. Certain stimuli cause the nipple to become erect as a result of contraction of the underlying smooth muscles. Keratinized stratified squamous epithelium forms the epidermis of the areola and nipple. It exhibits long, dermal papillae invading deep into the epidermal surface. The areola contains scattered sebaceous glands, sweat glands, and clustered openings of modified mammary glands (of Montgomery). These glands have a structure intermediate between sweat glands and true mammary glands and produce small elevations on the surface of the areola. The lactiferous ducts of the mammary gland open onto the nipple. The lactiferous duct is an extension of the lactiferous sinus, which is located beneath the skin of the areola.

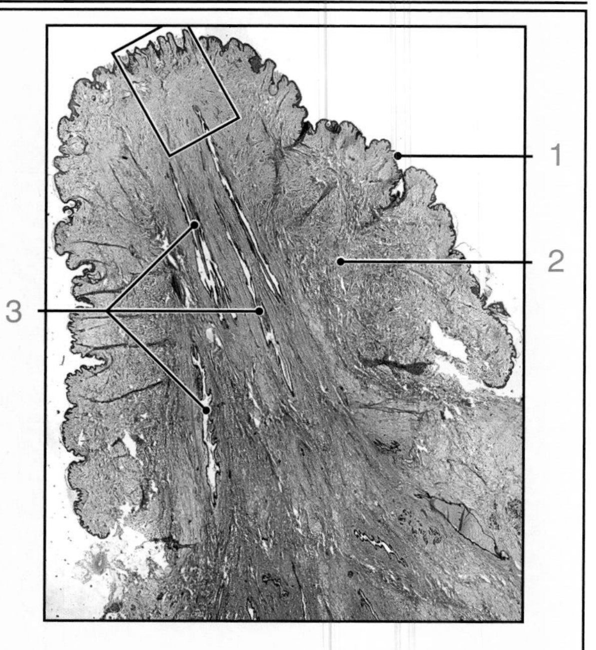

ORIENTATION MICROGRAPH: This low magnification micrograph shows a sagittal section through the nipple. The wrinkled contour of the nipple is lined by **stratified squamous epithelium** (1). The core of the nipple consists of **dense irregular connective tissue** (2), containing bundles of smooth muscle and elastic fibers, and the **lactiferous ducts** (3), which open at the tip of the nipple.

Nipple, human, H&E x45.

The **epidermis** (1) of the nipple consists of typical, keratinized stratified squamous epithelium. In this region it appears highly pigmented and wrinkled, due to long **dermal papillae** (2). The epithelium invaginates into **fissures** (3) that mark the opening into the lactiferous ducts. Near these openings, the lactiferous ducts are lined with stratified squamous epithelium. The epithelial lining of the duct shows a gradual transition from stratified squamous to a double layer of cuboidal cells in the **lactiferous sinus** (4) and finally to a single layer of columnar or cuboidal cells through the remainder of the duct system. **Keratinized cell debris** (5) is usually found plugging lactiferous duct openings. Large **sebaceous glands** (6) occupy the space just deep to the epidermis. Their holocrine secretion, the sebum, is discharged either directly onto the epidermal surface or, often, into openings of the **lactiferous ducts** (7). The core of the nipple consists of dense irregular connective tissue, containing smooth muscle bundles.

Nipple, human, H&E x180; inset x450.

This is a higher magnification of the boxed area, surrounding the lactiferous sinus, on the top micrograph. Bundles of **smooth muscle** (8) in longitudinal and cross-sectional profiles traverse the core of the dense irregular connective tissue. The wall of the **lactiferous sinus** (9) is wrinkled and lined by deeply stained, stratified cuboidal epithelium. Just beneath the epithelium, there is a thin layer of highly cellular **loose connective tissue** (10). The **inset** shows a higher magnification of the lactiferous sinus and its epithelium, consisting of two layers of cuboidal cells.

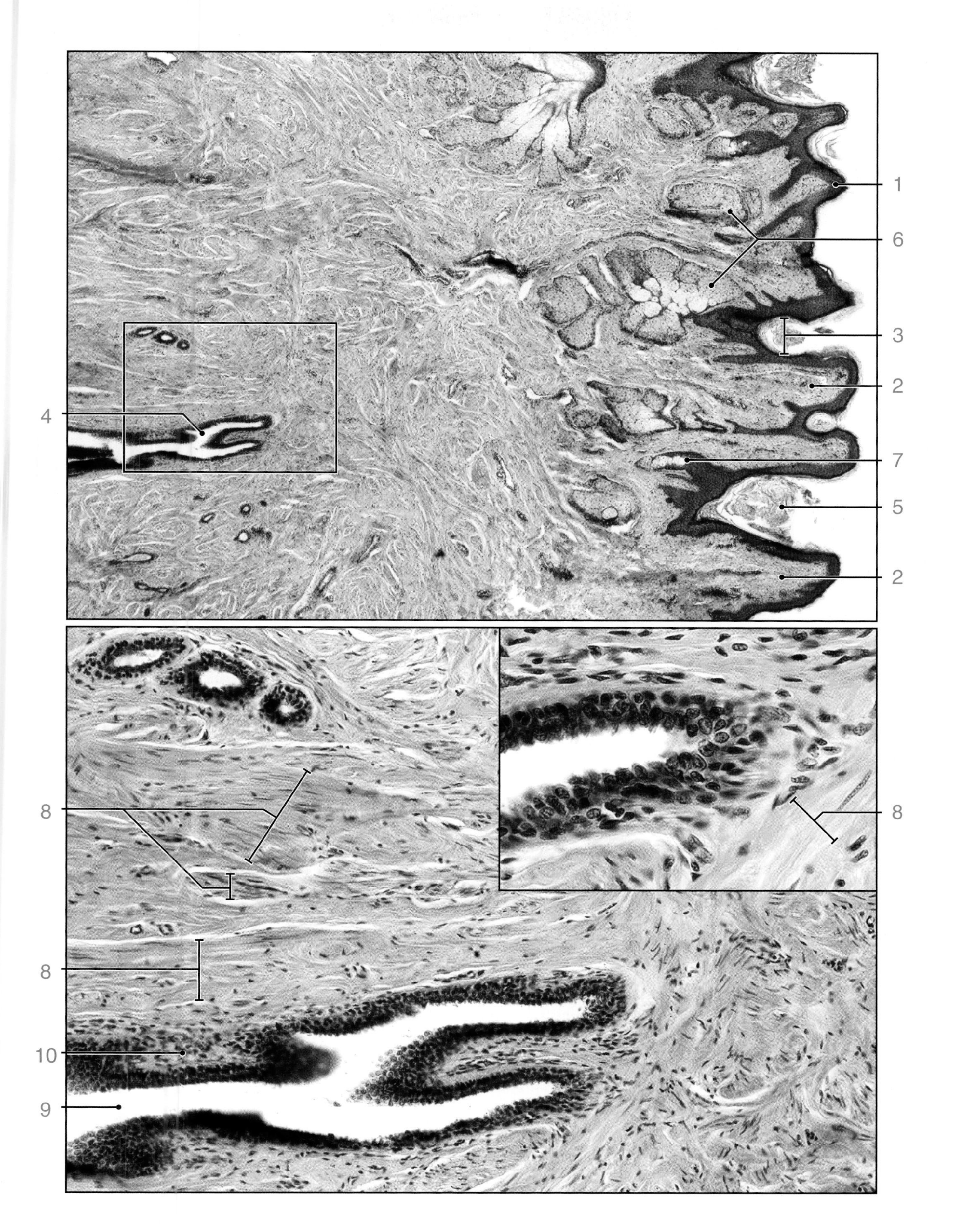

PLATE 151. NIPPLE

CHAPTER 20

Eye

PLATE 152. **GENERAL EYE STRUCTURE**

The human eye is a complex, spherical, sensory organ that collects and converts visual stimuli. It may be compared to a digital camera with an optical system to capture and focus light, a diaphragm to regulate the amount of light, and light detector (i.e., CCD) to capture and convert the image into electrical impulses. The wall of the eye consists of three concentric layers or coats: the retina, the inner layer; the vascular coat or uvea, the middle layer; and the corneoscleral coat, the outer fibrous layer. The cornea and lens concentrate and focus light on the retina. The iris, located between the cornea and lens, regulates the size of the pupil through which light enters the eye. Photoreceptor cells (rods and cones) in the retina detect the intensity (rods) and color (cones) of the light that reaches them and encode the various parameters for transmission to the brain via the optic nerve (cranial nerve II).

The eye measures about 25 mm in diameter. It is suspended in the bony orbit by six extrinsic striated muscles that control its movement. These extraocular muscles are coordinated so that both eyes move synchronously, with each moving symmetrically around its own axis. A thick layer of adipose tissue partially surrounds and cushions the eye as its moves within the orbit.

Drawing of a sectioned human eye, horizontal plane, modified from E. Sobotta.

The innermost layer of the eye, the **retina** (1), consists of a number of layers of cells. Among these are receptor cells (rods and cones), neurons (e.g., bipolar and ganglion cells), supporting cells, and a retinal pigment epithelium (see Plate 155). The photosensitive region of the retina is situated in the posterior three fifths of the eyeball. At the anterior boundary of the photosensitive region, the **ora serrata** (2), the retina becomes reduced in thickness. It continues forward to form a nonphotosensitive region of the retina that covers the inner surface of the **ciliary body** (3) and the **iris** (4). This anterior extension of the retina is highly pigmented, and the pigment (melanin) is evident as the black inner border of these structures.

The vascular coat or uvea, the middle layer of the eyeball, consists of the choroid, the ciliary body, and the iris. The choroid is a vascular layer; it is relatively thin and difficult to distinguish in the accompanying figure except by location. On this basis, the **choroid** (5) is identified as being just external to the pigmented layer of the retina. It is also highly pigmented; the choroidal pigment is evident as a discrete layer in several parts of the section.

Anterior to the ora serrata, the vascular coat is thickened; here, it is called the ciliary body (3). This structure (see Plate 156) contains the ciliary muscle, which brings about adjustments of the lens to focus light. The ciliary body also contains processes to which the zonular fibers are attached. These fibers function as suspensory ligaments of the **lens** (6). The iris (4) is the most anterior component of the chorioid layer and contains a central opening, the pupil.

The outermost layer of the eyeball, the corneoscleral layer, consists of the **sclera** (7) and the **cornea** (8). Both of these contain collagen fibers as their main structural element; however, the cornea is transparent, and the sclera is opaque. The extrinsic muscles of the eye insert into the sclera and effect movements of the eyeball. These are not included on this figure except for two small areas of a **muscle insertion** (9) located at lower and upper portions of the sclera. Posteriorly, the sclera is pierced by the emerging **optic nerve** (10). The region in the neural retina where nerve fibers converge to become part of the optic nerve is shaped into the **optic papilla** or **disc** (11). It is also called the blind spot because it is not sensitive to light. A shallow depression in the retina, lateral to the optic papilla, is the **fovea centralis** (12), the thinnest and most sensitive portion of the neural retina.

The lens is considered in Plate 154. Just posterior to the lens is the large cavity of the eye, the **vitreous cavity** (13), which is filled with a thick jellylike material, the vitreous body. Anterior to the lens are two additional chambers of the eye filled with aqueous humor, the **anterior** (14) and **posterior chambers** (15), separated by the iris.

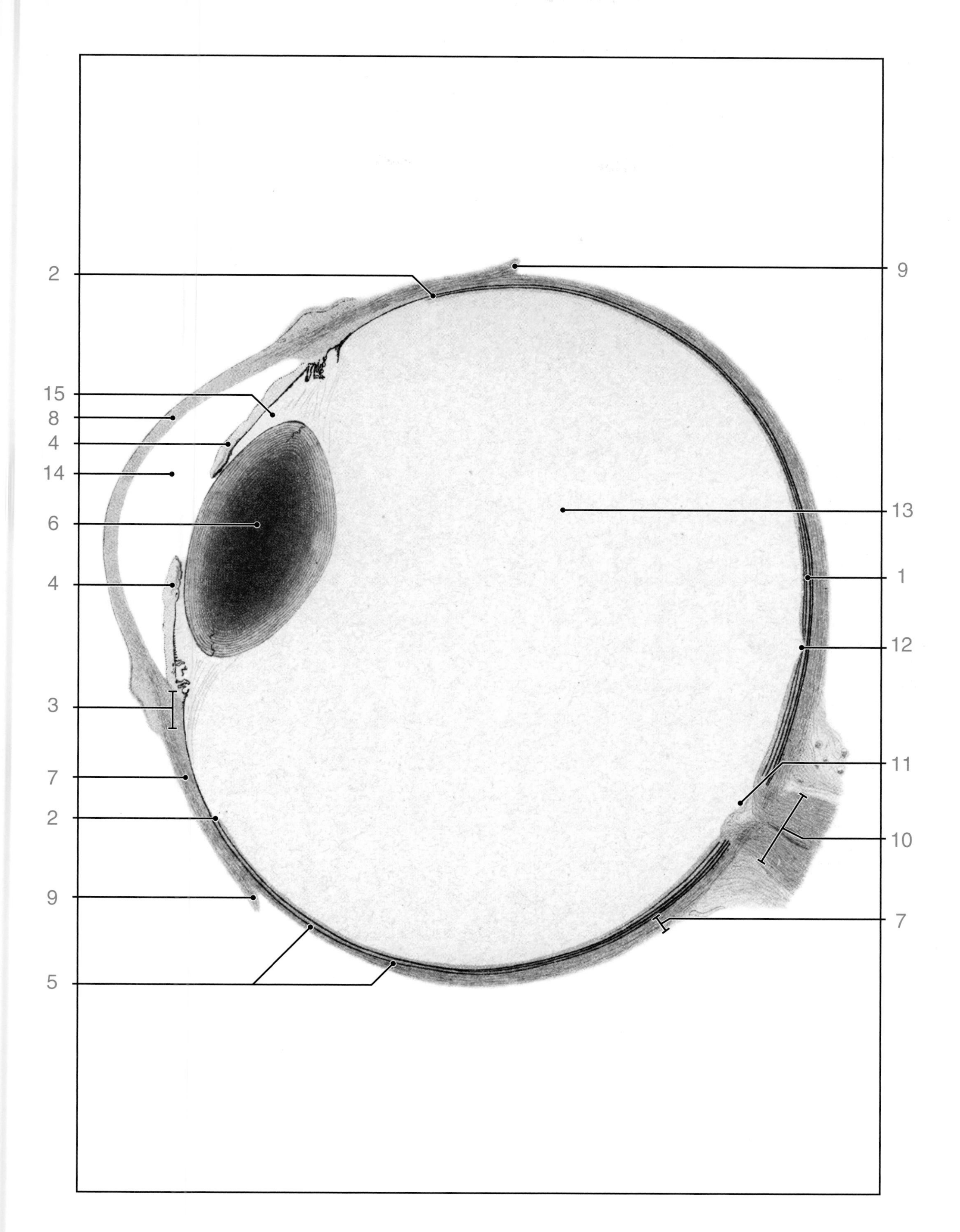

PLATE 152. GENERAL EYE STRUCTURE

PLATE 153. ANTERIOR SEGMENT OF THE EYE

The anterior segment is that part of the eye anterior to the ora serrata. It contains the anterior extension of the neural retina as well as two fluid-filled spaces, the anterior and posterior chambers, and the structures that define them. These structures are the cornea and sclera, the iris, the lens, and the ciliary body. The lens is suspended behind the iris and connected with the ciliary body by the suspensory ligament of the lens, made up of the zonular fibers. The posterior chamber is bounded posteriorly by the anterior surface of the lens and anteriorly by the posterior surface of the iris. The ciliary body forms the lateral boundary. Aqueous humor flows from the posterior chamber (where it is produced) through the pupil into the anterior chamber, which occupies the space between the cornea and the iris. The fluid from the anterior chamber is then drained into the canal of Schlemm.

ORIENTATION MICROGRAPH: This very low magnification micrograph identifies the structures that define the boundaries and contents of the anterior segment of the eye. They include the **cornea** (1), **sclera** (2), **iris** (3), **lens** (4), and **ciliary body** (5).

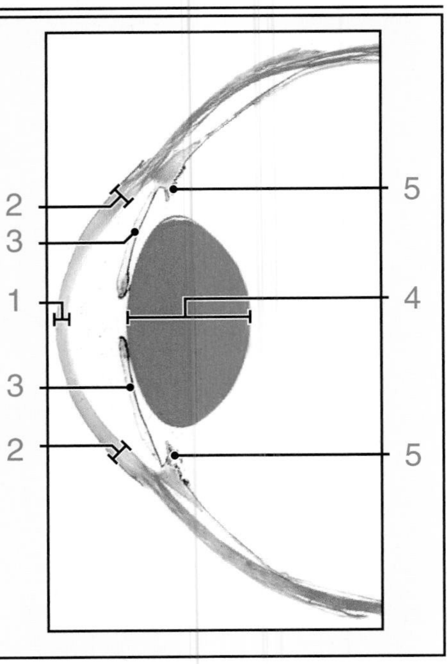

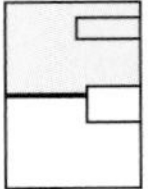

Eye, anterior segment, human, H&E x45; inset iris, pupillary margin x75.

A portion of the anterior segment of the eye shown in this figure includes parts of the **cornea** (1), **sclera** (2), **iris** (3), **ciliary body** (4), **anterior chamber** (5), **posterior chamber** (6), **lens** (7), and **zonular fibers** (8).

The relationship of the cornea to the sclera is illustrated clearly here. The junction between the two is marked by a change in staining, the substance of the cornea appearing lighter than that of the sclera. The **corneal epithelium** (9) is continuous with the **conjunctival epithelium** (10) that covers the sclera. Note that the epithelium thickens considerably at the corneoscleral junction and resembles that of the oral mucosa. The conjunctival epithelium is separated from the dense fibrous component of the sclera by a loose vascular connective tissue. Together, this connective tissue and the epithelium constitute the **conjunctiva** (11). The epithelial–connective tissue junction of the conjunctiva is irregular; in contrast, the undersurface of the corneal epithelium presents an even profile.

Just lateral to the junction of the cornea and sclera is the **canal of Schlemm** (12). This canal takes a circular route around the perimeter of the cornea. It communicates with the anterior chamber through a loose trabecular meshwork of tissue, the spaces of iridocorneal angle (of Fontana). The canal of Schlemm also communicates with episcleral veins. Through its communications, the canal of Schlemm provides a route for the fluid in the anterior chamber to reach the bloodstream.

The **inset** shows the edge of the iris. Note the heavy pigmentation of the posterior surface of the iris, which is covered by the same double-layered epithelium as the ciliary body and ciliary processes. In the ciliary epithelium, the outer layer is pigmented and the inner layer is nonpigmented. In the iris, both layers of the **iridial epithelium** (13) are heavily pigmented. A portion of the **sphincter pupillae muscle** (14) is seen above the epithelium.

Eye, ciliary body, human, H&E x90; inset ciliary process x350.

Immediately internal to the anterior margin of the **sclera** (15) is the **ciliary body** (16). The inner surface of ciliary body forms radially arranged, ridge-shaped elevations, the **ciliary processes** (17), to which the **zonular fibers** (18) are anchored. From the outside in, the components of the ciliary body are the **ciliary muscle** (19), the choriocapillary layer with surrounding connective tissue representing the **vascular coat in the ciliary body** (20), the **lamina vitrea** (21, **inset**), and the **ciliary epithelium** (22, **inset**). The ciliary epithelium, shown in the **inset**, consists of two layers, the **pigmented layer** (23) and the **nonpigmented layer** (24). The lamina vitrea is a continuation of the same layer of the choroid; here it serves as the basement membrane for the pigmented ciliary epithelial cells.

The ciliary muscle is arranged in three patterns. The outer layer is immediately internal to the sclera. These are the meridionally arranged fibers of Brücke. The outermost of these fibers continues posteriorly into the vascular layer and is referred to as the tensor muscle of the choroid. The middle layer is the radial group. It radiates from the region of the sclerocorneal junction into the ciliary body. The innermost layer of muscle cells is circularly arranged. They are seen in cross-section in this micrograph. The barely discernable **circular artery** (25) and the **circular vein** (26) for the iris, also cut in cross-section, are just anterior to the circular group of muscle cells.

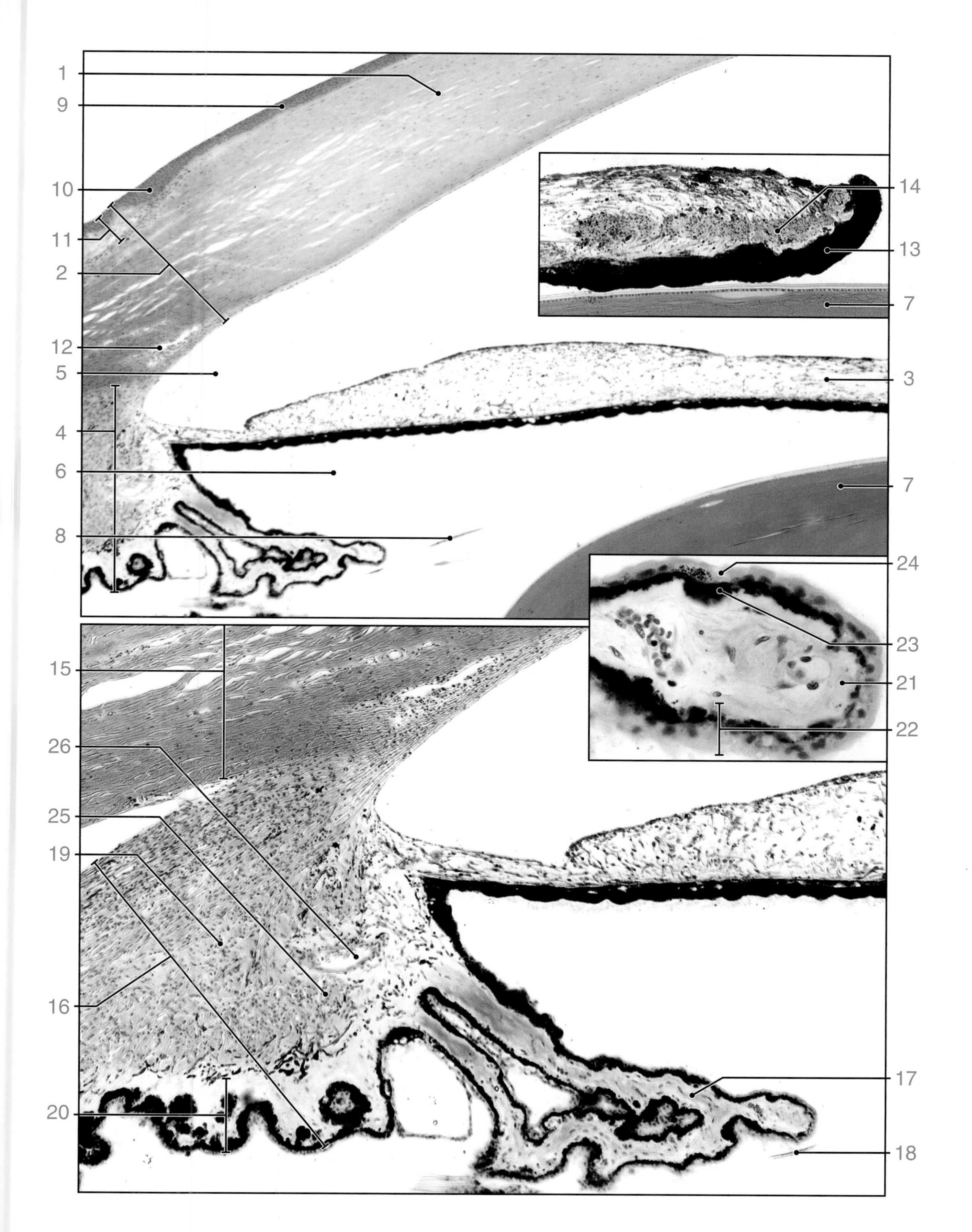

PLATE 153. ANTERIOR SEGMENT OF THE EYE

PLATE 154. **SCLERA, CORNEA, AND LENS**

The transparent cornea is the primary dioptric (refractive element) of the eye. Its anterior surface is covered with corneal epithelium, a nonkeratinized stratified squamous epithelium which rests on Bowman's membrane. The stroma of the cornea consists of alternating lamellae of parallel bundles of collagen fibrils and occasional fibroblasts (keratocytes). The fibrils in each lamella are extremely uniform in diameter and are uniformly spaced; fibrils in adjacent lamellae are arranged at approximately right angles to each other. This orthogonal array of highly regular fibrils is responsible for the transparency of the cornea. Similarly, the posterior surface of the lens is covered with a single layer of squamous cells, the corneal endothelium, which rest on a thickened basal lamina called Descemet's membrane. Nearly all of the metabolic exchanges of the avascular cornea occur across the corneal endothelium. Damage to this layer leads to corneal swelling and can produce temporary or permanent loss of transparency.

The lens is a transparent, avascular, biconvex epithelial structure suspended by the zonular fibers. Tension on these fibers keeps the lens flattened; reduced tension allows it to fatten, or accommodate, to bend light rays originating close to the eye to focus them on the retina.

Eye, sclera, human, H&E x130.

This low magnification micrograph shows the full thickness of the sclera just lateral to the corneoscleral junction, or limbus. To the left of the *arrow* is sclera; to the right is a small amount of corneal tissue. The **conjunctival epithelium** (1) is irregular in thickness and rests on a loose vascular connective tissue. Together, this epithelium and the underlying connective tissue represent the **conjunctiva** (2). The white opaque appearance of the sclera is due to the irregular, dense arrangement of the collagen fibers that make up the **scleral stroma** (3). The **canal of Schlemm** (4) is seen at the left, close to the inner surface of the sclera.

Eye, epithelium at the sclerocorneal junction, human, H&E x360.

This is a higher magnification micrograph showing the transition from the **corneal epithelium** (5) to the irregular and thicker **conjunctival epithelium** (6) covering the sclera. Note that **Bowman's membrane** (7), lying under the corneal epithelium, is just perceptible but disappears beneath the conjunctival epithelium.

Eye, sclera, canal of Schlemm, human, H&E x360.

This micrograph shows the **canal of Schlemm** (8) at higher magnification than the top left micrograph. The space is not an artifact, as evidenced by the presence of **endothelial lining cells** (9) facing the lumen.

Eye, cornea, human, H&E x175.

This low magnification micrograph shows the full thickness of the cornea and can be compared with the sclera shown in the top left micrograph. The **corneal epithelium** (10) presents a uniform thickness and the underlying **corneal stroma** (11) has a more homogenous appearance than the scleral stroma (3 in top left micrograph). The white spaces seen in this micrograph are artifacts. **Nuclei of the keratocytes** (12) of the corneal stroma lie between lamellae. The corneal epithelium (10) rests on a thickened basement membrane called **Bowman's membrane** (13). The posterior surface of the cornea is lined by a simple squamous epithelium called the **corneal endothelium** (14); its thick basement membrane is called **Descemet's membrane** (15).

Eye, corneal epithelium, human, H&E x360.

This micrograph is a higher magnification showing the **corneal epithelium** (16) with its squamous surface cells, the very thick homogeneous-appearing **Bowman's membrane** (17), and the underlying **corneal stroma** (18). Note that the stromal tissue has a homogeneous appearance, a reflection of the dense packing of its collagen fibrils. The flattened nuclei belong to the keratocytes.

Eye, corneal endothelium, human, H&E x360.

This micrograph shows the posterior surface of the cornea. Note the thick homogeneous **Descemet's membrane** (19) and the underlying simple squamous **corneal endothelium** (20).

Eye, lens, human, H&E x360.

This micrograph shows a portion of the lens near its equator. The lens consists entirely of epithelial cells surrounded by a homogeneous-appearing **lens capsule** (21) to which the zonula fibers attach. The lens capsule is a very thick basal lamina of the epithelial cells. A single layer of **cuboidal lens epithelial cells** (22) is present on the anterior surface of the lens, but at the lateral margin, they become extremely elongated, forming layers that extend toward the center of the lens. These elongated columns of epithelial cytoplasm are referred to as **lens fibers** (23). New cells are produced at the margin of the lens and displace the older cells inwardly. Eventually, the older cells lose their nuclei, as evidenced by the deeper portion of the cornea in this micrograph.

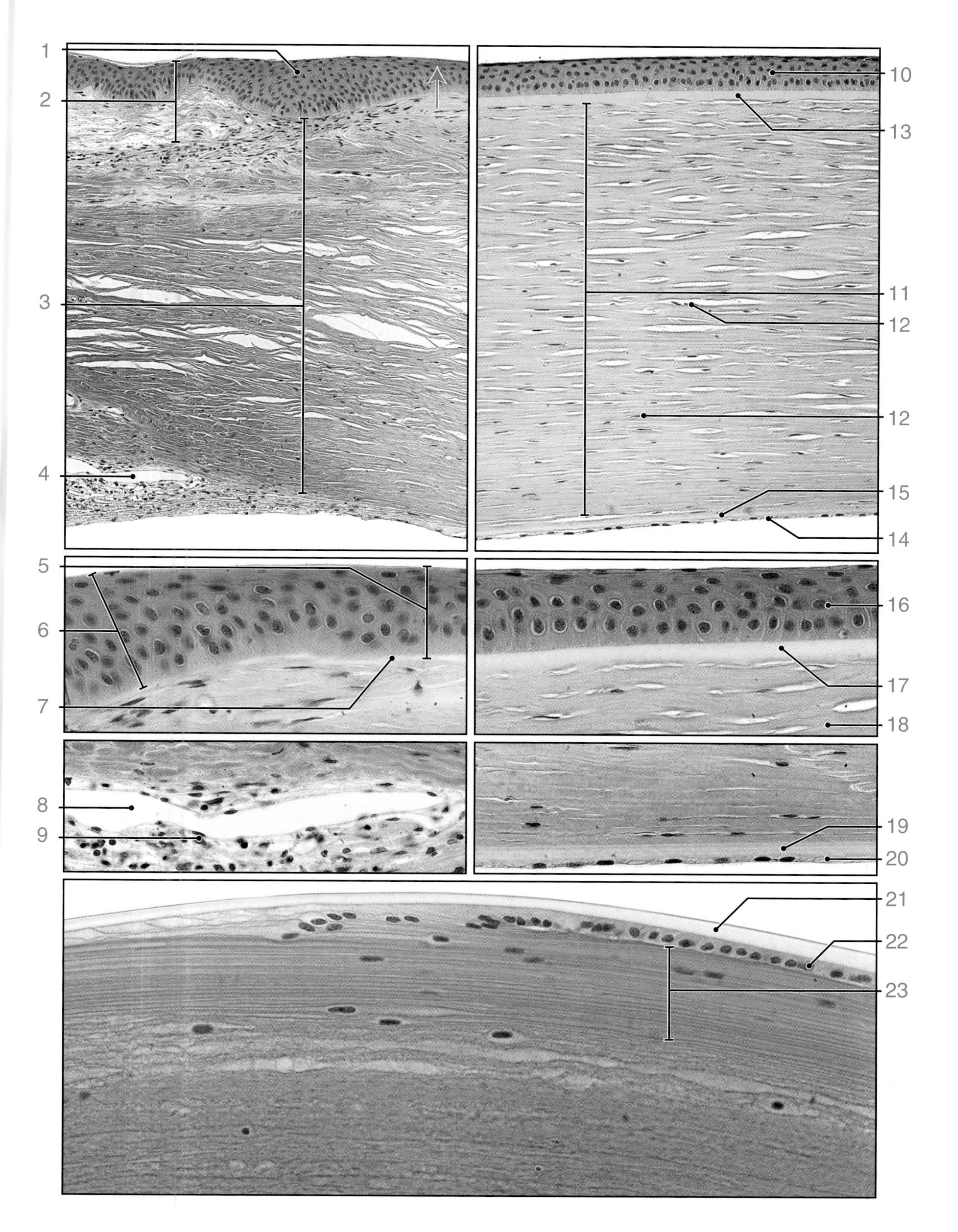

PLATE 154. SCLERA, CORNEA, AND LENS

PLATE 155. POSTERIOR SEGMENT OF THE EYE

The retina and optic nerve are projections of the forebrain. The fibrous cover of the optic nerve is an extension of the meninges of the brain. The neural retina is a multilayered structure consisting of photoreceptors (rods and cones); neurons, some of which are specialized as conducting or associating neurons; and supporting cells (Müller's cells). External to the neural retina is a layer of simple cuboidal retinal pigment epithelium. The Müller's cells are comparable to the neuroglia in the rest of the central nervous system. Processes of Müller's cells ramify through virtually the entire thickness of the retina. The inner limiting membrane is the basal lamina of Müller's cells; the outer limiting membrane is actually formed by the apical boundaries of Müller's cells and the apical junctions of Müller's cells. This level is also characterized by the presence of anchoring, cell-to-cell junctions (zonulae adherents) between processes of Müller's cells.

The neurons of the retina are arranged in three layers: a deep layer of rods and cones; an intermediate (inner nuclear) layer of bipolar, horizontal, and amacrine cells; and a superficial layer of ganglion cells. Nerve impulses originating in the rods and cones are transmitted to the inner nuclear layer and then to the ganglion cell layer. Synaptic connections occur in the outer plexiform layer (between the rods and cones and the neurons of inner nuclear layer) and the inner plexiform layer (between the neurons of inner nuclear layer and the ganglion cells), resulting in summation and neuronal integration. Finally, the ganglion cells' axons extend to the brain as components of the optic nerve.

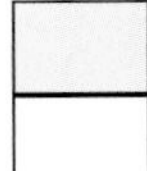

Eye, layers of the retina, human, H&E x325.
Based on structural features evident in histologic sections, the retina is divided into ten layers, beginning from the outside, as listed below and labeled on this micrograph:

Retinal pigment epithelium (1), the outermost layer of the retina;

Layer of rods and cones (2), the photoreceptor layer of the retina;

Outer limiting membrane (3), a line formed by the anchoring junctions between processes of Müller's cells;

Outer nuclear layer (4), containing nuclei of rods and cones;

Outer plexiform layer (5), containing neural processes and synapses of rods and cones with bipolar, amacrine, and horizontal cells;

Inner nuclear layer (6), containing nuclei of bipolar, horizontal, amacrine, and Müller's cells;

Inner plexiform layer (7), containing processes and synapses of bipolar, horizontal, amacrine, and ganglion cells;

Ganglion cells layer (8), containing cell bodies and nuclei of ganglion cells;

Layer of optic nerve fibers (9), containing axons of ganglion cells;

Inner limiting membrane (10), consisting of the basal lamina of Müller's cells.

This figure also shows the innermost layer of the **vascular coat**, or **choroid** (11), and a cell-free membrane, the **lamina vitrea** (12), also called Bruch's membrane. Immediately external to the Bruch's membrane is the choriocapillary layer of the vascular coat. These vessels supply the outer part of the retina.

Eye, retina and the optic nerve, human, H&E x65.
The site where the **optic nerve** (13) leaves the eyeball is called the **optic disc** (14). It is characteristically marked by a depression, as is evident here. Receptor cells are not present at the optic disc, and because it is not sensitive to light stimulation, it is sometimes referred to as the blind spot.

The fibers that give rise to the optic nerve originate in the retina, more specifically in the ganglion cell layer (see above). They traverse the sclera through a number of openings to form the optic nerve. The region of the sclera that contains these openings is called the **lamina cribrosa**, or **cribriform plate** (15). The optic nerve contains the central retinal artery and vein (not seen here) that also traverse the lamina cribrosa. Branches of the **central retinal artery** and **vein** (16) supply the inner portion of the retina.

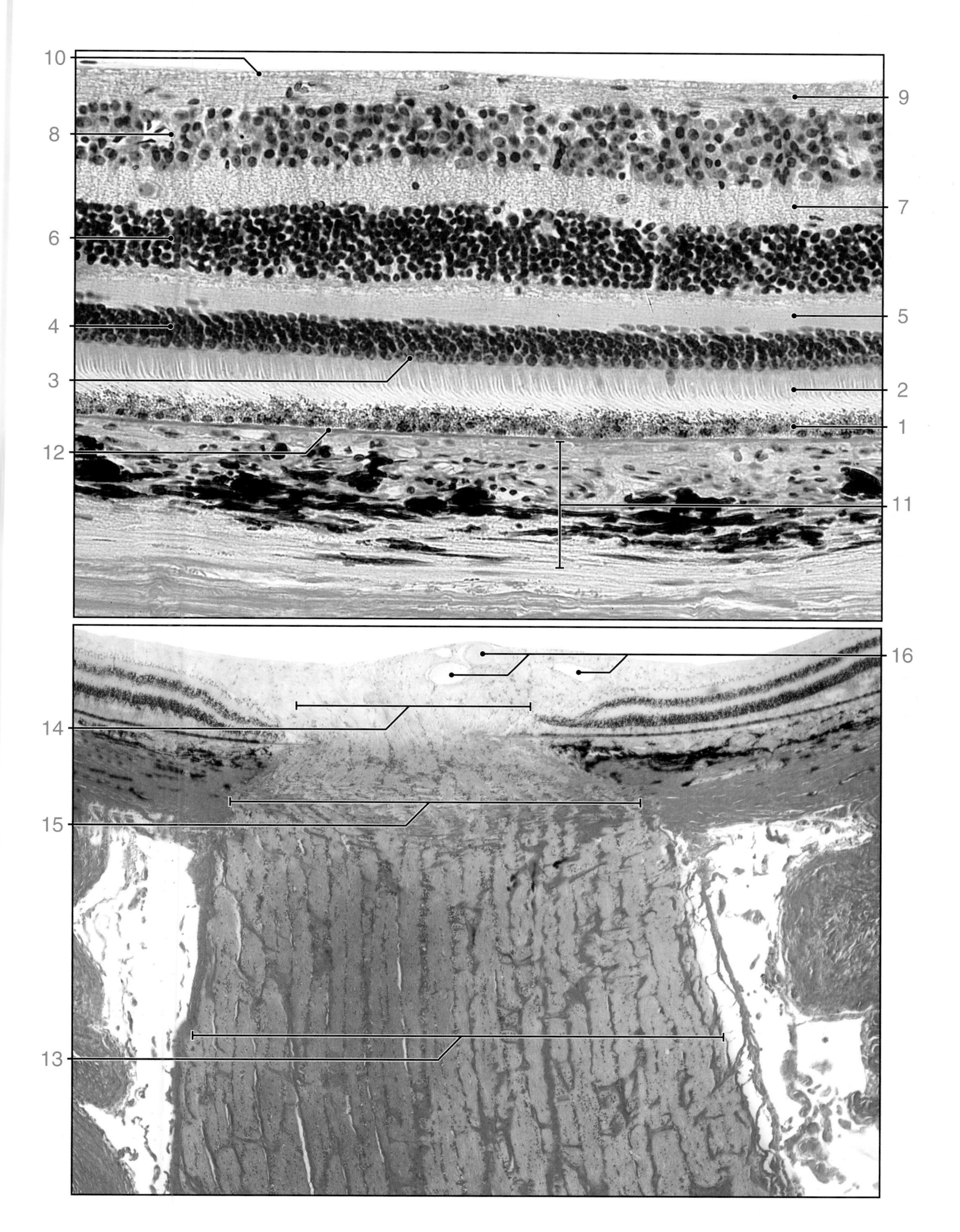

PLATE 155. POSTERIOR SEGMENT OF THE EYE

PLATE 156. DEVELOPMENT OF THE EYE

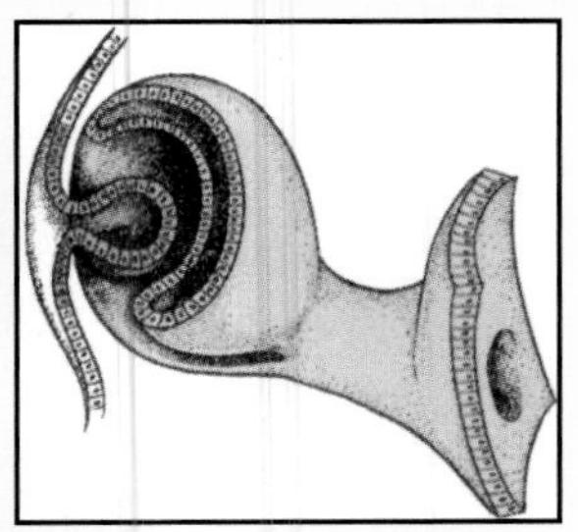

The eyes develop as evaginations from the developing forebrain. Even before the neural folds close to form the neural tube, the paired sites from which the optic vesicles will evaginate from the neural tube are visible as the optic sulci at the cranial end of the neurectoderm. Once the neural tube is formed, each optic vesicle grows laterally toward the basal lamina of the head ectoderm and its connection to the forebrain narrows to become the optic stalk. Where the optic vesicle approaches the head, the ectoderm thickens to form the lens placode. Both the lens placodes and the optic vesicles then invaginate; the lens placodes become the lens vesicles and the optic vesicles become the double-walled optic cup*s*. In the 5th week of development, the lens vesicle detaches from the surface ectoderm and comes to lie in the mouth of the optic cup. The region from where the lens vesicle detaches thickens again to give rise to the corneal epithelium. Mesenchymal cells then migrate into the acellular, extracellular matrix beneath the epithelium to give rise to the stroma and the corneal endothelium.

The outer layer of the optic cup forms a single layer of pigmented cells that will differentiate into the retinal pigment epithelium, the pigmented layer of the ciliary epithelium, and the anterior pigmented epithelial layer of the iris. The inner layer gives rise to the nine layers of the neural retina, the non-pigmented layer of the ciliary epithelium, and the posterior pigmented epithelial layer of the iris.

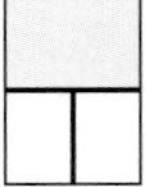

Developing eye, human (8 weeks), Masson Trichome x 220.

This low magnification micrograph shows a developing human eye at approximately eight weeks. All of the basic structures of the adult eye have now been formed, but the **eyelids** (1) are still closed and head **surface ectoderm** (2) still covers the future palpebral fissure. Developing **hair follicles** (3) of eyelashes are seen in one of the developing eyelids. There has been considerable shrinkage between the two layers of the optic cup, causing disruption of the continuity of the inner layer so that the developing **neural retina** (4) is widely separated from the **pigmented retinal epithelium layer** (5). The developing neural retina is also torn free from the anterior extensions of the **optic cup** (6), which are developing into the **ciliary body** (7) and **iris** (8). The **cornea** (9) is recognizable beneath the lids, and the developing **lens** (10) is still quite close to the posterior surface of the cornea. Many nuclei are still present in the developing **lens fibers** (11), even in the center of the lens. The developing **optic nerve** (12) is evident passing through the neural retina at the posterior pole of the eye. Small **blood vessels** (13), representing branches of the hyaloid artery and vein, can be seen in the **vitreous body** (14) and abutting the posterior surface of the developing lens.

Eye, anterior segment, human (8 weeks), Masson Trichome x 540.

This higher magnification micrograph shows the torn end of the neural retina beneath the developing **ciliary body** (15) and **iris** (16). The two-layered structure of the iridial and ciliary epithelia is evident here. The **corneal epithelium** (17) and **corneal endothelium** (18) are seen in apposition to the developing **corneal stroma** (19). Even at this magnification, the parallel organization of the corneal keratocytes and the stromal lamellae is evident. Numerous mesenchymal cells are present (not visible) in what will become the **anterior chamber** (20). The anterior epithelium and the equator of the lens are evident, and the turning in of the epithelial cells to form the **lens fibers** (21) is more obvious and much more active here than in a mature lens.

Eye, posterior segment, human (8 weeks), Masson Trichome x 540.

This higher magnification micrograph shows the developing **optic nerve** (22) leaving the eye through the **neural retina** (23) and **sclera** (24). While the thickness and cellularity of the developing neural retina are evident, it is not possible to distinguish clear layers in the structure. The accumulation of nuclei in the region of the **optic disc** (25) represents both proliferating glial cells, which will accompany and myelinate the axons of the optic nerve as they leave the eye, as well as some blood vessels entering and leaving the eye along the developing optic nerve. Small **blood vessels** (26) are distinguishable in the **vitreous body** (27).

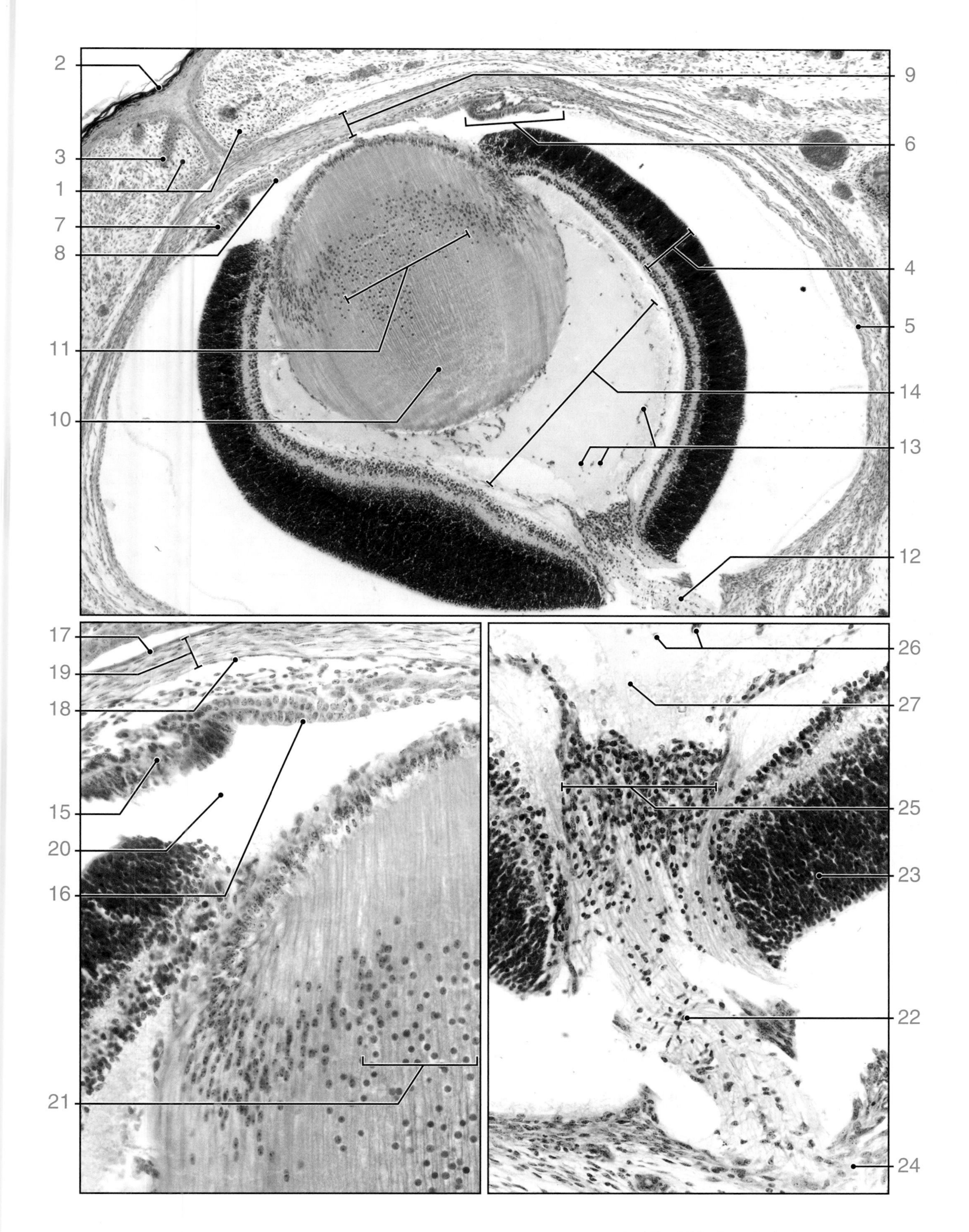

PLATE 156. DEVELOPMENT OF THE EYE

PLATE 157. **EYELID**

The upper and lower eyelids are modified folds of facial skin that protect the anterior surface of the eye. The skin of the eyelids is loose and elastic to accommodate their movement. Within each eyelid is a flexible support, the tarsal plate, consisting of dense, fibrous and elastic tissue and embedded tarsal glands (Meibomian glands). The lower border of the tarsal plate extends to the free margin of the eyelid, and its superior border serves for the attachment of smooth muscle fibers of the superior tarsal muscle (of Müller). The eyelids are covered posteriorly by a layer of palpebral conjunctiva. The orbicularis oculi muscle, a facial expression muscle, forms a thin oval sheet of circularly oriented skeletal muscle fibers overlying the tarsal plate. The connective tissue of the eyelid contains tendon fibers of the levator palpebrae superioris muscle, which opens the eyelid. In addition to eccrine sweat glands, which discharge their secretions directly onto the skin, the eyelid contains sebaceous glands (glands of Zeis) and apocrine glands of eyelashes (glands of Moll), both associated with eyelashes, and small, accessory lacrimal glands.

ORIENTATION MICROGRAPH: This is a section perpendicular to the surface of the eyelid. The outer surface of the lid is covered by **skin** (1); the inner surface, facing the eyeball, is covered by the **palpebral conjuntiva** (2). Adjacent to the palpebral conjunctiva are the **tarsal glands** (3), and more deeply located is the **orbicularis oculi muscle** (4). Portions of several **eyelashes** (5) are seen in the region of the conjunctiveocutaneous junction.

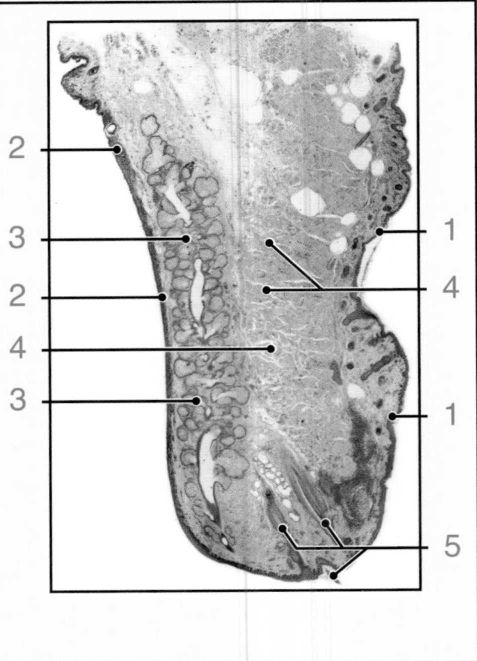

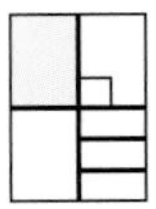

Eyelid and tarsal plate, posterior surface, human, picric acid, x75.

This micrograph of a sagittal section of the eyelid shows the **tarsal plate** (1) with embedded sebaceous glands, the **tarsal** or **Meibomian glands** (2). They consist of numerous alveoli that appear on this micrograph as yellow outgrowths from the straight **excretory duct** (3) that runs parallel to other such ducts in the plane of the tarsal plate. The sebaceous secretion of the tarsal glands leaves excretory ducts through their openings at the free border of the eyelid, located posterior to the row of eyelashes. The tarsal plate lies interior to the **plapebral conjunctiva** (4), which consists of a stratified columnar epithelium, containing numerous goblet cells. Striated muscle fibers of the **palpebral portion of the orbicularis oculi muscle** (5) are present interior to the tarsal plate. In this preparation, muscle and connective tissue stain yellow, and the epithelial cells of skin, conjunctiva, and glandular epithelium stain green.

Eyelid, anterior surface, human, picric acid x75; inset, Orbicularis oculi muscle, x160.

This micrograph shows the anterior part of the eyelid covered by **skin** (6) and the underlying **subcutaneous tissue** (7) containing **blood vessels** (8) and lymphatic vessels. Concentrically arranged striated muscle fibers of the **orbicularis oculi muscle** (9) form a thin sheet interior to the skin and underlying connective tissue. The **inset** shows the striated muscle fibers in cross-section at higher magnification.

Eyelid, eyelashes, human, picric acid x75.

The **eyelashes** (10) emerge from the most anterior edge of the lid margin, in front of the openings of the tarsal glands. They are short, stiff, curved hairs that emerge from **hair follicles** (11). Small, modified **sebaceous glands of eyelashes**, the **glands of Zeis** (12), empty their secretions into the follicle of the eyelash. Also, note the small sweat glands that begin as simple spiral tubular profiles, the **apocrine glands**, or **glands of Moll** (13). They also discharge their secretion into the hair follicles of the eyelashes.

Eyelid, palpebral conjunctiva, human, picric acid, x240.

This high magnification micrograph shows the **palpebral conjunctiva** (14) that consists of stratified columnar cells and goblet cells. The epithelium rests on a lamina propria composed of loose connective tissue. The conjunctiva, as well as the underlying connective tissue, is infiltrated by **lymphocytes** (15); the small, round, dense nuclei are characteristic of lymphocytes.

Eyelid, conjunctivocutaneous junction, human, picric acid x240.

This high magnification micrograph shows the **nonkeratinized stratified squamous epithelium** (16) at the conjunctivocutaneous junction. This junction is located at the free edge of the eyelid, usually between the ducts of the tarsal glands and the eyelashes.

Eyelid, palpebral skin, human, picric acid, x240.

This high magnification micrograph shows the palpebral skin, which has a **keratinized stratified squamous epithelium** (17).

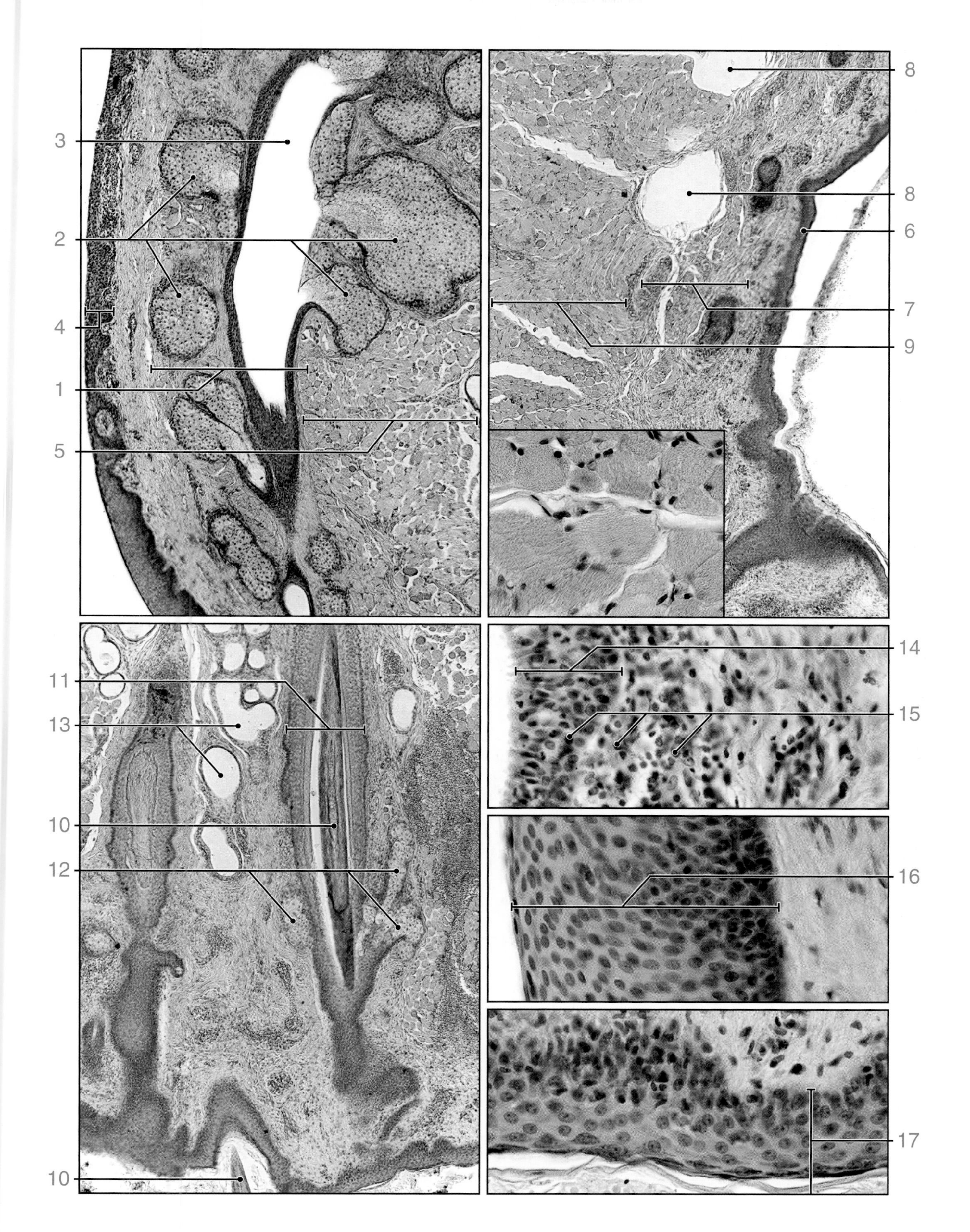

PLATE 157. EYELID

CHAPTER 21
Ear

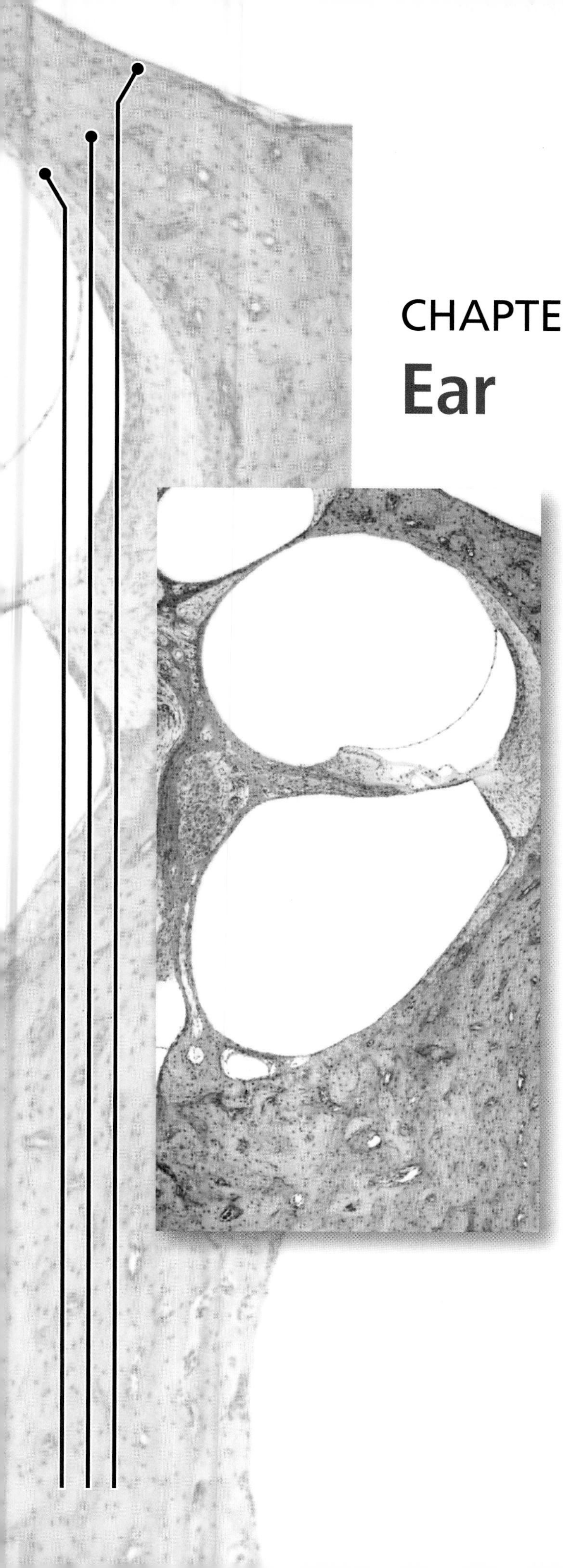

PLATE 158. EAR

The inner ear, located in the temporal bone, consists of a system of chambers and canals that contain a network of membranous channels. These are referred to, respectively, as the bony labyrinth and membranous labyrinth. In places, the membranous labyrinth forms the lining of the bony labyrinth; in other places, there is a separation of the two. Within the space lined by the membranous labyrinth is a watery fluid called endolymph. External to the membranous labyrinth, between the membranous and bony labyrinths, is additional fluid called perilymph.

The bony labyrinth is divided into three parts: cochlea, semicircular canals, and vestibule. The cochlea and semicircular canals contain membranous counterparts of the same shape; however, the membranous components of the vestibule are more complex in form, being composed of ducts and two chambers, the utricle and saccule. The cochlea contains the receptors for hearing, the organ of Corti; the semicircular canals contain the receptors for movements of the head; and the utricle and saccule contain receptors for positions of the head.

Ear, guinea pig, H&E, x20.

In this section through the inner ear, bone surrounds the entire inner cavity. Because of its labyrinthine character, sections of the inner ear appear as multiple separate chambers and ducts. These, however, are all interconnected (except that the perilymphatic and endolymphatic spaces remain separate). The largest chamber is the **vestibule** (1). The left side of this chamber (*arrow*) leads into the **cochlea** (2). Just below the *arrow* and to the right is the **oval ligament** (3), surrounding the base of the **stapes** (4). Both structures have been cut obliquely and are not seen in their entirety. The **facial nerve** (5) is in an osseous tunnel to the left of the oval ligament. The communication of the vestibule with one of the semicircular canals is marked by the *arrowhead*. At the upper right are cross sections of the membranous labyrinth passing through components of the **semicircular duct system** (6).

The cochlea is a spiral structure having the general shape of a cone. The specimen illustrated here makes three and a half turns (in humans, it makes two and three quarters turns). The section goes through the central axis of the cochlea. The cochlea consists of a bony stem called the **modiolus** (7). It contains the beginning of the **cochlear nerve** (8) and the **spiral ganglion** (9). Because of the plane of section and the spiral form of the cochlear tunnel, the tunnel is cut cross-wise in seven places (hence three and a half turns). Note the **christa ampullaris** (10). A more detailed examination of the cochlea and the organ of Corti is provided on Plate 159.

Ear, guinea pig, H&E, x225; inset x770.

This is a higher magnification of one of the semicircular canals and of the crista ampullaris (10 in the upper micrograph) within the canal seen in the lower right of the upper micrograph. The receptor for movement, the crista ampullaris (note its relationships in the upper figure), is present in each of the semicircular canals. The **epithelial surface** (11) of the crista consists of two cell types, sustentacular (supporting) cells and hair (receptor) cells. (Two types of hair cells can be distinguished with the electron microscope.) It is difficult to identify the hair and sustentacular cells based on specific characteristics; they can, however, be distinguished based on location. The **inset** shows that the **hair cells** (12) are situated in a more superficial location than the **sustentacular cells** (13). A gelatinous mass, the **cupula** (14), surmounts the epithelium of the crista ampullaris. Each receptor cell sends a hairlike projection deep into the substance of the cupula.

The epithelium rests on a loose, cellular **connective tissue** (15) that also contains the nerve fibers associated with the receptor cells. The nerve fibers are difficult to identify because they are not organized into a discrete bundle.

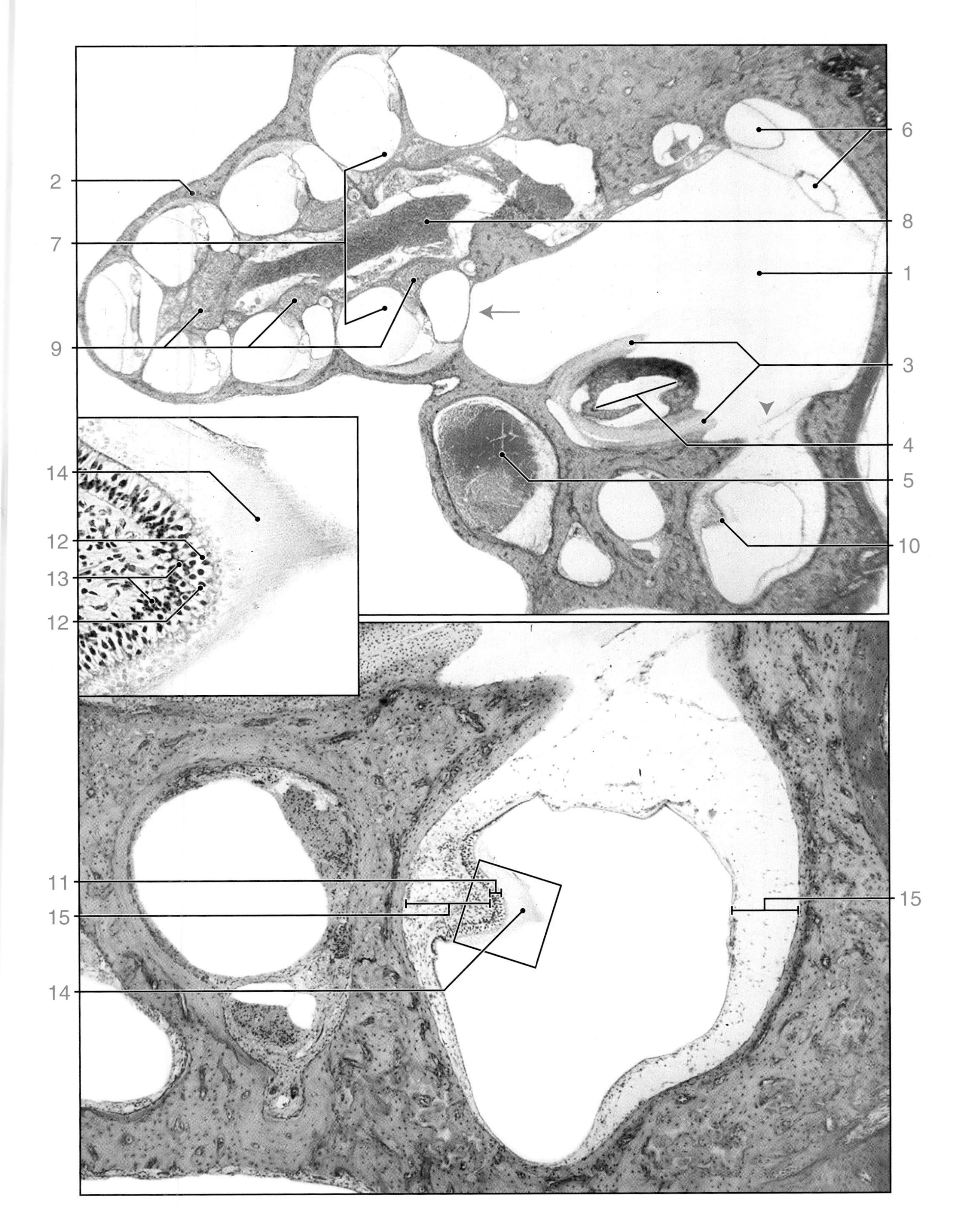

6
2
8
7
1
9
3
4
5
14
10
12
13
12
11
15
15
14

PLATE 159. **ORGAN OF CORTI**

The hair cell, a nonneuronal mechanoreceptor, is the common receptor cell of the vestibulocochlear system. Hair cells are epithelial cells that possess numerous stereocilia, which are modified microvilli also called sensory hairs. They convert mechanical energy to electrical energy that is transmitted via the vestibulocochlear nerve (cranial nerve VIII) to the brain. Hair cells are associated with afferent, as well as efferent, nerve endings. All hair cells have a common basis of receptor cell function that involves bending or flexing of their stereocilia. The specific means by which the stereocilia are bent varies from receptor to receptor, but in each case, stretching of the plasma membrane caused by the bending of the stereocilia generates transmembrane potential changes that are transmitted to the afferent nerve endings associated with each cell. Efferent nerve endings on the hair cells regulate their sensitivity.

Ear, guinea pig, H&E, x65; inset x380.

A section through one of the turns of the cochlea is shown here. The most important functional component of the cochlea is the organ of Corti, contained in the boxed area and shown at higher magnification in the micrograph below. Another structure in this micrograph is the **spiral ligament** (1), a thickening of the periosteum on the outer part of the tunnel. Two membranes, the **basilar membrane** (2) and the **vestibular membrane** (3), join with the spiral ligament and divide the cochlear tunnel into three parallel canals, namely, the **scala vestibuli** (4), the **scala tympani** (5), and the **cochlear duct** (6). Both the scala vestibuli and the scala tympani are perilymphatic spaces; these communicate at the apex of the cochlea. The cochlear duct, on the other hand, is the space of the membranous labyrinth and is filled with endolymph. It is thought that the endolymph is formed by the portion of the spiral ligament that faces the cochlear duct, the **stria vascularis** (7). The stria vascularis is highly vascularized and contains specialized "secretory" cells.

A shelf of bone, the **osseous spiral lamina** (8), extends from the modiolus to the basilar membrane. Branches of the **cochlear nerve** (9) travel along the spiral lamina to the modiolus, where the main trunk of the nerve is formed. The components of the cochlear nerve are bipolar neurons whose cell bodies are shown at higher magnification in the **inset**. The spiral lamina supports the elevation of cells, the **limbus spiralis** (10). The surface of the limbus is composed of columnar cells.

Ear, guinea pig, H&E, x180; inset x380.

The components of the organ of Corti, beginning at the **limbus spiralis** (11), are as follows: **inner border cells** (12); **inner phalangeal and hair cells** (13); **inner pillar cells** (14); (the sequence continues, repeating itself in reverse) **outer pillar cells** (15); **hair cells** (16) and **outer phalangeal cells** (17); and **outer border cells**, or **cells of Hensen** (18). Hair cells are receptor cells; the other cells are collectively referred to as *supporting cells*. The hair and outer phalangeal cells can be distinguished in this figure by their location (see **inset**) and because their nuclei are well aligned. Because the hair cells rest on the phalangeal cells, it can be concluded that the upper three nuclei belong to outer hair cells, whereas the lower three nuclei belong to outer phalangeal cells.

The supporting cells extend from the **basilar membrane** (19) to the surface of the organ of Corti (this is not evident here, but can be seen in the **inset**), where they form a **reticular membrane** (20). The free surface of the receptor cells fits into openings in the reticular membrane, and the "hairs" of these cells project toward, and make contact with, the **tectorial membrane** (21). The latter is a cuticular extension from the columnar cells of the limbus spiralis. In ideal preparations, nerve fibers can be traced from the hair cells to the **cochlear nerve** (22).

In their course from the basilar membrane to the reticular membrane, groups of supporting cells are separated from other groups by spaces that form spiral tunnels. These tunnels are named the **inner tunnel** (23), the **outer tunnel** (24), and the **internal spiral tunnel** (25). Beyond the supporting cells are two additional groups of cells, the **cells of Claudius** (26) and the **cells of Böttcher** (27).

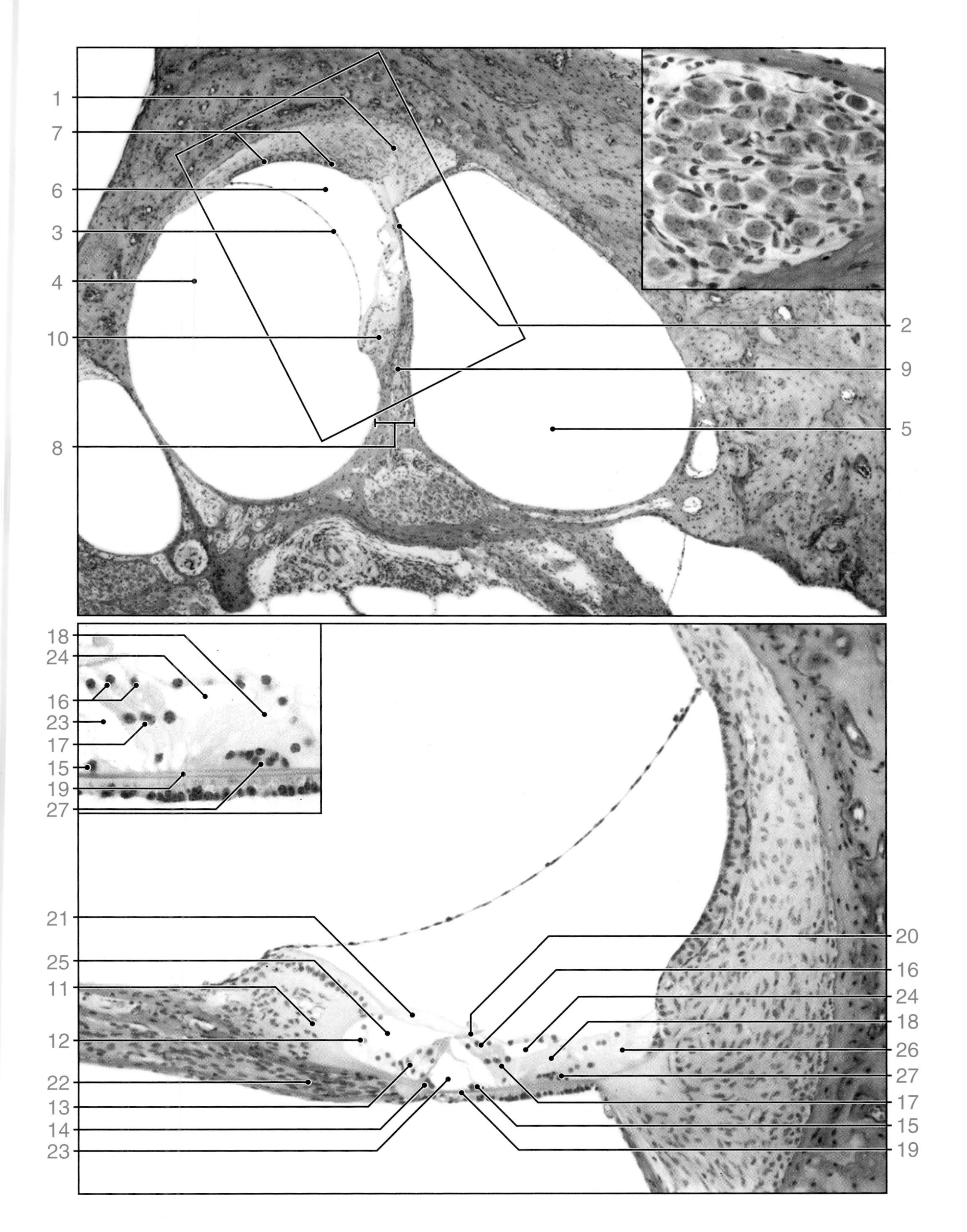

PLATE 159. ORGAN OF CORTI

Index